Applied Geoscience in Shale Exploration and Production

APPLIED GEOSCIENCE
IN SHALE EXPLORATION AND PRODUCTION

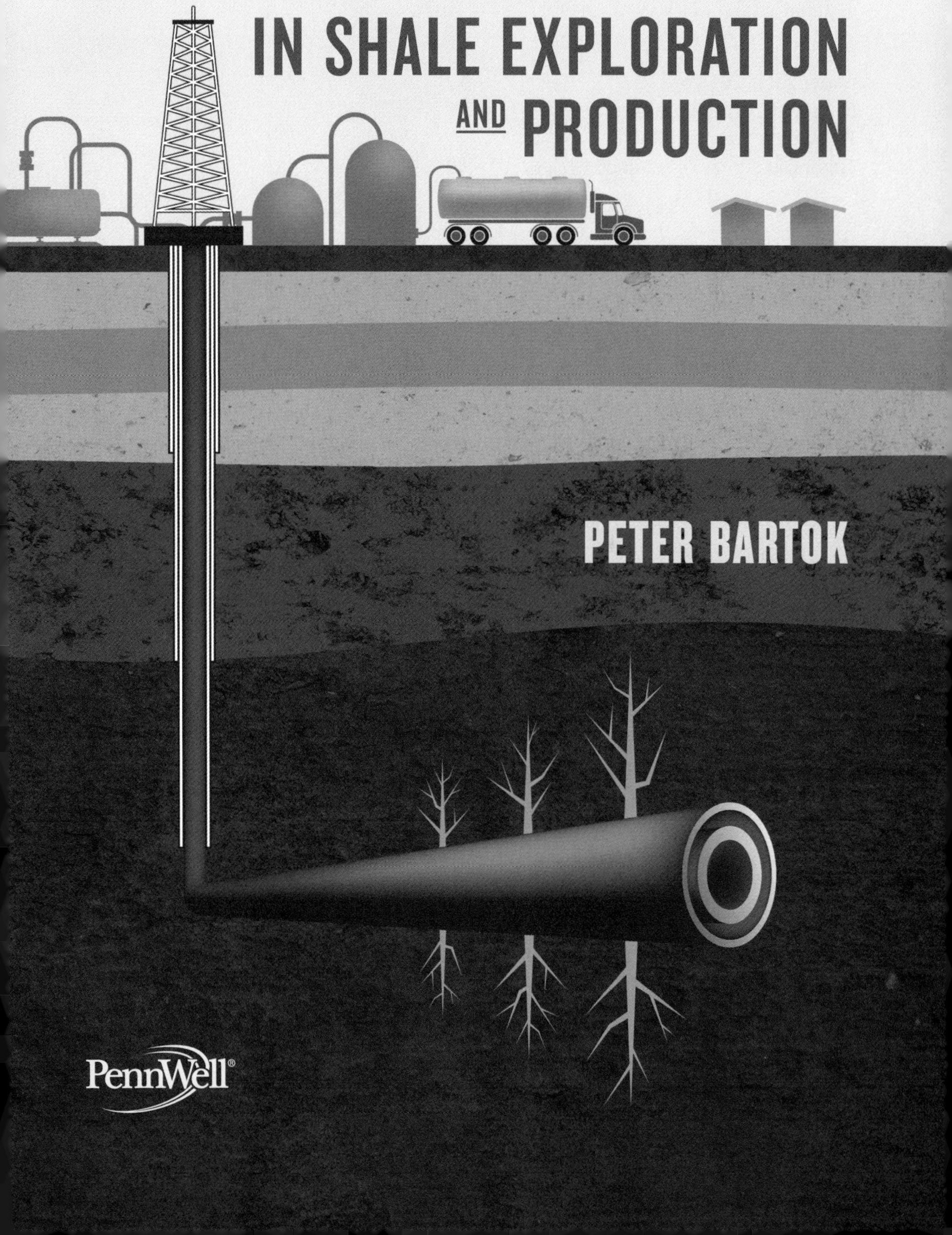

PETER BARTOK

PennWell®

Disclaimer

The recommendations, advice, descriptions, and methods in this book are presented solely for educational purposes. The author and publisher assume no liability whatsoever for any loss or damage that results from the use of any of the material in this book. Use of the material in this book is solely at the risk of the user.

PennWell Corporation
1421 South Sheridan Road
Tulsa, Oklahoma 74112-6600 USA

800.752.9764
+1.918.831.9421
sales@pennwell.com
www.FireEngineeringBooks.com
www.pennwellbooks.com
www.pennwell.com

Publisher: Matthew Dresher
Managing Editor: Mark Haugh
Production Manager: Sheila Brock
Production Editor: Tony Quinn
Cover Designer: Charles Thomas
Book Designer: Susan Ormston

Library of Congress Cataloging-in-Publication Data

Names: Bartok, Peter, 1947- author.
Title: Applied geoscience in shale exploration and production / Peter Bartok.
Description: Tulsa, Oklahoma : PennWell Corporation, [2017] | Includes bibliographical references and index.
Identifiers: LCCN 2017041173 | ISBN 9781593704070
Subjects: LCSH: Shale oils. | Petroleum--Prospecting.
Classification: LCC TP699 .B37 2017 | DDC 622/.3383--dc23

Printed in the United States of America

1 2 3 4 5 22 21 20 19 18

CONTENTS

PREFACE

In the early days of modern exploration, following World War II, a clear distinction was made between the various disciplines involved in the search for hydrocarbons. The geologists would initiate the process by gathering regional data and propose areas of interest. The geophysicist would soon follow with proposals for regional 2-D seismic lines. Following a structural interpretation of the data, the geology would be superimposed, the particular blocks acquired, and the engineering staff would take up the task of proposing the wells, drilling, and producing the hydrocarbons. Communication between the disciplines was sparse, and rarely would one question members of different disciplines. Often they were on different floors and rarely interacted. Today the opposite prevails, yet there is still some level of mistrust. With rare exceptions, most professionals lack the exposure to various disciplines until they reach senior management levels, and then they are rotated to supervise areas they are unfamiliar with.

The objective of this textbook is to bridge that gap by providing sufficient background in the various disciplines to have meaningful interaction among the parties involved.

I am indebted to the thoughtful readers who contributed time and effort to edit portions of the textbook. In particular I would like to thank Dr. Kurt Marfurt, professor at the University of Oklahoma in Norman, OK, and Dr. Yoginder Chugh, professor at Southern Illinois University in Carbondale, IL. Mr. Tom Venetis assisted in the editing of the text and the efforts are appreciated.

INTRODUCTION

Continuing advances in petroleum technology have altered the simplistic view proposed by the followers of Hubbert (1949). Hubbert assumed that, with conventional onshore drilling and shallow offshore shelf drilling, oil production worldwide would peak by the 1970s. Later, Hubbert modified this estimate to the year 2000. More recently, HSBC Bank predicted in 2016 that the world oil production prognosis would peak at 21 billion barrels per year in 2016 and begin a 7% decline per year, reaching a minimum in 2040 of 3.4 billion barrels per year (Fustier et al., 2016).

While timing for *peak oil* predictions proved popular with some, four key factors were not taken into consideration that modified the prediction of both the maximum peak production and when that peak would be achieved:

- Development of accessible economic three-dimensional seismic surveys and advanced processing techniques that significantly reduced drilling risks and extended existing fields
- Ability to drill into deep formations and in ultradeep water, which led to discoveries of major accumulations worldwide
- Reduction of oil demand by using renewable energy sources and making engines more energy efficient
- Technology to exploit unconventional resource shale oil and gas in many traditional basins that significantly increased oil production in several countries

The last of these four contributions is the focus of this book. What new technology yet to be discovered will further improve recovery? The challenge is to produce the hydrocarbons from both conventional and unconventional resources plays, particularly shales, at the lowest possible cost, in an environmentally sound manner—and thus maximize profits and minimize liabilities. The newly developed technologies for unconventionals can also be used in conventional oil and gas fields to improve production.

What is the history of developing gas from shales and what is its future? In 1825, the first known commercial use of shale gas was in Fredonia, New York, where the streetlights were fueled by gas from the Paleozoic Dunkirk shale. Modern shale gas development began in the Fort Worth Basin, Texas, where gas was sought in the Early Carboniferous Barnett shale. Although early investigations on various drilling methods began in the 1980s, it was not until the year 2000 that economic production was achieved through the introduction of multistage hydraulic fracturing and long-reach horizontal wells. Thus, large-scale commercial development of the Barnett shale became technologically feasible.

The Devonian Marcellus shale gas development in West Virginia and Pennsylvania followed. By 2005, development of the Marcellus shale play was well under way when interest began to spread to older sections, such as the Ordovician Utica shale. The interest quickly spread from West Virginia to Ohio. In all of the shale plays, the time involved from the initial exploration to the drilling and production phases has been extremely short. For example, in a matter of four years, from 2006 to 2010, 200 wells were drilled in the Eagle Ford trend, in South Texas (U.S. Energy Information Agency, 2013). At an average cost of $5–$10 million (MM) per well and subsequent field infrastructure, costs for field investment in excess of one billion U.S. dollars became commonplace. Even with the additional reserves from shale oil production, if we assume the current production and consumption rates, at best, oil may last until 2100. Better understanding of shales and development of new technologies to identify and produce shale resources will significantly lower costs.

Because each shale fairway is different, there is no pre-established workflow that fits all. Procedures and studies must be tailored to the needs of each individual project. However, it is important to carry out comparative studies on other basins to perfect the workflow for a specific area of interest, avoid mistakes, and diagnose critical elements. Basic principles exist that should be followed, and these will be detailed later in this chapter.

As in all economic cycles, supply and demand works in tandem to sustain a reasonable price scenario. It is critical to constantly review costs and improve the cost per barrel produced to ensure an adequate rate of return during economic downturns. The U.S. Department of Interior's Energy Information Administration (EIA) suggested that the recoverable hydrocarbon from shales is between 3% and 5% of the original oil in place, or stock tank original oil in place (STOOIP). The challenge is how to improve that recovery, especially as the geographic distribution of these shale resources is extensive (fig. 1–1).

For the Bakken, the minimum estimated ultimate recoverable per well must exceed 100,000 barrels of oil (BO) if the price of oil is $100 per barrel. A study by the EIA in the Bakken demonstrated that, of the 1,132 wells studied in the Williston Basin, 1,120 exceeded 20,000 BO. Lower oil prices would require correspondingly higher reserves that must be produced at higher rates. Although most studies have focused on the United States, there are over 40 countries worldwide with proven shale resource plays (fig. 1–2).

Shales, which are the focus of this book, are not the only type of unconventional play. *Unconventionals* include coal-bed methane, basin-centered gas, tight gas sands, heavy oil, and deep water frozen methane hydrates (clathrates). An additional subcategory is the *hybrid play*, composed of tight sands, or inter-bedded thin sands and source rock shales.

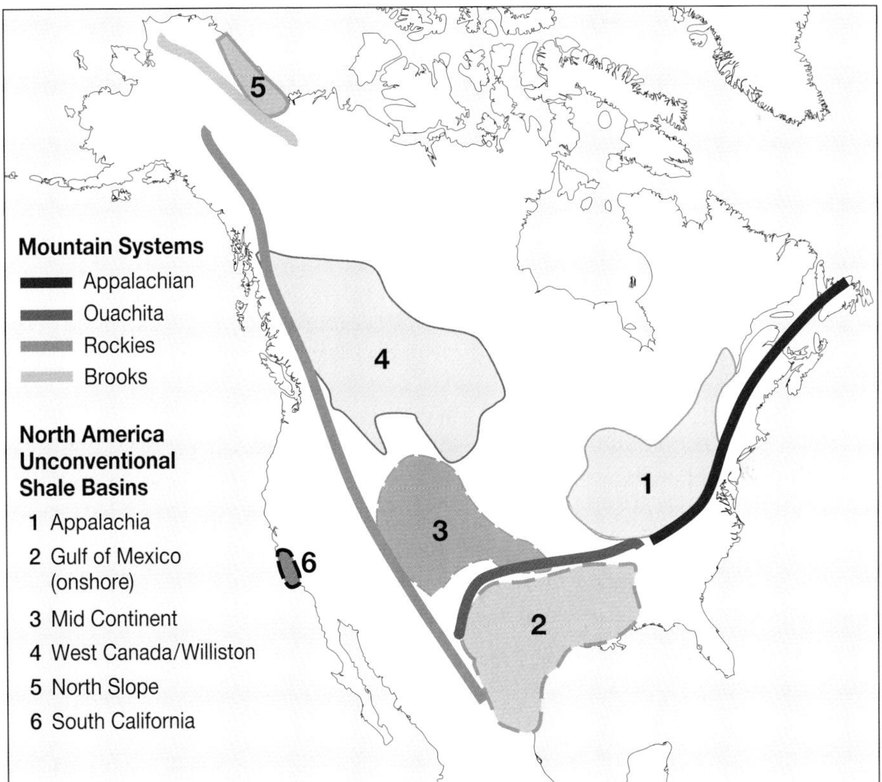

Fig. 1–1. North American shale resource play areas and major mountain ranges

Fig. I–2. Major world shale resources. The regions listed are as follow: (A) North America including (1) Western Canada and Williston Basin, (2) Michigan/Illinois, (3) Appalachia, and (4) Northern Gulf of Mexico; (B) South America including (5) Maracaibo/Llanos, (6) Parana, and (7) Neuquén; (C) Africa including (10) Saharan and (8) Sub-Saharan Africa; the (D) Middle East with special focus on (9) Saudi Arabia; (E) Europe including (11) Paris, (15) the Pannonian, (15) Poland, and (14) the Baltic region; (F) Russia including (16) the Caspian, (13) Timan/Pechora, and (12) West Siberia; (G) China including (17) East China, (18) Tarim, and (19) Junggar; (H) Australia including (20) Canning and (21) Bonaparte. Note that size of each basin does not indicate level of significance.

An unconventional play is a hydrocarbon reservoir that requires nonconventional techniques to obtain flow and extraction because of its physical or chemical conditions of entrapment. Unconventional resource plays have elements in common with conventional exploration and production of hydrocarbons. For shale resource plays and conventional plays, both require storage capacity in the form of porosity and permeability (either natural or induced), and both need adequate seals. The source rock is obvious since it is the target in any search of the shale resource play. The timing of generation is also important for both. Often, overall porosities and permeabilities in the shale resource plays are less than 10% and under 0.5 millidarcies.

What distinguishes a *shale play* from a conventional resource is that, in shales, the natural porosity is stored in nanopores; thus, connectivity and permeability are often measured in nanodarcies. Porosity cannot be measured directly in the subsurface except where core samples are taken. Porosities are usually estimated from logs on the basis of changes in density of the bulk rock or changes in the velocity that sound travels through the sediment. Permeabilities are estimated on the basis of empirical relationships with porosity corroborated with core data. When fluids are injected into the formation during well testing, the permeability can be estimated more accurately. Later chapters of this book (see chaps. 5 and 8) discuss these topics in detail.

Organic-rich shales are often not bound by structural traps. Strictly speaking, shale resource plays are a variation of the stratigraphic trap. However, they may be compartmentalized by faults or facies changes. They require hydraulic fracturing to improve connectivity of pores and thus enable production.

Shale resource plays must have adequate thermal conditions and contain proper seals even after stimulation; otherwise, overlying thief zones will draw any induced hydrocarbon production from the shale or specific producing interval. The thermal history must be appropriate for generation—neither too undercooked nor too overcooked. Because of the nanopore size, the ultimate target of the unconventional resource play is to produce *wet gas*, which is a gas rich in condensate oil under reservoir conditions and where appropriate to produce shale oil. The characteristics of the seal must also be studied. Often, the unfractured shale overlying the producing interval can provide the seal. *Seal capacity* is a function of capillary pressure, which is dependent on fluid densities and pore throat size (see chap. 8). By definition, oil or gas will flow from higher pressure to lower pressure, from deeper reservoirs to shallower reservoirs. An understanding of the hydrology of the reservoir fluids is fundamental.

Wells are drilled horizontally along the optimum reservoir section to improve access to the reservoir. The geosteering must be accurate and consistent. A gyroscope-based tool, located in the bottom-hole assembly (BHA) of the well and transmitted by use of either wireline or mud pulses, determines the position of the bit in the subsurface. The mode of production of the resource play is different than for conventional resources. Because they exhibit very low permeability, shales must be *fracked* (short for hydraulic fracturing).

Vertical wells are most often drilled as pilot wells to determine optimum steering guidelines for the subsequent horizontal wells and not as the principal producing well. Vertical wells are also used to determine basic rock properties from cores and the effects of stimulations such as minifracs and acidification. Preexisting vertical wells are often used to record microseismic signals and thus determine the direction, extent, and efficiency of the hydraulic frac. All available techniques should be applied to determine the best location for the wells, field development, and drilling and completion techniques.

Large amounts of water are needed at the onset of production to frac the sediment and stimulate the reservoir. In conventional reservoirs, water is used in a secondary recovery near the end of the field production cycle. Traditional reservoirs that encounter porous sandstones or carbonates, hydrocarbons can be produced with little or no additional stimulation. For unconventional resources, horizontal wells and hydraulic fracking are always recommended. The reasons for introducing water are its relatively low cost and the ability to recover a portion of the injected water after its use. As a general rule, only one-third of the water injected is recovered. Although significant water volumes are used in fracking, the amount is insufficient to act as a water drive. Once a development phase is initiated, additional formation water is produced together with the hydrocarbons. Their disposition is discussed in detail in chapter 9.

The drilling objective in the unconventional shale play is to penetrate a brittle horizon adjacent to the organic-rich shale, to propagate the induced fractures from the brittle interval into the more ductile shale and allow the flow of the hydrocarbons. Care must be taken to retain the seal integrity of the reservoir by avoiding over fracking and avoiding proximity to significant faults. Synthetic proppants and originally well-rounded and sorted fine sands are used to keep the fractures open.

One problem faced by all geoscientists in complex integrated studies is to ensure that the various languages spoken in each discipline are clearly understood. For example, a petroleum engineer might ask the geologist to generate an "RQD map" of the area of interest. This is a *rock quality distribution map* of the target horizon. For most geoscientists, even that description is insufficient. Specifically, what does the petroleum engineer mean by *rock quality*? The petroleum engineer requires an understanding of stress fields in the subsurface, as well as the rock mechanics of the target horizon, described further in chapter 8. The intensity of natural fractures is described as RQD. It is common to refrac an interval to stimulate additional production later in the production cycle. The possibility of refracking depends on the original completion method, the brittleness, and the RQD of the formation.

Properly spaced horizontal wells allow for every available segment of the source rock to be tapped and produced (fig. 1–3). Originally, it was thought that an ideal spacing distance keep the frac wings from adjacent wells from interfering with each other. As more closely spaced infill wells were drilled, it was observed that instead of interfering with production, the production in both wells increased. This seems counterintuitive until you consider how difficult a task it is to communicate with all the available pore space. Terms such as *coplanar* or *staggered multilateral* (fig. 1–4) are used to differentiate from simply stacked multilaterals to further enhance drainage. Other terms describe variations in drilling patterns. It is also possible to modify the orientation of wells that are stacked, to achieve changes on the frac wings (see chap. 8).

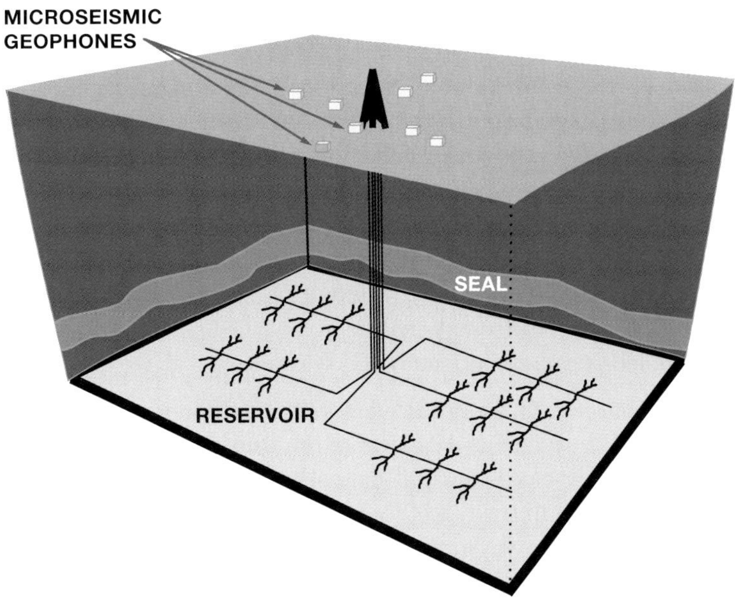

Fig. I–3. Schematic demonstrating the harvesting of hydrocarbons using horizontal wells

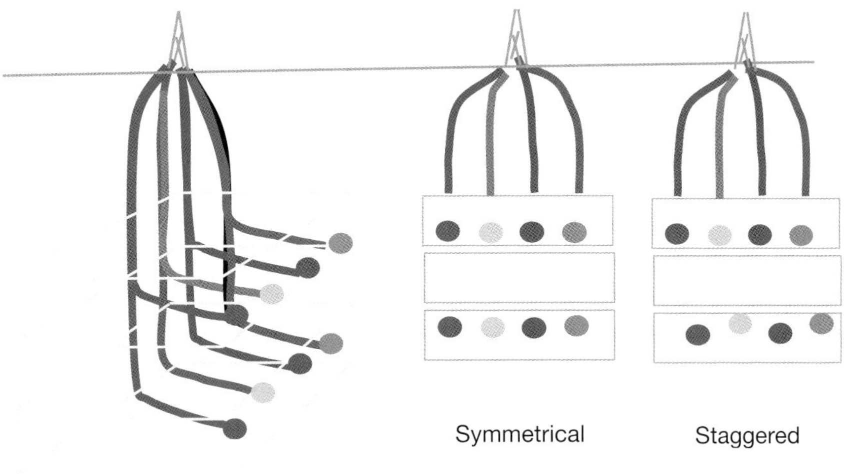

Fig. I–4. Examples of multistory horizontal wells

Investigations are under way to apply technology used in heavy oil production to the shale resource play, such as *steam-assisted gravity drainage* (SAGD). Here, one drills two horizontal wells, one beneath the other, and injects steam into the upper well and produces from the lower well. SAGD is a common practice in the Alberta heavy oil belt. Application of SAGD to shale resource plays with lower kerogen maturities can improve production. Alternate injection of gas, as either methane or nitrogen, can increase reservoir pressure or improve the ratio of free gas to adsorbed gas (see chap. 8). Free gas requires less energy to produce. In North Dakota, tests are being proposed to drill multiple stacked horizontal wells that penetrate the Three Forks Formation (underlying the Bakken target), possibly at three different levels called the *Three Forks benches*.

In the early phase of shale resource development, there was a misconception in industry and in the business community that the hydrocarbon source material had nearly a layer-cake distribution, with the total organic carbon (TOC) evenly distributed within the target formation. It became evident once significant development drilling was initiated that shale richness was unevenly distributed horizontally and can vary rapidly vertically. Investigations focused on defining the characteristics that identified the sweet spots, requiring in-depth understanding of the geology, geophysics, rock physics, and rock mechanics of the reservoir.

The inability to clearly define the richness and exact distribution of the source rock directly affects the volumetric calculation of the field potential. Overall volumetrics depend on estimating the volume of the pore space available both before and after fracking, as well as determining the volume of free gas, adsorbed gas, water saturation in the system, and organic carbon content of the shale. Detailed methods for evaluating volumetrics will be discussed in chapter 9. For now, the shale resource play is defined by facies variations, pore type, temperature, pressure boundaries, and estimates of original source rock richness and the quality of the kerogen (source rock type). These parameters are often difficult to define but must be estimated prior to entering the field development phase.

More modern drilling practices have extended the concept of horizontal drilling to multistory horizontal array drilling. This has allowed more intensive drilling programs, and it is not uncommon to drill more than 16 wells from a single pad. Typically each well requires two months to complete, and the drilling rig may be in place for almost three years. With subsequent refracking or additional stimulation and workovers, the rig may remain on site for several more years. Such multistory arrays provide significant cost savings and reduce the footprint left behind. Long lateral wells can extend up to 10,000 ft (3,000 m).

Can anything else be learned from unconventional resources and applied to conventional oil and gas exploration? The sciences of geosteering, multistage fracking, microseismic, logging, multicomponent seismic, and rock physics have opened new avenues of investigations previously considered only as academic exercises. Seismic processing has advanced to the point where attributes related to sweet spots can be identified and mapped. With the proliferation of four-dimensional seismic (i.e., repeated three-dimensional seismic over the life of the field), the progressive depletion of the reservoirs can also be mapped, thereby improving the harvesting of the hydrocarbon resource.

STATEMENT OF OBJECTIVE

In the assessment of shale resource plays, it becomes evident that a multidisciplinary approach is critical, as well as for advanced conventional development. The first rule of an integrated study is that all participants must learn to speak the different languages used by each profession and understand the subtle cultural differences specific to each discipline. The second rule is that all team participants must be familiar with the techniques used by the various disciplines, to optimize their integration.

An example of language differences is the scale factor. A sedimentologist can describe massive sandstone bedding on a decimeter scale, whereas to the geophysicist a massive sandstone is a package 10–20 m thick; moreover, to the engineer, the massive sandstone is a continuous sand unit with no discernable internal barriers and with similar measurable reservoir parameters. The proliferation of technical language and techniques in the fields of geoscience and engineering and the subsequent use of contractions can lead to confusion and ultimately foster distrust. For example, an engineer might overhear the following conversation between a geophysicist and a petrophysicist: "My AVO signature is clearly class III and suggests 'a,' 'b,' and 'c.'" The petrophysicist might reply, "But my LMR plot, based on the well logs, suggests that for the depth range in question, it is more likely to expect a class II AVO, and that may modify your interpretation." In this example, the engineer needs to understand the impact on the amount of free gas in the reservoir, a topic that was never mentioned in this conversation between the two geoscientists, who are essentially speaking different languages.

One objective of this book is to bridge these cultural barriers across disciplines and improve communication among various industry professionals. A second objective is to ensure that sufficient knowledge is available to readers to facilitate conversations with their counterparts in the various disciplines and to contribute to the discussion of all aspects characterizing a reservoir and affecting its management. It is also important to enable professionals to read the literature from the various disciplines on related topics and determine their importance and relevance to their current study.

The present text is not intended as an in-depth discussion of the specific topics mentioned, as each requires much more training to achieve proficiency before applying the techniques discussed. Instead, the goal is to provide the general groundwork that allows the reader to understand how information is acquired and applied, as well as some of the limitations of this information. Note that this publication assumes that the author is supplying information but is not attempting to render engineering or geoscience professional services. If such services are required, the assistance of an appropriate professional should be sought. At the same time, an understanding of how the data are obtained provides a degree of confidence to the interpreter.

Two factors dominate both conventional and resource plays: first, costs are high; and second, the margin of error is extremely small. The ultimate goal is to maximize the rate of return on the investment. For this reason, once a producing trend is discovered, the engineering aspects of the play take on a dominant role. However, it is the geoscientist who can provide the necessary means to define and pursue the sweet spots in the region and thus increase the chances for profitability. Engineers and geoscientists both play a vital role.

The lifetime of the average shale resource well is quite short. To determine the life span of a field, the engineers generate production decline curves for each well, discussed in more detail in chapter 9. A way to visualize the decline curve is that if the initial production is 100 within the first two years, the decline may be 60% (i.e., 40% the first year and 20% the second year). The tail end of the decline curve is an exponential curve with the remaining 40% of the reserves produced over 8–10 years.

What is an optimal initial work flow for a shale resource assessment? The most logical answer is to follow the geologic process that established the resource in the first place and then determine the most critical conditions for the specific reservoir under evaluation, including its predictability and mode of production. As discussed by Jenkins (2017), each of the four basic phases of exploration and development of a field—exploration, appraisal, demonstration and development—must be carried out in a systematic manner under strict actuarial controls. Similar to Maslow's personal hierarchy of needs, in geoscience the basic exploration and production (E&P) hierarchy of needs (fig. 1–5) is as follows:

- Compiling historical data on the region
- Carrying out basin studies including mega-scale tectonostratigraphic analyses
- Preparing regional maps

- Establishing a documentation path to avoid repeating previous steps to allow the focus to be on data pertinent to the main objective while allowing new data to be introduced.

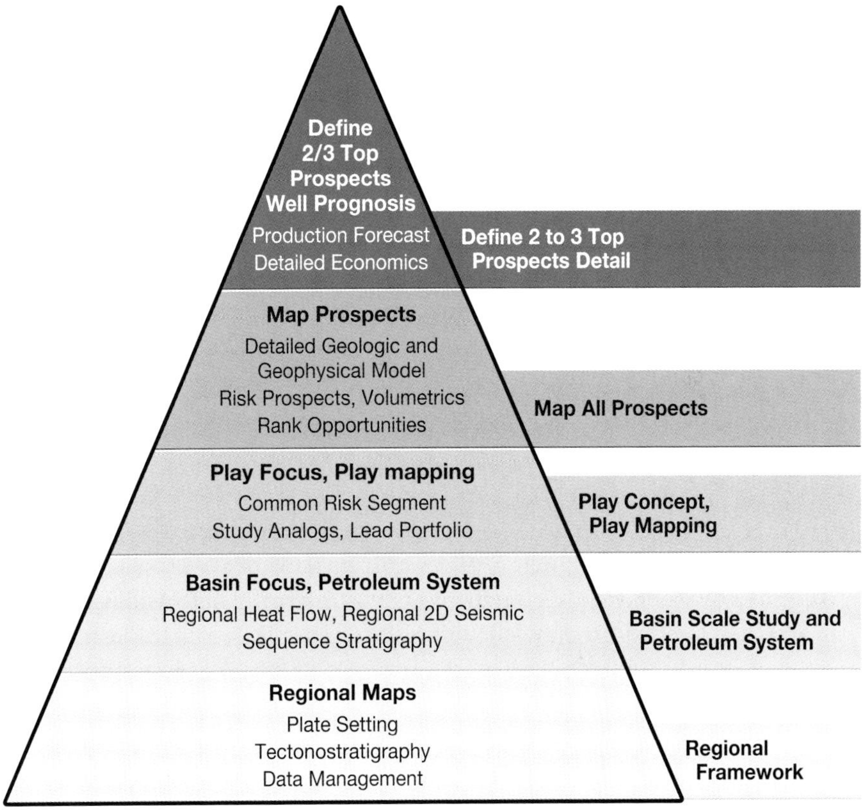

Fig. I–5. Pyramid of need—from exploration concept to production

The second stage is the *play concept*. Here, the focus is on defining the petroleum system of the basin by using the following techniques:

- Performing detailed sequence stratigraphy
- Developing the geologic history of the basin through time by generating *gross depositional environment* (GDE) maps
- Defining the criteria to be used in the *common risk segment* (CRS) analysis and applying them on a basin scale, giving a high grade to more promising areas of the basin

Defining leads follows naturally from the CRS analysis and should include more detailed analyses of the following:

- Describing the mineral constituents themselves—in particular, the clay minerals that comprise the shale and how they behave
- Determining their diagenetic overprint
- Explaining the petrophysical log response
- Analyzing the target organic material—namely, the type of organic matter and the style of kerogen network
- Establishing how pores are created in the sediment, including the organic material, even at micro scales
- Estimating the physical behavior of fluids inside the pores when they are subjected to subsurface reservoir pressures and temperatures before and after fracking
- Establishing techniques to predict the sweet spots (What is required to produce the shale resource? Is it maturity? Is it TOC? Is it overpressure? If the rock must be fracked, how should the fracking be executed?)
- Evaluating how seismic can assist in the shale property prediction
- Last, listing all factors that have an impact on commerciality

The *prospect* has specific characteristics. It is the precursor to drilling the pilot well and therefore requires the following:

- Extremely precise depth information
- A prognosis of the reservoir and fluid conditions in the subsurface
- Definition of potential volumetrics to enable ranking of the drilling sequence
- Estimation of the number and orientation of wells to be drilled
- A drilling plan that includes all the necessary details on coring, logging, and stimulation plans
- Estimating drilling and development costs

In addition to the scheme outlined in figure 1–5, it is essential to prepare a decision tree and an exit plan. Failures will always occur. It is important to define the criteria to separate a geologic concept failure from an operational problem. The second can be corrected, whereas the first requires a return to basics.

To achieve complete understanding of the shale resource potential, one first begins with a description and definition of the shale from an analytical standpoint, understanding its distribution through the application of stochastic methods and by study of the diagenetic overprint of its constituent particles. The optimum environment of deposition requires data collection and analysis, followed by application of predictive tools to determine both the extent of resources and how the fairways can be established and mapped. Once the resource is detected, then one can establish the optimum extraction method.

This book is organized according to the general work flow described above. The initial section defines the shale, its diagenesis, and how to quantify the constituent particles (chaps. 2 and 3) and later describes its environment of deposition (chap. 4). In addition, the porosity and physical properties of the shales must be established on the basis of cores, and the analysis of the samples must then be integrated with the petrophysical properties (chap. 5). Geophysics and geochemistry provide the tools for extrapolating the results laterally to reduce risks (chaps. 6 and 7). Finally, the engineering, reserve calculation, and project economics provide the necessary conditions for production of the resource (chaps. 8 and 9).

Critical negative consequences to either the environment and or aquifers must be addressed. A significant amount of criticism has been levied against fracking because of the potential for aquifer contamination, surface seepage, and even detectable seismic events. Nevertheless, fracking itself is not a likely cause. Great care is taken to set proper casing and liners to isolate the production casing. The integrity of seals is preserved as much as possible, and major faults are avoided to counteract thief zones. However, as a consequence of producing the shale resource, large volumes of water with added chemicals must be reinjected into the subsurface area. Additional water is produced from highly water-saturated reservoirs and must also be disposed of, and this is where the problem lies. The water is normally injected into highly porous reservoirs without regard to proximity to faults. The volumes of water can be quite large, in excess of 10,000 barrels of water (BW) per day, and if faults are present, the water can easily migrate through the faults. More careful study must be made of the recipient formations and consequent leakage. It has been well documented that large volumes of water injected into or in close proximity to an active fault plane will lubricate the fault plane and activate the fault. Some areas are more sensitive than others. The higher-risk areas should be properly mitigated. Either water should be recycled, or great care must be taken during the reinjection process. Detailed study of aquifers is highly recommended. These studies are also vital in the selection of adequate casing program to ensure isolation of the aquifers.

REFERENCES

Fustier, Kim, Gordon Gray, Christoffer Gundersen, and Thomas Hilboldt. 2016. "Global oil supply: Will mature field declines drive the next supply crunch?" HSBC Global Research. petrole.blog.lemonde.fr/files/2017/01/HSBC-peak-oil-report-2017.pdf.

Hubbert, M. King. 1949. "Energy from fossil fuels." *Science*. 109: 103–109.

Jenkins, Creties. 2017. "Appraising and developing your mudrocks: How to avoid squandering billions of dollars next time." Paper presented at Houston Geological Society Applied Geoscience Conference, The Woodlands, TX, March 7–8.

U.S. Energy Information Agency. 2013. "Technically recoverable shale oil and shale gas resources: An assessment of 137 shale formations in 41 countries outside the United States." Open-file report. Washington, DC: U.S. Department of Energy. www.eia.gov/analysis/studies/worldshalegas/archive/2013/pdf/fullreport_2013.pdf.

2

SHALES, CLAY MINERALOGY, AND ASSOCIATED FEATURES (SURFACE AND SUBSURFACE)

TAKEAWAY

- Definition of shale based on grain size and mineralogy and significance
- Response of shale to gamma-ray log
- Clay minerals—mineral composition and significance, with focus on smectite/illite structure
- Mineral composition–based definition of brittleness
- Diagenesis of smectite and relationship to overpressure and free gas
- Porosity, permeability, and organic content response in shales
- Tools used in the study of shales and their applications: XRD, XRF, SEM, TEM, CEC, DTA, EDAX, QEMSCAN, IBM-SEM, FTIR, CT scan, and Langmuir isotherm

GENERAL PRINCIPLES

All professionals involved in the study of the shale resource play require an accurate definition of the shale, including its physical and chemical properties. At the same time, it is important to understand how the data are obtained to determine their accuracy and limitations of the definition. In addition, establish what questions to ask to improve the definition of these attributes.

There are several distinct ways of classifying shales and determining their properties. To the sedimentologist, grain size (<3.9 mm) and mineralogy are critical and equally important. Shales are mainly composed of clay minerals known as *phyllosilicates*, the most common of which are kaolinite, smectite, illite, and chlorite. However, to petrophysicists and engineers, the definition of shale is modified to suit their needs. Katahara (2008) has suggested that aside from the mineralogical characteristics of interest to petrophysicists, shales also have nanopores and permeabilities measured in nanodarcies. For the engineer, the brittleness and chemical properties are critical.

The gamma-ray tool is used for initial detection of shale present in the subsurface. The total shale volume in the rock can be estimated by reading the gamma-ray response along the borehole at the level of interest and comparing it to shales in close proximity to the studied interval. The maximum deflection of the gamma-ray tool is considered to be a *clean* shale, and the ratio between the *diluted* shale and the clean shale is used to compute the shale volume, V_{sh}. Note that the gamma-ray response relationship described above applies only to non–organic-rich shales. In organic-rich shales, the chemical structure of the kerogen often traps radioactive materials, yielding high gamma-ray readings; thus, these are often referred to as *hot shales*. When the gamma-ray log is used in combination with sonic, density, and neutron porosity logs, many additional attributes of the shale can also be determined, including porosity, density, and sonic travel time. This is usually the initial step in defining the presence of shale and estimating several of its properties. Cores are subsequently taken over the zone of interest to ensure appropriate calibration with the logs because the samples can be carefully analyzed with multiple techniques that will be described in this chapter. Additional details are explained in chapters 4 and 5.

Because the principal radioactive isotopes that yield a gamma-ray response in non–organic-rich shales are known, tools have been designed to differentiate constituent clay minerals. Studies have shown that the principal radioactive components—thorium, uranium, and potassium—can be related directly to the dominant clay type: kaolinite, smectite, illite, mixed-layer illite/smectite, or chlorite. The tool used is known as a *spectral log* and is described in more detail in chapter 5. The ability to separate the clay fraction into its mineral constituents is vital to define the seal characteristics of a reservoir and to determine the physical rock properties of the reservoir to be hydrofracked.

To the engineer, shales are further classified by their physical and chemical properties. Because all wells are drilled with drilling mud, the potential for interaction between the drilling chemicals and the formation can result in serious wellbore stability problems. Furthermore, the mechanical properties of shales, such as load bearing and water content, are significant to the engineer.

The distinction between *mudstone* and shale is also fundamental. Mudstones show considerable elasticity over a wide range of stress conditions, whereas shales are more fissile, brittle, and often strongly laminated. Their behavior under stress is quite different. Often, in overpressure conditions, "shales" can behave as mudstones. In addition, some of the constituent clay minerals can have extremely large surface areas onto which water, cations, and organic material can be adsorbed. Adsorption occurs when, as a consequence of electrostatic forces, ions or molecules attach to clay mineral surfaces and organic materials without chemical bonding. The topic of adsorption is discussed later in this chapter and is of major concern since adsorbed gas is more difficult to mobilize and hence exerts an impact on the reservoir conditions and production.

Grain size distribution

Grain size distribution is obtained, to determine the size fraction in the sample with a sand shale mixture, by moderately grinding a rock sample to break up the intergrain cements, weighing the sample, and then passing the sample through several sieves using a Ro-Tap apparatus. The equipment shakes and taps to collect the various sample grain sizes. Each size range in the sample is measured in discrete intervals starting at 0.0625 mm (or 1/16 mm), which is the lower cutoff for sand, and 0.004 mm (or 1/256 mm), which is the cutoff for clay. The term ϕ was introduced by Krumbein in the 1930s as a modification to the original Wentworth grain size scale and is currently assigned to each sieve size. Each half-grain size is assigned a larger number (i.e., 1/2 mm is 1, 1/4 mm is 2, etc.). The smallest conventional scale is usually 1/256 mm (sieve 8). Below that value, all particles are considered clay size regardless of composition (fig. 2–1). Special mesh sizes are available for finer clay separation. On completion of the process, the sample in each mesh size is weighed, and its relative weight percent is determined. Plotting the cumulative weight percent versus size (Visher, 1969) allows for statistical determinations of the grain size distribution used in the classification of sandstones and shales and related to their environments of deposition. It is also important to determine the composition of the various size fractions.

The clay fraction is subsequently placed in a large cylinder of distilled deionized water, and a deflocculent is added. The sample is then stirred and allowed to settle. A hydrometer can be used to determine the density of the water and, thereby, the relative presence of the specific clay size fraction. The results are based on empirical studies examining the settling velocities of various clay grain size fractions and can be summarized by Stoke's law:

$$V = D^2 g \frac{\rho_1 - \rho_2}{18\mu_v} \tag{2.1}$$

where V is the settling velocity, D is the particle diameter, g is the acceleration of gravity, ρ is the density of the suspension before and after, and μ_v is the viscosity of the fluid. With water at 30°C, the formula is simplified as follows:

$$V = kD^2 \tag{2.2}$$

where k = 11,214.

The cylinder and hydrometer are gauged to measure grams/liter. Sand particles settle within 40 seconds and silt within 2 hours. For a more detailed discussion, the reader is directed to the standard test method of the American Society for Testing and Materials (1985). For carbonate shale mixtures, the classification is mainly optical at both a macroscopic and a microscopic level. This is due to the cementing effect of the carbonates.

The hydrometer method assumes spherical settling and therefore provides a general guide but lacks accuracy owing to the platy nature of clays. A more precise method for grain size distribution is provided by x-ray crystallography and scanning electron microscopy (SEM). These topics are elaborated further later in this chapter. Establishing the presence or absence of sand in the sample has a bearing on the compaction of the shale, brittleness, and other physical properties, as well as on the diagenesis of the sample. If carbonates are mixed with the sample, the dry sample is weighed, washed in a weak hydrochloric acid (HCl) solution, dried, and reweighed to establish carbonate content. In addition, an uncovered slide is spread with a warm epoxy or resin, and the cuttings samples are spread on the slide. Half the slide is dipped in alizarin red solution and analyzed optically. All calcite grains will stain red, and noncalcite will remain white. A point count determines the percent of calcite in the sample. If dolomite is suspected in the sample, a portion of the slide is dipped in potassium ferricyanide, which stains dolomite blue. Estimating the carbonate content allows determination of the brittleness of the sample and has implications with respect to the log response. Limestone and dolomite are very dense minerals (2.65–3.0 g/cm^3) and therefore affect porosity determination. At reservoir levels, shales often have densities ranging from 2.4 to 2.65 g/cm^3.

The content of the organic material is an essential element of the shale resource play. For decades, shale samples have been treated with hydrofluoric acid (HF) to extract pollen grains and conodonts (discussed in detail in chap. 3). The same procedures can be used to extract the organic material by dissolving all the silica-based minerals. Weighing the residue after the previous treatments provides a general organic content of the sample. An additional tool to measure organic content in a sample is to heat the sample and measure the carbon dioxide (CO_2) produced.

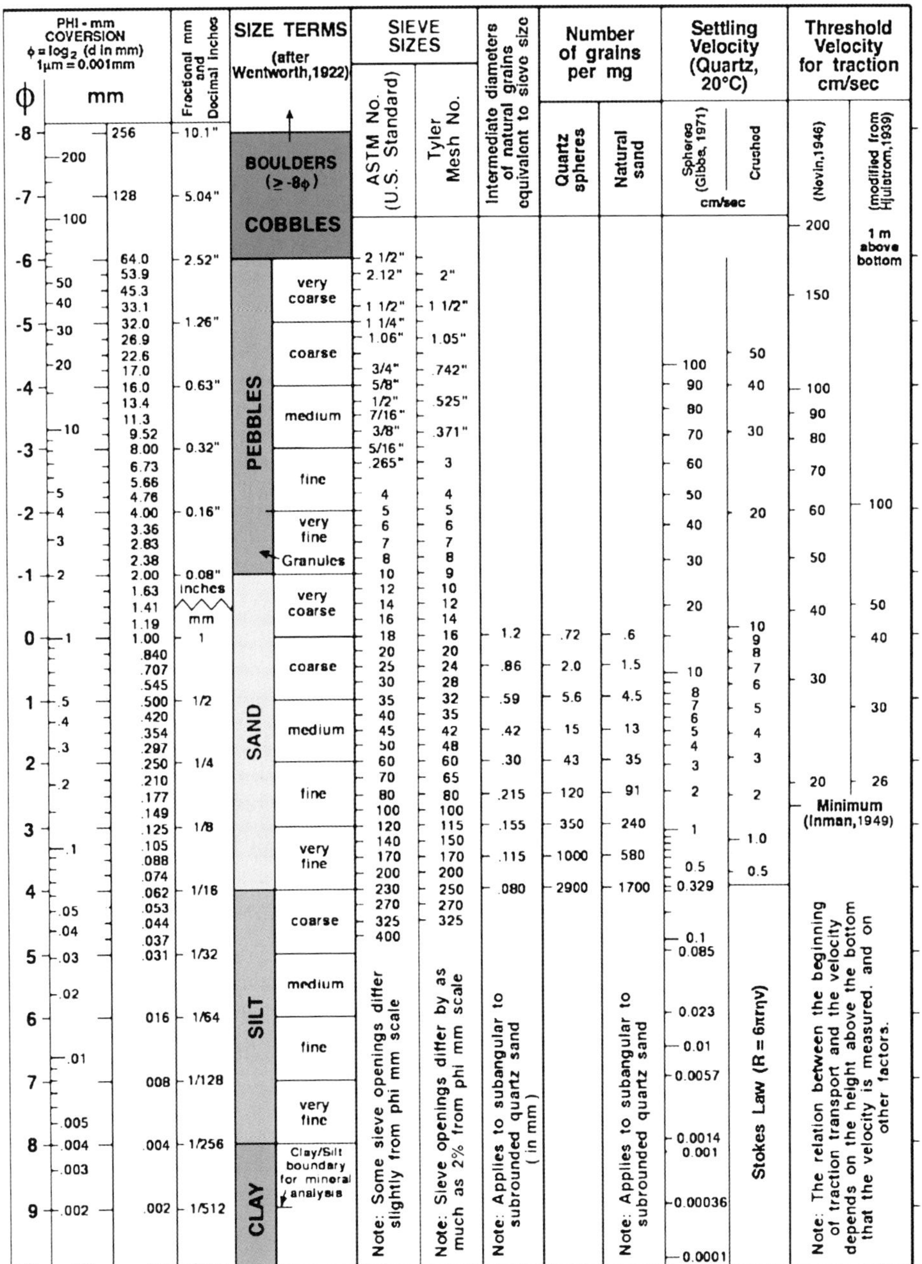

Fig. 2–1. Standard grain size chart (Data from U.S. Geological Survey open-file report; Poppe et al., 2013)

Core data

Core information is fundamental for the unconventional resource play. Aside from the multiple measurements described later in this chapter and in chapter 8, the geological and paleontological description provides the basic control for the well log calibration, definition of genetic units, and correlations that would otherwise not be understood. As an example, figure 2–2 shows a typical Utica/Point Pleasant core. The dark shale is organic rich. Interbedded in the shale are layers of brachiopod packstones that would otherwise not be detected and are critical for defining the environment of deposition. In other regions, local layers of pyrite or ash beds need to be identified and calibrated with the logs in terms of frequency and distribution.

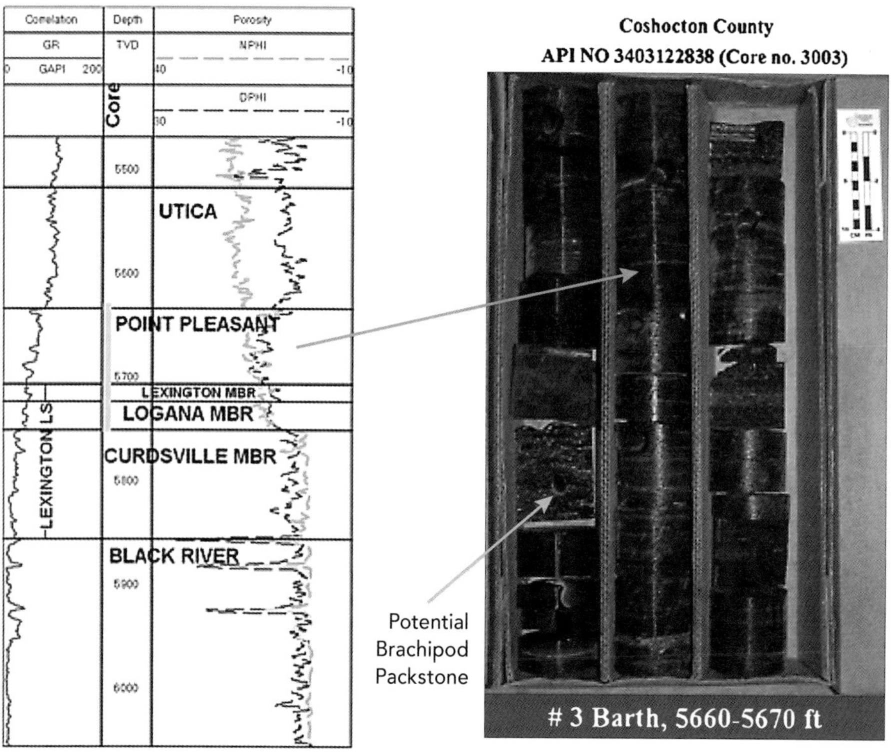

Fig. 2–2. Core data from the Ordovician Utica/Point Pleasant Formation, Ohio (Reproduced with kind permission of Shale Experts; Riley and Baranoski, 2011)

Shale mineralogy

The second definition of shales is based on composition. Surprisingly, this definition is not as robust as one might expect. Shales are rarely pure clay minerals. In fact, clay minerals themselves can be either allochem or authigenic (autochem). The term *allochem* is more commonly applied to carbonates, and it is used to distinguish carbonates transported into the environment from those produced in situ, which are considered *autochem*. However, the terminology applies very well to shales, and allochem is analogous to detrital clays. It is important to establish if the clay fraction is an allochem or an autochem. Usually, the autochem has a significantly smaller grain size and can be separated by the hydrometer method described previously. The autochem in shales refers to diagenetically generated or biologically derived clays. Shales can also contain silica particles as either detrital quartz grains or fossil precipitates (e.g., sponge spicules and radiolaria). Moreover, they can contain varying amounts of carbonate, either as pure calcium carbonate ($CaCO_3$) from foraminiferal tests (shells) or mixed with autochem clay minerals, often originating from algae and bioturbation. Several tools are used to establish the mineral composition of the shale including x-ray crystallography, x-ray fluorescence, electron microprobe, calibrated well logs (to a certain extent), and other techniques discussed later in this chapter. The overall composition of resource shales can be plotted on a ternary diagram (fig. 2–3) and compared to other shale resource plays. Importantly, as figure 2–3 shows, there are no unconventional shale plays with a very high clay fraction.

Ternary plots are available for the composition of most of the world's shales. Several of the major North America shales are shown in figure 2–3. There is a wide range of compositions, and no unique composition provides a better candidate shale resource play. Each has its own benefits and deterrents. The most important aspects of studying analogue basins are to determine the differences and similarities between your current study area and those of other basins and to establish the most appropriate tools to define the specific problem in your area of study.

The higher the clay content is, the more ductile the shale will be—and therefore more difficult to frac. The presence of carbonates has significant impact on diagenesis and compaction of the shale. Understanding the composition of the shale is crucial in unraveling basin history, investigating overpressure, and designing proper completion techniques. New classification of shales are being proposed to combine aspects of composition. Terms like *siliceous-mudstones* and *mixed clay poor mudstones* have been proposed by Donovan (2017).

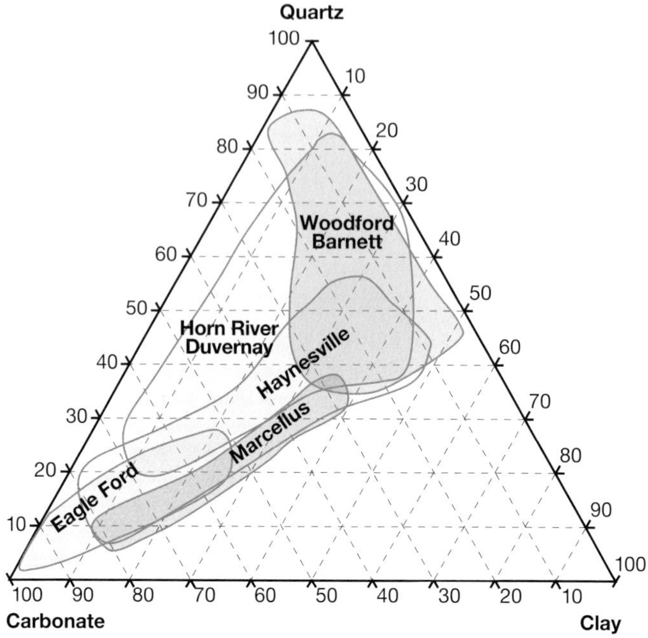

Fig. 2–3. Ternary diagram of the composition of common shale resource plays in the United States (Based in part on data from Colorado School of Mines)

There are two definitions for brittleness of sediments. The first is based on mineralogy and the second on rock physics parameters. We will discuss the second definition in chapter 6. The mineralogy definition of brittleness, or *B*% (Mavko et al., 2009), is as follows:

$$B\% = \frac{Q + \text{calcite}}{Q + \text{calcite} + \text{carbon} + \text{clay}} \tag{2.3}$$

where *Q* is silica or quartz, hence the importance of the mineralogical analysis. The term brittleness should be used as a relative term within the area of study. Two similar composition shales from different regions may fracture differently even if the B% is the same.

The most common clay minerals are kaolinite, chlorite, smectite, illite, and mixed-layer illite/smectite. Of these four minerals, only the smectite group is considered an expanding clay under specific conditions and plays a key role in affecting reservoir conditions in the subsurface. The genesis of all the clay minerals that may be present must be considered to understand shales. Kaolinite is more abundant near continental source areas because it is primarily derived from weathered K-feldspars. Allochem types (detrital-sourced clay minerals) such as chlorite, smectite, and illite are commonly associated with either a lower-grade metamorphic provenance or weathering of igneous, metamorphic, or older sedimentary terrains. The higher temperature ranges of the provenance rocks allow formation of the complex phyllosilicates. The clay minerals can be transported into different environments of deposition. Clay minerals are common in deltaic and coastal sediments and less common in carbonate environments since most organisms cannot thrive in muddy environments (except where sourced from carbonate pellet mudstones and bioturbation). Because smectite is not entirely dependent on feldspar weathering, its distribution is widespread into slope and basinal environments of deposition. Clay minerals also form as autochem or diagenetic products in situ. Both illite and chlorite—and, to a lesser degree, kaolinite—can form as authigenic minerals. Only autochem illite is reviewed in the present discussion because it is very common and has a major impact on reservoirs (Bettison-Varga and Mackinnon, 1997). Note, however, that the kaolinite family of minerals also form diagenetically and can be derived from smectite (Buhmann and Grubb, 1991). Before we can describe crystals and clay minerals, their basic structure requires review.

Basic crystallography and its application to clay minerals

Crystal faces are described in three-dimensional space. A coordinate system is used to describe the attitude of each crystal face. Miller, in the 1830s, was first to catalogue crystals, and the system bears his name, the *Miller index*. The indices are comprised of three numbers (*h*,*k*,*l*), with (000) assigned to the origin; 111 would correspond to the complementary crystal face diagonal across a cube or isometric crystal such as halite (fig. 2–4). In other words, the face on the cube is one unit cell away from the origin in the *h*,*k*,*l* direction. For clay minerals, the most important face is 001. It represents the dominant plane of cleavage for clays (i.e., flat plane).

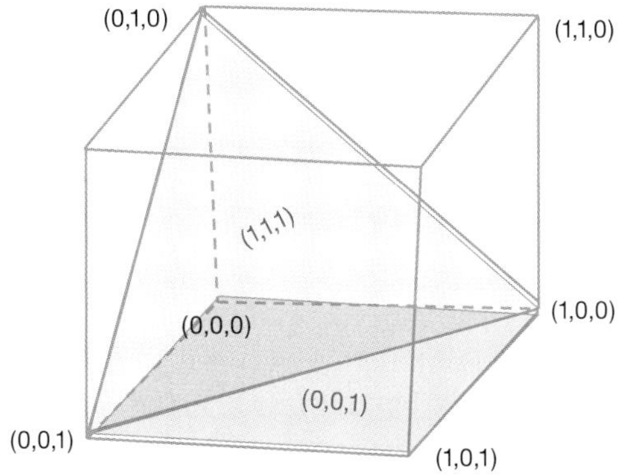

Fig. 2–4. Basic cube for Miller index classification showing crystal planes (111) and (001)

STRUCTURE OF CLAY MINERALS

Clay minerals belong to a family of silicates known as phyllosilicates. These are thinly layered crystals exemplified by micas. The mica minerals also elaborate on several of the physical attributes of phyllosilicates. They readily cleave along a specific plane (001) (fig. 2–4), suggesting the presence of space between the laminae. Shales behave in a similar manner, and the degree of cleavage plays a role in their fissility. The basic structure of clay minerals is a series of ring structure each comprised of six silicon dioxide (SiO_2) tetrahedrons (t) underlain by a layered octahedral complex (o) of $AlMg(OH)_2$. The simplest clay mineral is kaolinite, $AlSi_3O_5(OH)_4$. Note the similarity of the composition with its parent mineral orthoclase, $KAlSi_3O_8$. In an arkosic sand source, as orthoclase alters to kaolinite, potassium ions are liberated. This will become important when studying the alteration of smectite. The octahedral composition can change and as a consequence have different chemical and physical properties. In dioctahedral smectite all octahedral cations are comprised trivalent aluminum and iron. In trioctahedral smectite the dominant octahedral cations are bivalent iron and magnesium. The effects are described in chapter 9.

In the case of kaolinite, the crystal structure is composed of one set of *t-o* layers overlain by a similar set. A small gap is maintained between the two lowers that is the result of dominant oxygen atoms facing each other and their electrostatic repulsion. Even though the octahedral external layer is a hydroxide (OH) structure, the oxygen dominates owing to its large ionic radius. However, the basic charges are opposite between the facing layers; hence, the structure has limited expandability. We will elaborate on this topic in chapter 8.

The distance between the top of the upper tetrahedral layer and the top of next lower tetrahedral layer defines the crystal unit cell size. In kaolinite, that distance is typically close to 7 angstroms (Å; 1 Å = 0.1 nm), and in x-ray crystallography, it is referred to as the *d* spacing. The density of kaolinite is 2.62 g/cm^3. Note that this very close to the density of sandstone (2.65 g/cm^3). The same crystal structure does not apply to smectite. Its crystal structure is a *t-o-t* couplet (fig. 2–5). In other words, a similar *t-o-t* layer underlies the upper *t-o-t* layer. The two opposing faces of *t-o-t* define the crystal lattice of a smectite mineral, and the *d* spacing of the crystal lattice is the distance from the top *t-o-t* to the top of the next *t-o-t* layer. The *d* spacing for smectite is 14–15 Å (fig. 2–5). In addition, because the two opposing layers have the same electric charge, they repel each other (recall that electrostatic forces can either attract or repel). As a consequence, the crystal can readily expand along the interface. The binding agents in the crystal lattice are cations, water, organic material, or a combination thereof.

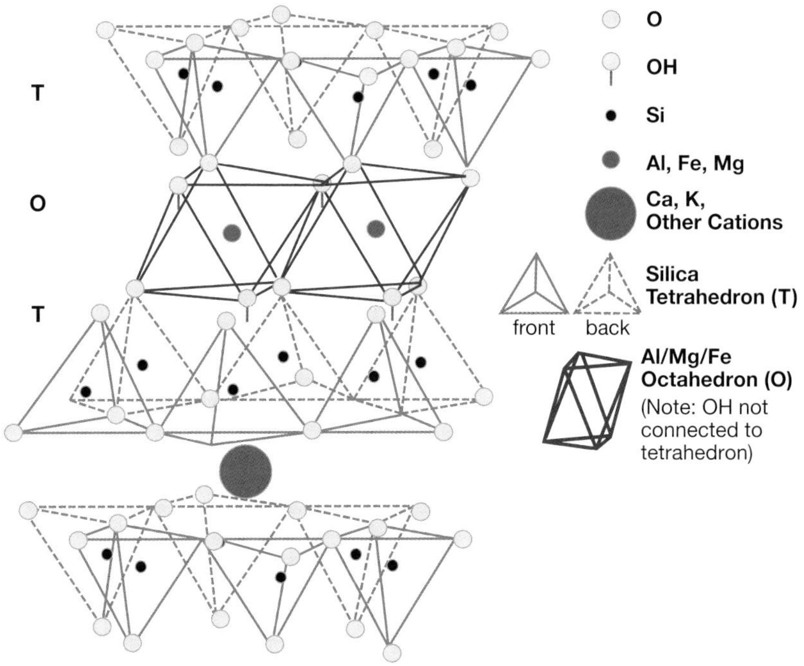

Fig. 2–5. Crystal structure of montmorillonite, a smectite showing typical *t-o-t* layers

An example of how the expanded smectite layer functions is carbonless paper. Ink is microencapsulated in a protein or other agent and inserted into an expanded smectite slurry. By creation of the proper cation exchange environment, the encapsulated ink droplet enters the smectite interstitial space, and a thin coating of the slurry is placed on the back of the paper and dried. Once pressure is applied, the smectite crystal collapses, and the ink is expelled. In medicine, clays are also used for their adsorbing effects resulting from van der Waals forces, electrostatic attraction. An example is kaopectate, which desorbs kaolinite; once digested, it is able to adsorb viral agents. Kaolinite is used because of its low adsorption coefficient. When a stronger adsorption agent is needed, the clay mineral either smectite or attapulgite (closely related to smectite) are used. The medical literature on using dioctahedral or trioctahedral smectite is extensive due to the chemical differences. The density of smectite is 2.01 g/cm^3 and reflects its open structure. Differentiating smectite and illite is discussed in the section describing their diagenesis and x-ray response. As will be seen in the discussion on seismic, the reflection coefficient (contrast) of a signal is strongly dependent on density. A kaolinite-rich shale overlying a sandstone will yield a low reflection coefficient and differ from a smectite-dominant shale. For a more in-depth discussion of clay mineralogy, refer to the work of Eslinger and Pevear (1988). Also note that an illite rich shale will yield a harder shale than a smectite rich shale.

Kaolinite and chlorite phyllosilicates are often present in shales, but both play a minor role in shale resource assessment. Kaolinite is a simple *t-o* layer tightly packed and does not expand. Chlorite has a *t-o-t* arrangement and is differentiated from the smectite group by the presence of a gibbsite ($Al(OH)_3$), brucite ($Mg(OH)_2$), or goethite (FeO(OH)) layer between the *t-o-t* layers, resulting in closure of the crystal lattice and therefore having no expansion.

One important factor to remember about the clay minerals is that they are not charge neutral. Water and hydrocarbons attach to quartz grains owing to surface tension. In clays and organic material, they are attached by electrostatic forces. This topic will be reviewed in chapter 8. Water in shales is either free water or bound water.

Diagenesis of smectite and illite

Under specific temperature conditions, smectite converts to illite. Illite has the same basic structure as smectite except that in the illite crystal the interstitial space between the *t-o-t* layers is occupied by a large potassium cation. This results in closure of the layers and tightening of the structure. A second very important effect of the alteration of smectite to illite is the water expelled during the conversion (Bjørlykke, 1998). Several important aspects describe the expelled water:

- The water is fresh. In wells where the illitization is reaching its peak, water salinities may be as low as 15,000 ppm chlorides versus an ambient salinity of the brine often as high as 60,000 ppm chlorides. Note that freshwater is a leaching agent, particularly if carbonates and feldspars are present.
- Because the shale is its own seal, the excess water creates overpressure. The relationship between illitization and overpressure is well documented (Jeans, 1994).
- The illitization process occurs at the same temperature as early hydrocarbon generation. The carboxylic acids produced by the cracking of the kerogen combined with the generated CO_2 (Surdam et al., 1984) are readily dissolved in the fresher water and aid in the leaching process of fractures and pores, thus becoming a pathway generator for the soon-to-follow hydrocarbons. Carboxylic acids will be reviewed in chapter 8.
- The density increase in the illite dominant shale also affects its interval velocity. This has major implications in the seismic modeling of basins undergoing illitization (Hart et al., 2013).

- The excess water results in unloading of the shale and a consequent decrease in interval velocity.

Clay diagenesis is an involved process. The focus of the present chapter is to define those elements of clay diagenesis that are critical to the shale resource play. None is more important than the alteration of smectite to illite. In brief, the process entails the replacement of most of the cations between the *t-o-t* layers of smectite with a potassium cation by a process known as *cation exchange*. The resulting *d* space for illite is 10 Å, and the density can increase up to 2.75 g/cm^3. This difference is of great importance in estimating the porosity in the shales, seismic response, and several other important physical phenomena described later in this chapter. The potassium cation is the perfect fit for the intercrystalline lattice space since the ionic radius adjusts and fits perfectly in the opening of the tetrahedral ring at the base of the smectite lattice. As a result, the entire clay structure collapses around the potassium ion. This results in a closure of the layers and a tightening of the structure, which becomes more compact and loses its capacity to expand. The greatest degree of compaction of shales occurs during illitization. Beyond how the shale compacts, the following questions regarding smectite-to-illite conversion require investigation:

1. What is the temperature relationship for the reaction?
2. How are pressure requirements and pressure consequences associated with the reaction?
3. What are the effects of illitization on gas adsorption?
4. What is the potential impact on the seal capacity of the shale?
5. What are potential sources for the potassium in the system and consequences if insufficient potassium is available?

As stated previously, the smectite-to-illite reaction can be treated as an equilibrium equation because, at the appropriate temperature and pressure, the reaction takes place very quickly. The actual reaction has been described by Rask et al. (1997) as either a solid-state reaction or a solution-precipitation reaction (fig. 2–6). The kinetic constraints of the reaction have been discussed in detail by Berger et al. (1997). The first reaction does not produce any additional water and requires more potassium than the latter. Because dewatering of the shales is a common phenomenon of illitization, only the latter reaction is discussed. However, both are likely to occur. In the solution-precipitation example, the equation described by the authors is as follows:

$$\begin{gathered} 1.57\ KNaCa_2Mg_4Fe_4Al_{14}Si_{38}O_{100}(OH)_{20} + 10\ H_2O + 3.93\ K \rightarrow \\ K_{5.5}Mg_2Fe_{1.5}Al_{22}Si_{35}O_{100}(OH)_{20} + 1.57\ Na + 3.14\ Ca + 4.28\ Mg + \\ 4.78\ Fe + 24.66\ Si + 570 + 11.4\ OH + 15.7\ H_2O \end{gathered} \tag{2.4}$$

Equation (2.4) is a common sample formula for the conversion of smectite to illite (reproduced with the kind permission of the Clay Mineral Society; Rask et al., 1997) and shows the presence of an additional 15.7 moles of water and 24.66 moles of free silica.

Note that this reaction produces 36% less illite than the original smectite by volume. It also frees calcium, sodium, magnesium, silica, oxygen, hydroxyl radicals, and water. The reaction is inefficient because a portion of the smectite will convert to kaolinite and chlorite. In equation (2.4), for every 10 moles of water used in the alteration of smectite, 15.7 moles of water are produced during illitization. Freshwater is a significant solvent and can result in diagenesis of minerals and organic matter (Bjørlykke, 1998). The activity coefficient of the aqueous solution is a function of the ionic strength of the solution and follows the basic laws of thermodynamics. The water chemistry available during illitization will be discussed again in conjunction with organic geochemistry. During illitization, if there are any carbonates near the shale body entering the illitization phase, free magnesium ions will result in dolomitization of the limestones (McHargue and Price, 1982). The generation of dolomite in the shale increases the brittleness of the shale and is therefore important to consider. This is the case in the southwestern Williston Basin, where dolomitization within the Bakken Formation plays an important role.

The dewatering of the clay is of great importance because if good seals are present above the shale, the formation becomes overpressured (Jeans, 1994). In fact, the preservation of the overpressure itself requires improvement of the seal capacity. In addition, if water is allowed to migrate out of the shale, significant compaction may occur. Hydrous pyrolysis in a potassium chloride (KCl) solution has been used in laboratories to generate illite from smectite (Roaldset et al., 1998). Therefore, the temperature conditions for the reaction are well understood. The purpose of the experiment is to establish the activation energy of the reaction and evaluate the chemistry of the reaction. Roberson and Lahann (1981) have pointed out that once the physical and chemical conditions for illitization have been met, the reaction rate is very fast and allows for a simple equilibrium equation to occur. The basic principle of the experiment is to provide a proper hydrous solution for the reaction to take place. As the sample is heating to a specific stage, a portion is collected and analyzed by x-ray diffraction (XRD). The experiment is repeated until there is no longer any additional transformation. The results of the experiment have shown that the highest activation energy for the smectite-to-illite conversion occurs between 80°C and 100°C. This is coincident with the temperature required in order to generate hydrocarbons and several other diagenetic reactions. Therefore, it is of great importance to determine when this temperature was reached and estimate what the reservoir conditions were at that time.

Furthermore, by use of transmission electron microscopy (TEM), the actual alteration process can be imaged (fig. 2–6) and noted to be a structural change at the atomic level and not only a dissolution reprecipitation process. The process is massive and not layered. Interbedded smectite has been observed where the illitization is incomplete and provides what is known as mixed layer.

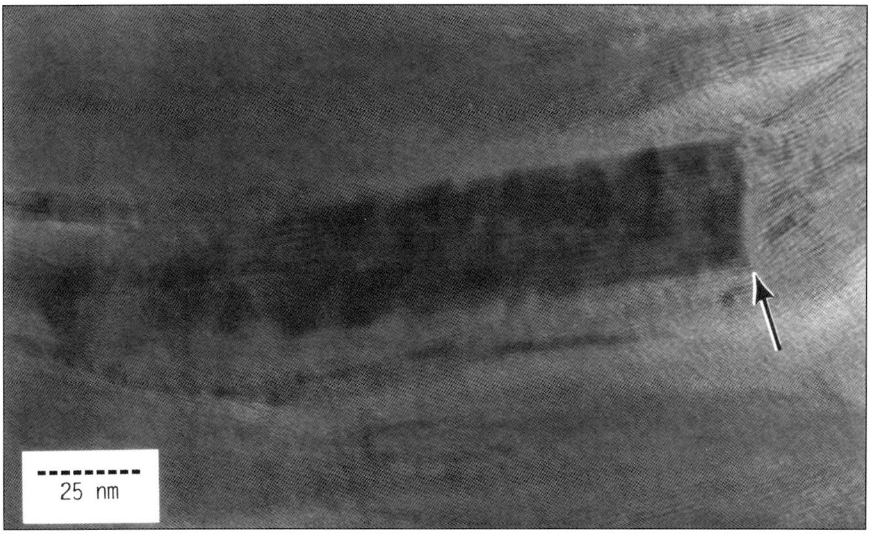

Fig. 2–6. Illite/smectite TEM slice of a Gulf of Mexico mudstone showing a structural change at the atomic level. Photomicrograph shows illite (dark) in the center and smectite (light) on the right (Reproduced with kind permission of the Clay Mineral Society; Rask et al., 1997)

Whitney's (1990) summary of the chemical reaction has led to an improved understanding of the role of the activation energy for the reaction (Shi, 2009). The literature on the subject is extensive because the reaction can be simulated under laboratory conditions and has major implications on reservoir conditions. The temperature relationship in the illitization process is likely one of the most important aspects related to the shale resource play owing to the close relationship to oil generation. Interestingly, potassium is also trapped in organic plant material. This is the reason for the addition of potash-rich fertilizer to soils. Since plant material became a dominant organic component in the ecosystem during the Late Paleozoic, it is possible that its absence during Early Paleozoic reduced amounts of potassium in the environment and resulted in the apparently slower conversion rates observed from smectite to illite in contemporaneous sediments.

The chemical reaction of illitization of smectites is also pH dependent. In young siliciclastic sediments, this is usually not a major problem. However, if carbonates are involved in solution, they negatively affect the transformation. Because carbonate sediments tend to have a lower concentration of potassium ions needed for the reaction to take place and act as buffer solutions in the chemical reaction, the smectite-to-illite alteration is slowed. Moreover, the solution pH also affects silica precipitation. As indicted in equation (2.3), for every 1.57 moles of smectite available in solution, 24.66 moles of Si are liberated. At temperature of 100–150°C and pH 6.0–7.0, silica precipitates as a cementing agent. This indurates the sediments and has an impact on the brittleness of the shale. Silica precipitation is discussed further in chapter 3.

Shale compaction is very well documented, and Rieke and Chilingarian (1974) dedicated an entire book to the topic. To understand the mechanics of shale compaction requires knowledge of pore pressure, water expulsion due to diagenesis and overburden stress. These topics will be expanded in chapter 8. For this section, it is necessary simply to provide a range. It is apparent that if compaction is evident, the consequence is a reduction in void space. This reduction is achieved by decreasing the pore size and the number of pores. The compaction may be as high as 25% to 30%. For this reason, shales that have undergone significant compaction may be less viable as a resource play. Compaction history is important in determining the detailed burial history of the sediment, discussed in chapter 7.

Overpressure is closely related to compaction and is a complex phenomenon resulting from several causes. Tectonics certainly plays a role. However, in relatively undeformed strata, as is common in shale resource plays, it appears that illitization plays a major role in the development of overpressure. Various authors have reviewed the relationship between illitization and overpressure. Chopra and Huffman (2006) and O'Connor et al. (2011) have analyzed various shales in different basins (fig. 2–7) and compared the density of the shales against velocity and depth. The unloading effect is due in part to fluid expansion during overpressure. Note, however, that local conditions may influence the required conditions for the overpressure to develop. The light gray line in figure 2–8 corresponds to the Gulf of Mexico; the dark gray line area, although not explicitly identified by the original authors, is most likely from the West Canada Basin. The close association between illitization, overpressure, and oil generation is evident from figure 2–8. The geochemical aspects of the generation of hydrocarbons will be reviewed in chapter 7. The important point is that the effects need to be calibrated on the basis of local conditions. Observed differences may be due to chemical composition differences between the various shales.

The smectite-to-illite conversion also affects the surface area on the clay mineral. Surface area is directly proportional to the adsorption capacity of the clay. Because surface area includes both internal and external surface area, as the interstitial space collapses, the internal surface is significantly reduced after illitization. As discussed previously, smectite crystals have interlattice space, as well as large external surface area (fig. 2–9 and table 2–1). Crystals, like kaolinite and chlorite, have only external surface area. The method used to determine the surface area is precisely based on its adsorption capacity and several variations of the analysis. Most sample laboratories have proprietary techniques to carry out these studies. Often complex organic compounds are used.

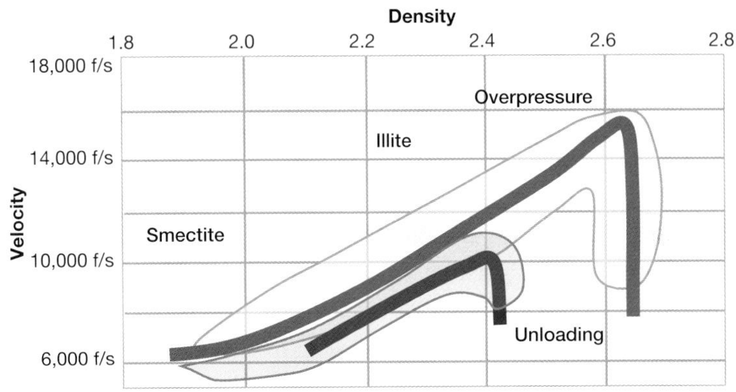

Fig. 2–7. Two example plots of velocity (ft/s) versus density (g/cm^3) showing the effect of illitization on possible variations in composition. The light gray line data are from O'Connor et al. (2011), and the dark gray line data are from Chopra and Huffman (2006). The same patterns are shown but under different conditions.

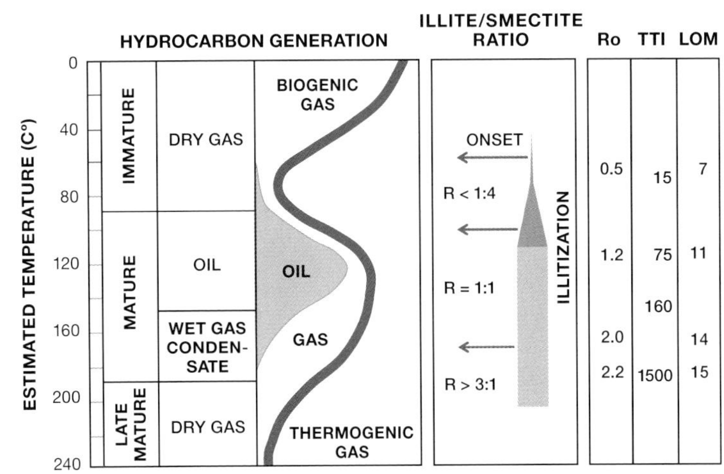

Fig. 2–8. Illitization related to the timing of hydrocarbon generation. Other maturation indicators, discussed in later chapters, include time-temperature index (TTI), vitrinite reflectance (R_o), and level of organic maturity (LOM). Dominant illitization occurs at 80°C–100°C. (Drawing adapted from Waples, 1980; Tissot and Welte, 1984; Burtner and Warner, 1986; and Pollastro, 1993)

Table 2–1. Surface area values in clays

	Surface Area (m^2/g)	
Clay	**Internal**	**External**
Smectite	750	50
Illite	5	25
Kaolinite	0	15
Chlorite	0	15

For cations to be trapped in the interlattice space, the available separation must be in excess of 1.5 Å (fig. 2–9), which is equal to the minimum molecular radii required in order to trap a water or methane molecule (Kammeyer and Whitman, 1972). The adsorbed water in smectite is one or two molecules thick. Because water is a polar molecule, it enhances the cation-trapping capacity within the clay interstitial space. Adsorption also plays a role with capillary pressure and will be discussed in detail in chapter 8.

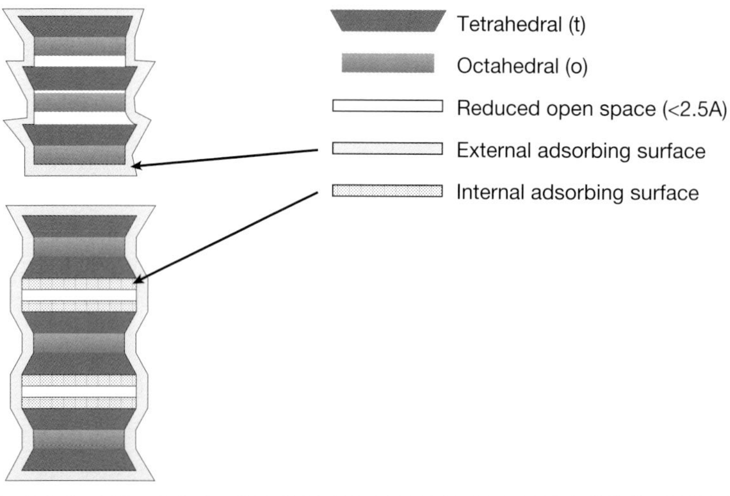

Fig. 2–9. Identification of external and interior surface area where cations can be adsorbed. The upper diagram is for a kaolinite structure, and the lower is for smectite.

In addition, trapping potassium cations in the illite matrix allows determination of the age of illitization. The technique studies the relationship between the parent radioactive isotope ^{40}K and the daughter product ^{40}Ar (Faul, 1966). The relationship is an exponential equation. The basic principle is based on the half-life of ^{40}K, which is 1.25 billion years. From the time the ^{40}K isotope is trapped in the illite until it is analyzed, the ratio of the parent to the daughter product, given the known half-life, will provide information on the timing of illitization. This method has been used from Precambrian rocks to recent sediments with less than 9 Ma margin of error for the oldest sediments and 1 Ma for young sediments. For the purpose of the burial history of the sediment, the accuracy is sufficient to be applied to the model. Burial history is described in chapter 7.

The objective of the burial history is to establish the timing for oil and gas generation in the hydrocarbon source rock shale. Where gas reservoir leakage is a concern, it may be important to explore in areas where maturation was relatively recent. As noted previously, allochem (detrital) illites have larger grain size than do their autochem counterparts. Therefore, before the age is determined, the sample is separated by grain size using the hydrometer method or fine sieves. It is evident from the two discussions that the presence of significant smectite in the shale resource play is of paramount importance to the storage capacity of adsorbed gas, especially when it is combined with the organic material present in the system. Illite crystallinity is the focus of further research to assist in developing a maturity index based on its structure. The Kubler index for both air-dried and glycolated illite was developed at the University of Neuchâtel, Switzerland, and its results have been compared with other maturation indices (Jaboyedoff et al., 2001). Analysis of clays is discussed in greater detail later in this chapter.

The ultimate goal of all analytical techniques applied to the study of any shale resource is to accurately describe its mineralogy, analyze the texture, determine the size of the particles involved, evaluate the pore space, and calculate the conductivity between the pores (permeability). In addition, once the sediment has been classified, its chemical properties should be used to predict potential drilling problems and how the hydrocarbons might respond to the various reactions that take place in the shale during diagenesis and drilling. A common tool used for the purpose of analyzing the mineralogy is x-ray diffraction. Other tools will also be presented to improve the determination of the mineralogy and pore space.

X-RAY DIFFRACTION

X-ray diffraction (XRD) is among the most accurate determinants of clay minerals. The clay sample is smeared on a glass slide before running the sample. XRD works on the principle that an x-ray beam is focused on a crystal such that, at a critical incidence angle, the beam focused on the dominant crystal plane will generate a high-intensity reflection. As the x-ray beam is reflected from parallel beds, the constructive interference resulting from the summation of reflectors and the small phase change of the x-ray wave associated with the distance between the parallel layers can be *deconvolved* to establish the distance between the layers, which yields the *d* spacing. A combined waveform signal results from either constructive or destructive interference. Because the source of signal has a constant wavelength, the lag time between the two (δ) can be deconvolved. The peaks are a function of the wavelength (λ):

$$\delta = n\lambda \tag{2.5}$$

where n is an integer when the waves are in phase and there is constructive interference. The angle 2θ is obtained from the XRD tool as it rotates the sample.

According to Snell's law, the incidence angle equals the reflection angle. Thus, in figure 2–10, angle *ABF* = *CBE*, *CBE* = *EBD*, and *AOB* = *CBE*. Therefore, angle *CBD* is 2θ. Line *AD* represents the direct ray path, and line *BC* represents the reflected ray path. The relationship between the two is the 2θ angle. The path length between the two waves that interfere is *ABC*. Therefore, δ = *ABC* or 2*AB*. From the given construction, the d spacing is line *OB*. Therefore, line $AB = d \sin\theta$, and according to Bragg's law,

$$n\lambda = 2d \sin\theta = \text{line } ABC \tag{2.6}$$

Because the source wavelength is known and an integer of 1 is used in the analysis, a direct relationship can be established between the 2θ angle and the d spacing (figs. 2–11 and 2–12).

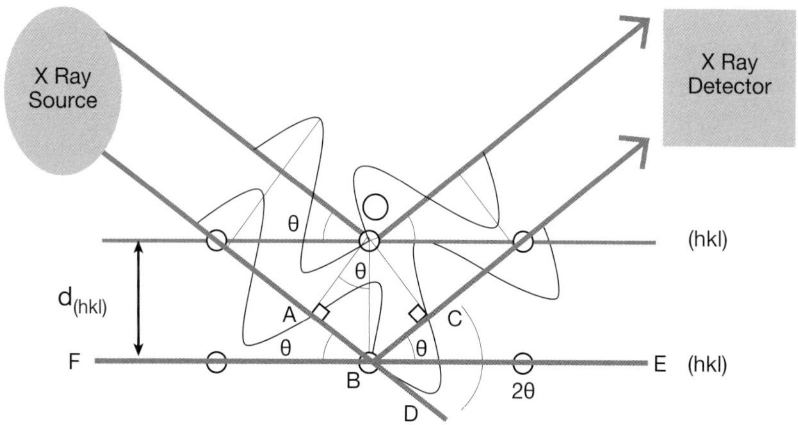

Fig. 2–10. Schematic of the x-ray path reflecting off two parallel crystal planes to define the mineral *d* spacing (line *OB*).

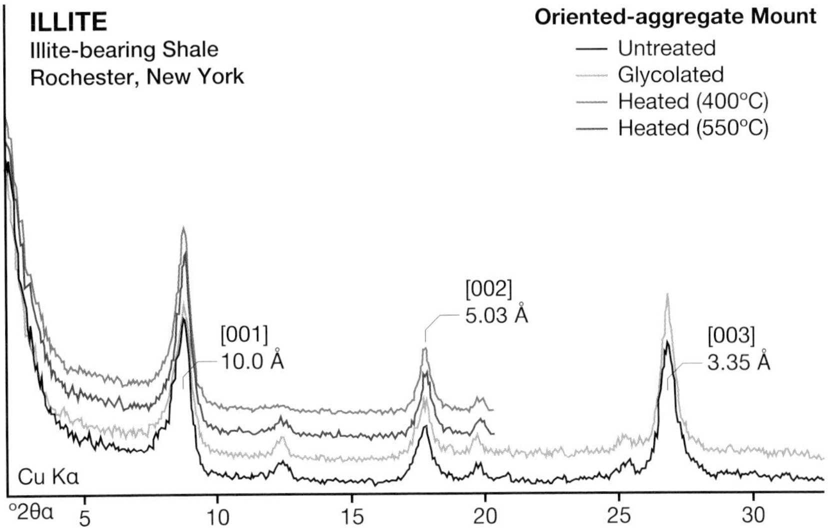

Fig. 2-11. Along the *x*-axis are the angle 2θ values, and the *y*-axis gives the scintillation counts, in counts per second (cps) at the specified angle. Note the reference to the Miller index. There is also no expansion in the illite clay sample (discussed later in this chapter). (Reproduced with permission from U.S. Geological Survey open-file report; Poppe et al., 2001)

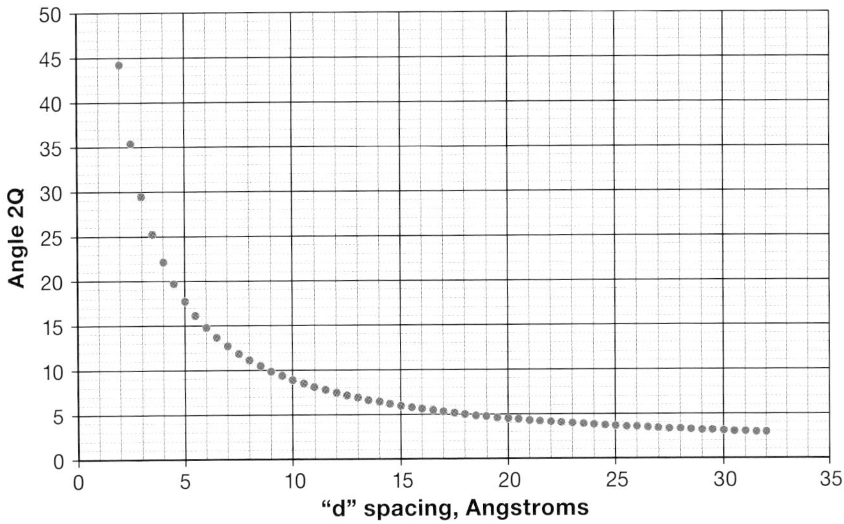

Fig. 2-12. 2θ vs. *d* spacing, in angstrom, of clay minerals, assuming a Cu Kα source with a wavelength of 0.15418 nm (Data from U.S. Geological Survey open-file report 01-041; Poppe et al., 2001)

The plot in figure 2–12 provides a direct conversion of 2θ angle to *d* spacing assuming a copper (Cu) Kα source with a wavelength of 0.15418 nm (1.5418 Å). The calibration tests for the XRD results should be requested in order to ensure the validity of the results. Remember that normally shales are smeared on the slide to ensure the proper orientation of the sample.

The resulting plot of angle 2θ versus cps (counts per second on the scintillation counter) can be applied to the graph (fig. 2–12) and used to convert 2θ to *d* spacing in angstroms. (For simplicity, usually only the 2θ angle is used in publications.) The area under the peak (fig. 2–13) of the (001) plane is an indication of relative abundance of the mineral. If the area under the illite (001) were twice as large as the area under the smectite (001), then the illite-to-smectite ratio would be 2:1 (Norrish and Taylor, 1962). Note that there is often a secondary peak at 2θ of 15.5–16 that corresponds to the mixed layer of smectite/illite; when present, this secondary peak should also be considered in the final ratio.

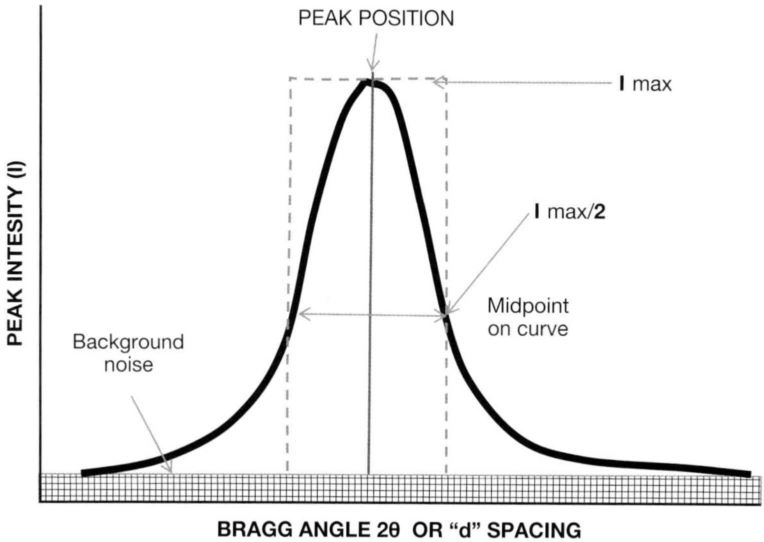

Fig. 2–13. Analysis of the XRD curve to determine volume ratio and crystal size. The same curve analysis technique is used for gas chromatography (see chap. 7).

The peak width has been used as a guide to determine crystal size since the early 1900s. The relationship, known as the *Scherrer equation* (Drits et al., 1997), is as follows:

$$B(2\theta) = \frac{K\lambda}{L\cos\theta} \tag{2.7}$$

where $B(2\theta)$ is the width along the 2θ axis, L is the crystal size, λ is the source wavelength, and K is a constant for the specific mineral shape. This constant needs to be determined for the minerals under study and for most cases approximates $K = 1$. Equation (2.7) serves as a guide and should be compared to SEM studies.

A second analysis that is performed on the clay sample is to soak the sample in ethylene glycol and rerun the XRD (fig. 2–14). The new sample is referred to as *glycolated* or *treated*. The illite peak does not change (fig. 2–11), but the 2θ on the smectite peak is skewed to the left, indicating that the smectite lattice has expanded. The lower the 2θ is, the higher will be the *d* spacing. Only smectite expands its cell size.

The lower 2θ indicates a larger *d* spacing in figure 2–14. The heated sample shows a tendency toward the illite peak, as expected at 2θ of 9° (fig. 2–14). The identification of the individual minerals and mixtures are determined using a catalog compiled by the International Centre for Diffraction Data (ICDD). All readings are carried out after the tool has been calibrated with standard samples. Proper guidelines must be followed to resolve the complexities of the XRD analysis. The calculations are relatively simple (Poppe et al., 2001).

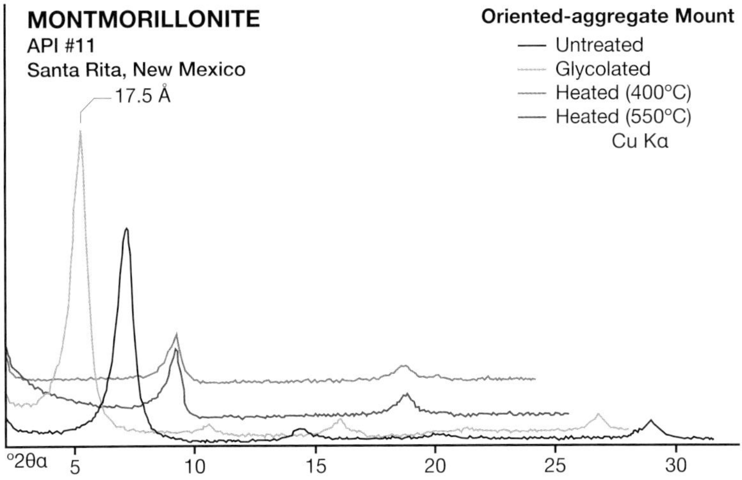

Fig. 2–14. Glycolated smectite clay sample. Note shift in smectite peak after glycol treatment. Also note that heated smectite collapses to an illite structure. (Reproduced with permission from U.S. Geological Survey open-file report; Poppe et al., 2001)

X-RAY FLUORESCENCE

X-ray fluorescence (XRF) analysis is commonly carried out on core samples. The theory behind XRF is intricate. The basic principle is to analyze the electron energy levels for individual atoms. Atomic nuclei have electron orbitals designated as the *K*, *L*, and *M* rings. When an x-ray beam is focused on a sample and collides with the *K* orbital, its energy level is slightly raised. As it returns to its stable state, it emits an x-ray at a particular wavelength that is unique to the affected element (fig. 2–15).

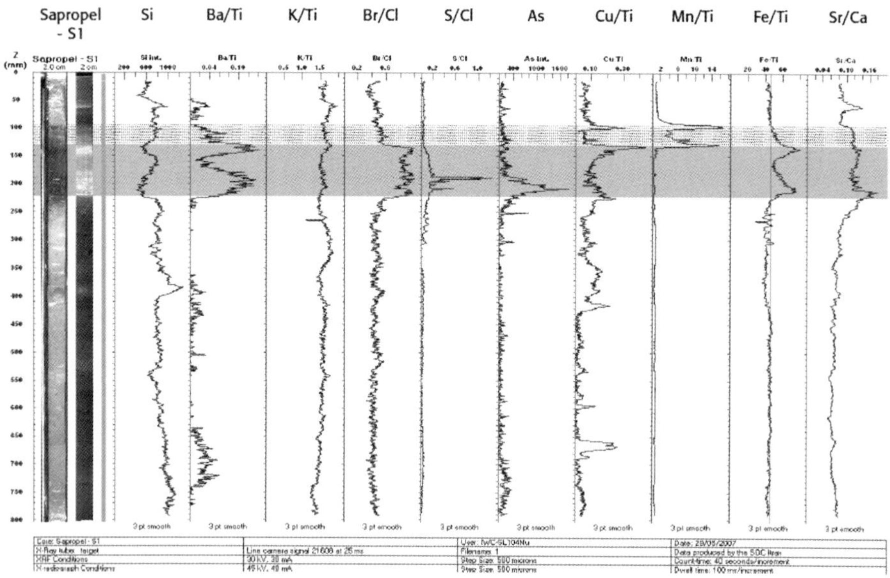

Fig. 2–15. Application of XRF for single-element analysis on a core sample (Reproduced with kind permission of I. Croudace; Croudace and Rothwell, 2010, 2015).

Because the energy required for the change in orbital is unique to the atom, the emitted fluorescence can indicate the element involved. The tool is a comparative measurement. Known certified references must always be used to calibrate the tool. It is very accurate, and today there are several portable XRF tools that are the size of a handheld electric drill. This miniaturization has led to a significant expansion for the use of the tool. For example, Croudace and Rothwell (2015) have applied the tool on a core (fig. 2–15). The tool is usually set to detect a specific element such as potassium content. As such, it readily demonstrates where the concentration increases and allows for an estimation of the depth at which illitization attains a maximum. The presence of other elements, such as chromium or titanium, can assist in defining the presence of volcanic ash in the sediment. Nayak and Singh (2007) have compared XRD, XRF, and Fourier transform infrared (FTIR) spectroscopy in differentiating components of the clay mineral. There are some issues when the tool is applied to cuttings (Chatellier et al., 2011). As will be discussed in chapter 8, the electrostatic charge on the smectite crystal varies as a function of the cation in the octahedral structure. It may be dioctahedral (Al,Fe^{+3}) or trioctahedral (Fe^{+2},Mg). Montmorillonite is generally diocatahedral.

CATION EXCHANGE CAPACITY

The same clay samples used in the previous studies can and should be analyzed to calculate their cation exchange capacity (CEC). When carried out in conjunction with each other, CEC and the previous analyses described reinforce one another. CEC analysis is routinely performed on all shales drilled because it is fundamental for the driller to understand how the drilling mud will respond to the shale under subsurface conditions (Burrafato and Miano, 1993).

Organic matter has the highest CEC values; hence, a humus-rich soil is the most effective medium from plant growth. The reason the organic matter has such a high CEC will be discussed further in chapter 7 (because it requires an introduction to organic chemistry). In brief, the trapping process is the same as that of clays. Once again, smectites far exceed other clay minerals in terms of CEC values owing to their structure and high expansion index. The oxygen compounds provide the strong negative charge adsorbing the cations and alkanes. CEC values of clay minerals (in mEq/100 g) are as follows (Drever, 1982):

- Smectites, 80–150
- Illite, 10–40
- Kaolinite, 1–10
- Chlorite, 1–10

A high CEC value in sample suggests high content of organic material or smectite (or both) in the sample. The common method used follows the American Petroleum Institute (API) protocol and involves titrating methylene blue into a treated specified sample. This is used to determine both the CEC and the surface area of the sample (Hang and Brindley, 1970). The CEC between the formation water and the drilling mud provides the response observed in the spontaneous potential curve logged in the well.

LANGMUIR ISOTHERM

Recall that CEC is a function of the surface area of material described previously. The process of trapping cations on a clay surface is adsorption. The adsorption index of shale resource play is fundamental for the storage and trapping of hydrocarbons—so much so that special studies are carried out to determine the adsorption coefficient known as the *Langmuir coefficient* of the shale.

The *Langmuir isotherm* can be directly related to the CEC. Importantly, the same Langmuir isotherm will be used to analyze adsorption of gas to organic matter (see chap. 7). Adsorption is very well understood and is the basis for charcoal filtering in water purification. The charcoal grains are desorbed. As fluids flow through the treated filter, contaminants are adsorbed until the filter reaches full capacity, at which point it is changed. Several universities have specialized research institutes dedicated to the topic, including the Langmuir Research Institute at Harvard University. There is even a journal solely dedicated to Langmuir issues published by the American Chemical Society.

The Langmuir isotherm is the volume of adsorbed gas in a shale. The method commonly used to determine the Langmuir isotherm applies varying pressures to the sample in the presence of gas. The sample is pulverized to ensure maximum surface area, then heated (with low heat) to remove any existing adsorbed gas. The process is known as *desorption* (fig. 2–16). If the sample contained any free gas, it would have been removed while it was pulverized. Methane is introduced into the pressure chamber containing the clean shale sample, and the pressure is increased while the temperature is maintained constant. The excess gas is removed and quantified. The difference is the adsorbed gas, measured in standard cubic feet of gas per ton of sample. The original free gas in the sample is obtained by immediately sealing the sample in wax when the core is taken, and subsequently the free gas in sample can be quantified in a laboratory. Note that the same process affects the sample during burial. As the methane begins to migrate through the shale and pressure is applied, there is a progressive increase in adsorbed gas until the sample is saturated.

In figure 2–16, the thick gray line is the Langmuir isotherm, the thick black line represents the maximum possible gas, and the thin line represents the amount of free gas. The free gas increases with pressure because any free gas in crystal lattice will be driven out by pressure. The core data can be plotted on a standard well log, and a Langmuir curve can be compared to other petrophysical parameters (Alexander et al., 2011). In relatively low-maturity reservoirs carbon isotope analysis of free gas and adsorbed gas should be the same. Otherwise they indicate migration of the free gas.

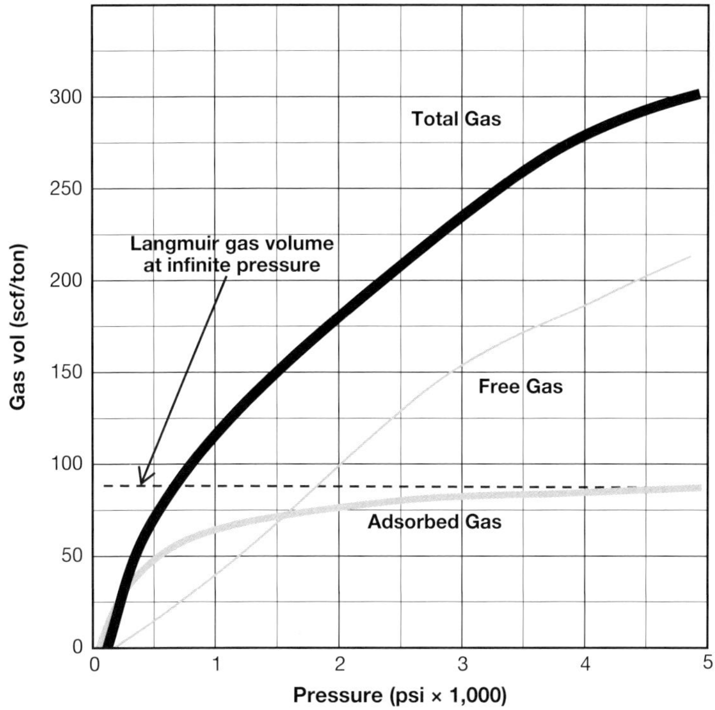

Fig. 2-16. Langmuir isotherm (Modified from Alexander et al., 2011)

DIFFERENTIAL THERMAL ANALYSIS

Differential thermal analysis (DTA) has been used for clay identification for over 60 years (Rowland, 1952). The theory behind this classification of the clays is that there are two chemical reactions commonly associated with heated clays, each of which has a different thermodynamic and kinetic reaction. The first is a relatively low-temperature endothermic reaction associated with dewatering. The second occurs at high temperatures and is an exothermic reaction that changes the chemical structure of the compound.

When a clay sample is progressively heated, endothermic and exothermic reactions are observed at unique temperatures for specific clays. Once again, miniaturization has allowed DTA to be applied in the field. Since 1947, portable DTA has been in use. Parker et al. (1956) demonstrated the usefulness of the DTA tool in differentiating varieties of smectites. With proper calibration, DTA can improve modeling of the illitization process.

DTA studies in clay minerals were initiated during the early part of the 20th century, with a major push to apply DTA to clay minerals by Grim in the 1940s (Grim, 1968). For many years, DTA was used as a descriptive tool; more recently, it has been combined with gravimetric studies to determine if the reactions were associated with weight loss. More modern techniques have significantly improved on the tools. An important contribution was compiled by Smykatz-Kloss and Warne (1991) and Yariv (2003). Their studies combined DTA and thermogravimetric analysis (TGA) with mass spectrometers and infrared spectrometers. When the three tools are combined they provide a means to differentiate the evolved gas generated in the heating process. TGA allows for the separation of weight-loss reactions and phase transition. Gas chromatography allows for an analysis of the compounds released from the adsorbed clay surface. DTA can be carried out in both an oxygen atmosphere and an inert gas atmosphere such as nitrogen. Further research is recommended to better understand variations in kerogens in a given basin and improve the definition of temperature-related critical reaction products.

SCANNING ELECTRON MICROSCOPY, TRANSMISSION ELECTRON MICROSCOPY, AND ELECTRON MICROPROBE ANALYSES

Scanning electron microscopy (SEM) and transmission electron microscopy (TEM) are both important tools in understanding shales. As noted previously, by definition, clay implies particle sizes less than 1/256 mm, equivalent to 4 micrometers (μm), or 4 microns. The size of clay particles is generally below optical microscopic resolution. However, SEM can resolve images down to 0.1 mm.

SEM focuses a beam of electrons (instead of a light beam in an optical microscope) onto the surface of a sample. The electron beam is reflected, collected, and transformed into an image (figs. 2–17 to 2–21). In the case of TEM, a beam of electrons is transmitted through an ultrathin sheet of the studied material, and the transmitted beam is then collected and analyzed (fig. 2–6). Both are widely applied in shale studies. The major difference between the two is that SEM allows for a three-dimensional view of the object under investigation. This allows the interpreter to assess the relationship of the various objects under the field of view, as well as the space between them. TEM represents only a two-dimensional perspective of the object under study. However, as shown in figure 2–6, TEM allows study of the manner in which the transformation of smectite to illite occurs.

Energy-dispersant x-ray spectroscopy (EDX or EDAX) is used in conjunction with SEM. A sample is bombarded by an electron beam, focused on a small target at a single point, causing the electrons in the specific atom to be elevated to a higher energy state. Once the beam is relaxed, the electron returns to its original energy state but in the process emits a unique x-ray wavelength signal that identifies the element (in a similar fashion as the XRF tool). This technique allows for a more detailed analysis of the internal chemistry of the object under investigation (figs. 2–17 and 2–18). SEM combined with EDAX has been applied to geological studies for over 50 years.

Bridging illite forms as authigenic illite (fig. 2–21). When present, it can significantly reduce permeability in fracture systems. SEM can also detect characteristics of the organic material in the sample. Organic material has the unique characteristic that its porosity increases with maturity, thereby creating additional storage space; this is most likely due to hydrocarbon expulsion (Löhr et al., 2015), which will be discussed in chapter 7. Loucks et al. (2009) have reported an excellent comparison between the size of the organic material nanopore and the size of a methane molecule (fig. 2–22). The efficiency of the nanopores as storage space for gas is evident. The SEM image may be enhanced using the ion beam milling (IBM) technique or the more advanced focused ion beam milling (FIB) described in the next section.

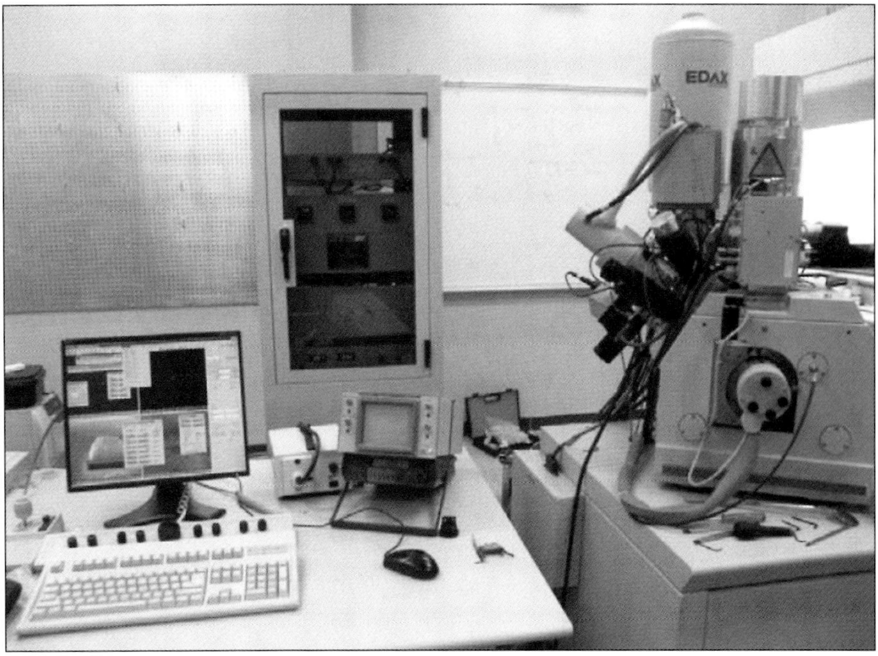

Fig. 2–17. SEM/EDAX unit at University of Houston nanotechnology laboratory, combined with focused ion beam (FIB) milling (Photograph published with permission from the University of Houston)

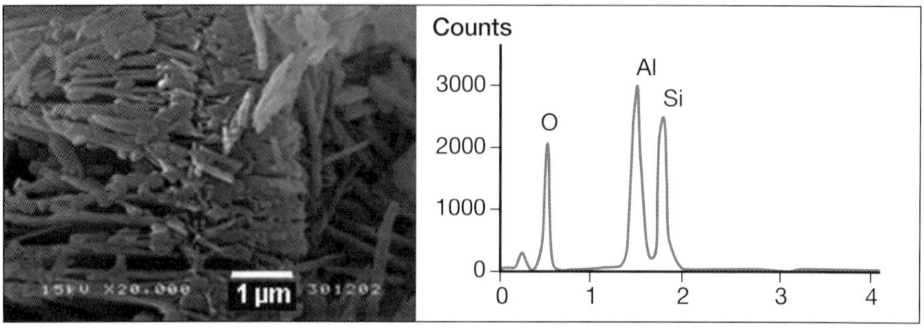

Fig. 2–18. SEM image of a kaolinite crystal with an EDX composition plot at a specific clay particle. The composition of kaolinite is $Al_2Si_2O_5(OH)_4$. (Reproduced with kind permission of the Clay Mineral Society; Worasith et al., 2011)

Fig. 2–19. Typical illite appearance on SEM image showing fibrous habitat (Reproduced with kind permission of Weatherford Inc.)

Fig. 2-20. Typical smectite appearance on SEM images. Note the authigenic quartz (faceted crystal) overgrowth into the pore space, as well as the large smooth plates and open texture. (Reproduced with the kind permission of Weatherford Inc.)

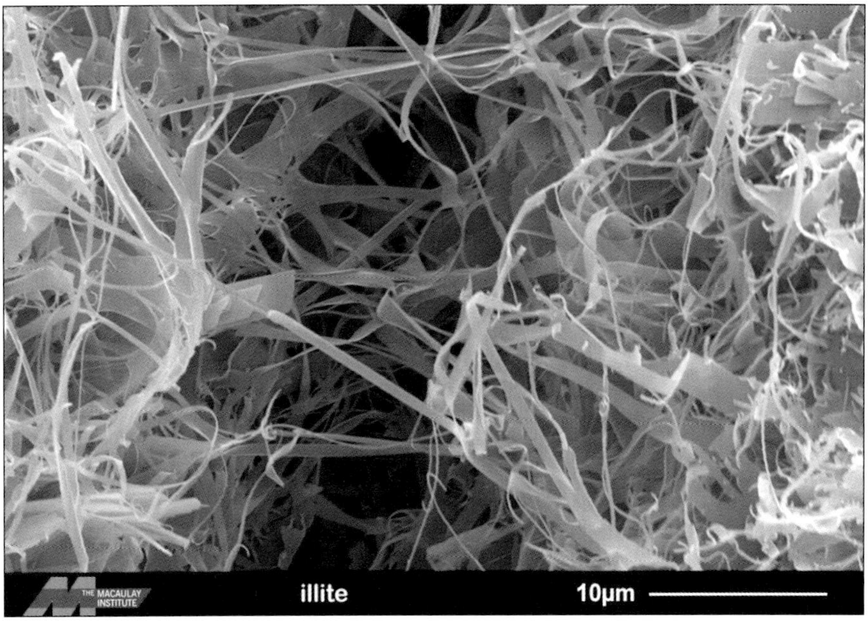

Fig. 2-21. Bridging illite, Rotliegendes sandstone, North Sea (Reproduced from the images of the Clay archive of the Mineralogical Society of Great Britain and Ireland and the Clay Mineral Society, www.minersoc.org/gallery.php?id=2; McHardy et al., 1982.)

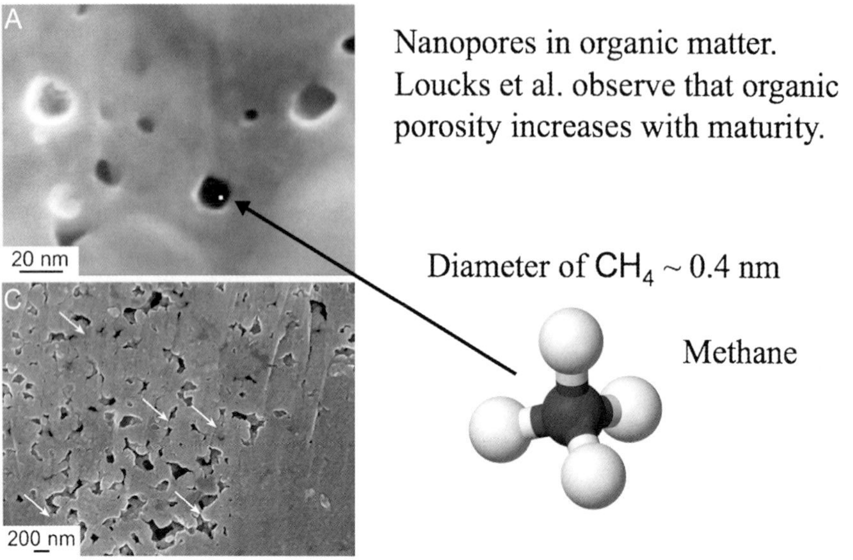

Fig. 2–22. SEM image relating nanopore and methane molecule sizes (Reproduced with kind permission of the Bureau of Economic Geology of Texas; Loucks et al., 2009)

Traditional SEM has been enhanced by an electron microprobe attachment (i.e., EDAX) that reads continuous data across a transect of the sample. On the basis of the formula of common minerals, the results from SEM/EDAX analyses can be readily translated into mineralogical compositions. This is the basis for QEMSCAN (fig. 2–23), which is currently in common use by many laboratories involved in shale assessment. Other similar techniques are available and may use different names so as not to infringe on patents.

Once again, the objective is to ensure proper classification of the shale and improve the petrophysical evaluation of the shale under investigation. As Sliwinski et al. (2010) have noted, QEMSCAN should be corroborated by selective XRD analyses. QEMSCAN's advantage is that it is automated and consistent.

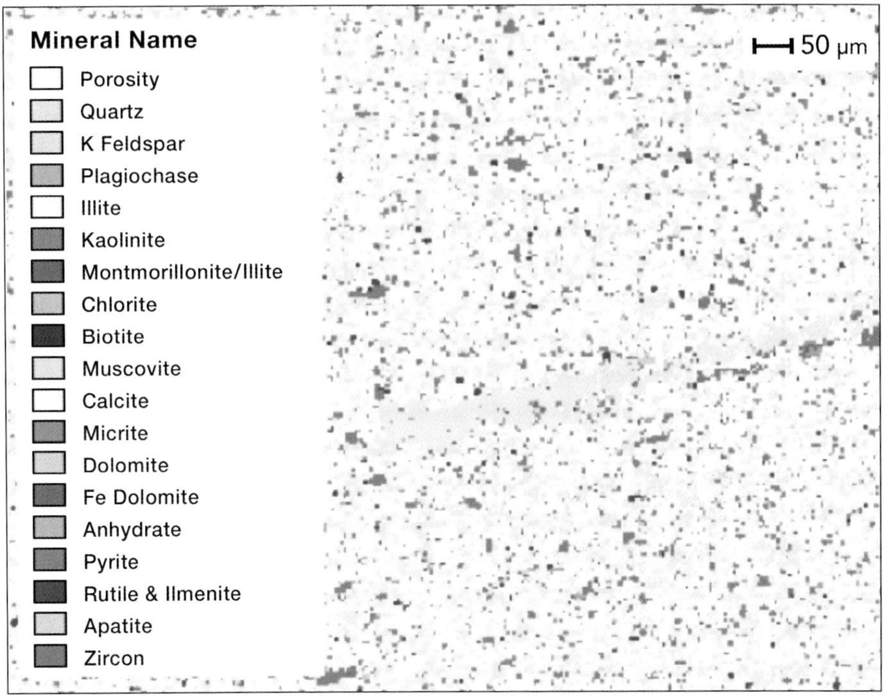

Fig. 2–23. QEMSCAN image shows color-coded mineral assemblage interpretation from the Cretaceous Colorado shale (Reproduced with kind permission of M. Power, SGS Canada Limited; Sliwinski et al., 2010)

ION BEAM MILLING WITH SEM AND FOCUSED ION BEAM

The technique of ion beam milling (IBM) with SEM (IBM-SEM)—or its more modern version, the focused ion beam (FIB)—has revolutionized the analysis and distribution of pore space in shales. A small, 5 mm cube can be analyzed using IBM-SEM; as the surface is milled by the ion beam, it is scanned as a SEM image to provide the basis for mapping the pore space (tomography). The method follows a similar technique as SEM; however, instead of energy-dispersant x-rays or electrons, IBM focuses an ion beam at an angle to act as a milling agent (commonly argon gas). Then, SEM focuses from above to capture images, and as the sample is sliced, the images of each slice are combined to provide a three-dimensional image of the minerals and the pore space. Because the result is a digital image, it can differentiate elements, group common elements, and color-code them (fig. 2–24). For example, the pores can then be separated by the contrast and the minerals removed such that only pores are shown in the image; this is vital to see the pore throat network and compare the observations with the petrophysical log response. Tortuosity can also be modeled by emulating the flow of gas through the pore network. The engineers can compare the effectiveness of stimulation on the variations of the pore geometry.

In all reservoir sequences, one language barrier that must be overcome is the description of pore space. Carbonates represent a good example of the difficulty in describing pores (Choquette and Pray, 1970). Shell used a pore classification that was numeric and sequential in nature (Reijers and Bartok, 1985). A similar problem arose with shale porosity. Slatt and O'Brien (2011) have described a series of pore types for the Barnett and Woodford shales, based on their source and cause. These include *porous floccules*, having a collapsed-cardhouse structure; *organo-porosity*, nanometer-scale pores within organic structures, generally isolated; *fecal pellet pores*, grain-size pellets with original preserved intraparticle micropores; *fossil fragment porosity*, intraskeletal pores that have remained open by virtue of sheltering; *intraparticle grain pores* (or *intragrain porosity*) with better chance of interpore communication; and *microchannel* and *microfracture*, nanometer- and micrometer-scale pathways that can lead to communication, which some have labeled as a *kerogen network*. This classification allows professionals studying the reservoir to have a better understanding of behavior of the fluids in the system. Importantly, within a company, there should be consistency in the pore descriptions to enable comparative studies to be carried out across a field reservoir system and between fields. Once the pore type is understood and the reservoir is fracked, an understanding of the pore distribution and connectivity will improve prediction of production from the interval.

The sample preparation for IBM-SEM is similar to that for SEM: cleaning, cutting, polishing, and epoxy impregnation for softer or more porous material. It is important that the surface be smooth at the micron level.

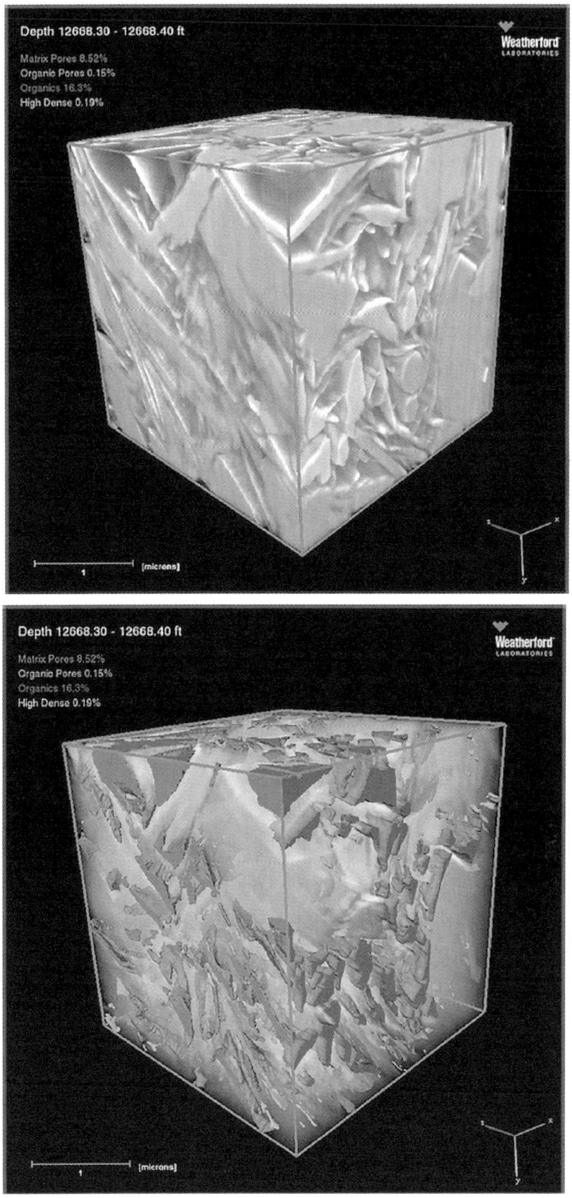

A.

B.

Fig. 2–24. IBM-SEM sample. SEM image shows pore distribution in dark gray. (Reproduced with kind permission of Weatherford Inc.; Rine et al., 2013; Slatt and O'Brien, 2011)

FOURIER TRANSFORM INFRARED SPECTROSCOPY

Fourier transform infrared (FTIR) spectroscopy is carried out for mineral identification, as well as for determination of organic composition. As with XRF, the analysis is dependent on the quantized vibrational energy of the atomic states (Chen et al., 2015). When the photon is absorbed by the molecule, the atomic energy state is elevated, resulting in the generation of detectable frequencies as the bonds vibrate. Although FTIR spectroscopy has been applied to coal ranking for several decades, only recently has it been applied to shale resource assessment. Aside from mineral characterization, FTIR spectroscopy is a valuable tool to determine thermal maturity of the shales. Because it can analyze micro samples, the techniques can be applied to discriminate various paleoclimate conditions in the environments of deposition. Sample preparation for FTIR studies is critical, and where the sample is analyzed, the preparation procedure should be reviewed with care. The standard preparation is to grind the material with 25 mg of potassium bromide (KBr) for an initial 30 seconds, then to grind the sample further for another 30 seconds with 125 mg of KBr. The sample is pressed into a pellet and analyzed. KBr is used because it is transparent to infrared. Chen et al. (2015) have outlined the details of sample preparation. The most important observation is that the various species or functional groups can be summarized by applying the Beer-Lambert law:

$$W = \frac{AM}{\varepsilon I \rho} \tag{2.8}$$

where w is the weight percent of interest, A is the absorbance factor (based on the relationship between the incident and transmitted light and sample thickness), ε is the normal absorptivity, I is the thickness of the sample, and ρ is the density of the sample. The result of equation (2.8) is given in *wavenumber.* In the language of frequency, wavenumber is the inverse of wavelength.

In coal macerals, FTIR spectroscopy clearly distinguishes liptinite, inertinite, and vitrinite. In the sample in figure 2–25, the kaolinite signal can easily be correlated among several wells.

As will be seen in chapter 7, conventional maturity reflectance must be measured on vitrinite only. Even more important, if the spectrometer is properly calibrated, FTIR spectroscopy can distinguish variations in kerogens. Moreover, distinction of macerals is vital when determining reflectance of only vitrinite and avoid including other reflectance values from other macerals. Because FTIR spectroscopy has been graduated to coal rank and in turn to vitrinite reflectance, it is an invaluable tool to determine maturation of macerals where vitrinite is absent. This is particularly true for shales from the pre-Carboniferous (Lis et al., 2005).

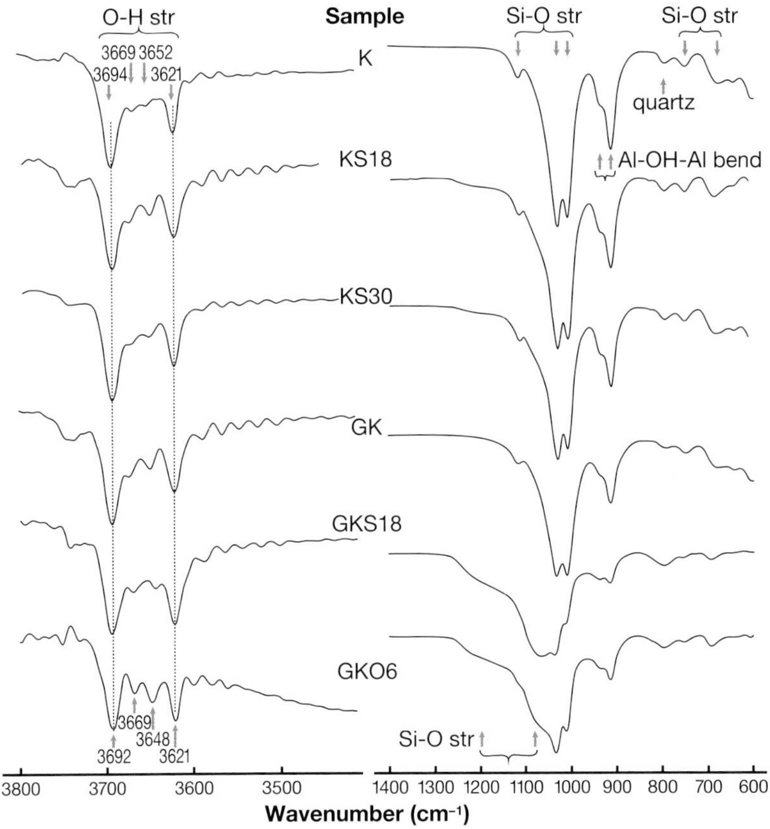

Fig. 2–25. FTIR plot for a Kaolinite sample, demonstrating consistency in response among various samples in the region (Worasith et al., 2011), which therefore allows clay types to be determined. Percent transmittance is measured along the *y*-axis. (Reproduced with kind permission of the Clay Mineral Society, publisher of *Clays and Clay Minerals*.)

COMPUTED TOMOGRAPHY AND SPECTRAL GAMMA RAY LOG

One of the problems faced when describing shale resources is the common physical appearance known as *poker-chip fractures*. These fractures are due to the expansion of gas as the sample is brought to the surface. Because of their appearance, it becomes difficult to assess the properties of the shale. However, they can be reconstituted and placed in a CT scanner. Because of the continuous x-ray scanning of the core sample using CT, a three-dimensional view of the sample can be partially reconstructed, and many of its internal characteristics can be determined.

Today, CT of core samples is a conventional procedure and is always carried out prior to slabbing the core (figs. 2–26 and 2–27). At the same time, a core *spectral gamma-ray log*, also known as a *spectralog*, is recorded. The latter tool is described in detail in chapter 5. The spectral gamma-ray log serves as a correlation tool to match the core with the subsurface gamma ray while providing valuable information on the clay mineralogy of the sample. Micro-CT scanners are now available that allow for very detailed analysis at the 2 mm level. This allows for an analysis of the microporosity in the shale.

Fig. 2–26. Poker-chip fracture pattern often observed in organic-rich shale cores at Shell 21 Texel well, Bakken Formation, North Dakota

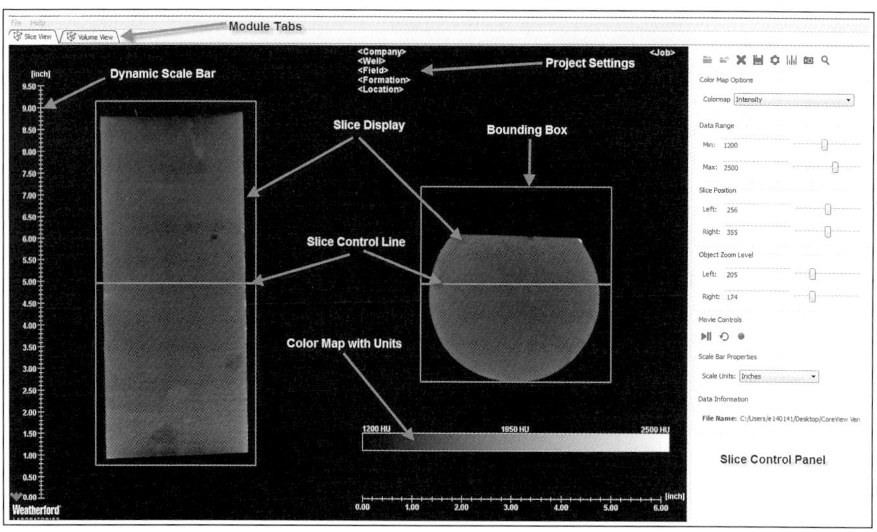

Fig. 2–27. Weatherford's presentation of CT results. (Reproduced with kind permission of Weatherford Inc.)

IMBIBITION STUDIES

The traditional method to evaluate permeability in rock samples that naturally have low porosity and permeability is by analyzing the imbibition characteristics of the sample. The basic method is to crush the sample and remove all fluids. Then, under pressure, mercury is introduced into the sample, and the equivalent water saturation is determined at each step. The volume of fluid that can be introduced is measured, as well as the velocity of the flow; it is the velocity of the flow, using Darcy's equation, that allows one to convert the velocity to permeability. The process is reversed and repeated several times.

One of the problems with this method is that it does not consider the wettability of the shale. *Water-wet* systems promote production, and *oil-wet* systems inhibit production. As described previously, shales adsorb water owing to van der Waals forces. The water is irreducible and will not affect production. When the sample is dried, all water, including irreducible water, is removed, and the permeability study ignores its presence. In shale studies, the minifrac is considered the most accurate determinant of permeability, and a more detailed description of this method will be given in chapter 8.

POROSITY, PERMEABILITY, AND WATER SATURATION IN SHALES

One key objective of any reservoir study is to determine the porosity and permeability of the reservoir as accurately as possible. This is to forecast producibility and estimate the ultimate recoverable reserves. The eventual goal is to relate porosity and permeability. As described already, shales tend to fragment if they have fissility.

Several tests can be run by contractors. Porosity can be measured on crushed samples with either helium or mercury porosimeters. Mercury is used because, similar to oil, it does not affect wettability of the shales. Water saturation and capillary pressure use adaptations of the Dean-Stark apparatus, which basically distills the water from the sample and, as it precipitates, measures its volume. Permeability is more difficult to measure. Aside from the imbibition studies, pressure decay can also be used. A crushed shale sample is subjected to injected gas (Lasswell, 2013), and the results are tabulated as permeability values. Two basic techniques are used: applying constant gas pressure and observing the decline; and apply pressure pulses and observing the declines. The latter is suggested as more efficient for nanodarcy permeabilities (according to Core Laboratories general procedures).

The end goal of porosity and permeability studies is to develop a relationship between the two properties that can be applied in the area of study. Porosity-permeability crossplots with nanopermeabilities indicate significant independence between the two (Blasingame, 2008). The table relating porosity and permeability must be tailored for the specific area involved in the study.

The problem with these analyses is that they need to consider the pore throat size. The IBM-SEM analysis provides this valuable information. The Winland equation (AAPG Wiki) is commonly used to ensure the best fit:

$$\log \mathrm{R35} = 0.732 - 0.588(\log K_{air}) - 0.864(\log \phi) \quad (2.9)$$

where R35 is a pore throat at which the water saturation is 35%, used as a reference point; K_{air} is the uncorrected permeability with air; and ϕ is the porosity.

In general, when compared with other porosity and permeability data, equation (2.9) (the Winland equation) has significant scatter and therefore serves only as a first approximation. Additional information will be discussed in the chapter 8 (see "Capillary Pressure, Wettability, and Pore Size").

SUMMARY

The problems of shale definition and determining the physical and chemical attributes of the shale require detailed analysis using multiple redundant techniques. Shale grain size and composition are of importance because they help define whether the specific shale will tend to be brittle or ductile.

Several analytical techniques allow for accurate determination of the clay mineral composition. The most common are XRD for crystal structure and XRF for elemental composition. Other tools include SEM, EDAX (again for composition), QEMSCAN (a more advanced EDAX for mineral composition), IBM-SEM or FIB-SEM (ion-milled samples for nanopore studies), CT, and Langmuir isotherm (to determine adsorbed gas content). Other techniques have also been discussed.

Clear understanding is needed by all involved as to what questions must be answered and what tools are available to assist in that effort. Mineralogy, porosity, pore type, permeability, and density are only a few of the questions that require study.

One of the clay minerals studied in detail is illite, as well as the smectite-to-illite conversion. There is a direct link between overpressure and this conversion.

REFERENCES

AAPG Wiki. "Rock quality." wiki.aapg.org/Rock_quality.

Alexander, T., J. Baihly, C. Boyer, B. C. G. Waters, V. Jochen, J. Le Calvez, R. Lewis, C. Miller, J. Thaeler, and B. E. Toelle. 2011. "Shale gas revolution." *Oil Field Review (Schlumberger).* 23: 40–54.

American Society for Testing and Materials. 1985. "Standard test method for particle-size analysis of soils D 422-63 (1972)." In *Annual Book of ASTM Standards 04.08*. Philadelphia: American Society for Testing and Materials, 117–127.

Berger, Gilles, Jean-Claude Lacharpagne, Bruce Velde, Daniel Beaufort, and Bruno Lanson. 1997. "Kinetic constraints on illitization reactions and the effects of organic diagenesis in sandstone/shale sequences." *Applied Geochemistry.* 12: 23–35.

Bettison-Varga, Lori, and Ian D. R. Mackinnon. 1997. "The role of randomly mixed-layered chlorite/smectite in the transformation of smectite to chlorite." *Clay and Clay Minerals*. 45: 506–516.

Bjørlykke, K. 1998. "Clay mineral diagenesis in sedimentary basins—a key to the prediction of rock properties: Examples from the North Sea Basin." *Clay Minerals*. 33: 15–34.

Blasingame, Thomas Alwin. 2008. "The characteristic flow behavior of low-permeability reservoir systems." Paper presented at SPE Unconventional Reservoirs Conference. Keystone, Colorado, February 10–12. dx.doi.org/10.2118/114168-MS.

Buhmann, C., and P. L. C. Grubb. 1991. "A kaolin-smectite interstratification sequence from red to black complex." *Clay Minerals*. 26: 343–358.

Burrafato, G., and F. Miano. 1993. "Determination of the cation exchange capacity of clays by surface tension measurements." *Clay Minerals*. 28: 475–481.

Burtner, Roger L., and Maurice A. Warner. 1986. "Relationship between illite/smectite diagenesis and hydrocarbon generation in Lower Cretaceous Mowry and Skull Creek shales of the northern Rocky Mountain area." *Clays and Clay Minerals*. 34: 390–402.

Chatellier, Jean-Yves, Erik Quartero, Mark Urban, Marianne Molgat, Arnaud Deconinck, and Pierre Francus. 2011. "Power and limitations of x-ray fluorescence from cuttings: A test in the Utica and Lorraine shales from Quebec." Paper presented at AAPG Annual Convention and Exhibition, Houston, Texas, April 10–13. *Search and Discovery* article 40766. www.searchanddiscovery.com/pdfz/documents/2011/40766chatellier/ndx_chatellier.pdf.html.

Chen, Yanyan, Caineng Zou, Maria Mastalerz, Suyun Hu, Carley Gasaway, and Xiaowan Tao. 2015. "Applications of micro-Fourier transform infrared spectroscopy (FTIR) in the geological sciences—a review." *International Journal of Molecular Sciences*. 16: 30223–30250.

Chopra, Satinder, and Alan Huffman. 2006. "Velocity determination for pore pressure prediction." *CSEG Recorder*. 31. csegrecorder.com/articles/view/velocity-determination-for-pore-pressure-prediction.

Choquette, Phillip W., and Lloyd C. Pray. 1970. "Geological nomenclature and classification of porosity in sedimentary carbonates." *AAPG Bulletin*. 54: 207–250.

Croudace, Ian W., and R. Guy Rothwell. 2010. "Micro-XRF sediment core scanners: Important new tools for environmental and earth sciences." *Spectroscopy Europe*. 22: 6–13.

———. 2015. "ItraxPlot: An intuitive flexible program for rapidly visualising Itrax data." In *Micro-XRF Studies of Sediment Cores*. Dordrecht, The Netherlands: Springer, 613–624.

Drever, James I. 1982. *The Geochemistry of Natural Waters*. Englewood Cliffs, NJ: Prentice Hall.

Drits, Victor, Jan Srodon, and D. D. Eberl. 1997. "XRD measurement of mean crystallite thickness of illite and illite/smectite: Reappraisal of the Kubler index and the Scherrer equation." *Clays and Clay Minerals*. 45: 461–475.

Donovan, Art, J. Evenick, and L. Banfield. 2017. "An organofacies-based mudstone classification for unconventional tight rocks and source rock plays." Unconventional Research Conference. Austin, TX, July 24–26, 2017.

Eslinger, Eric, and David R. Pevear. 1988. "Clay minerals for petroleum geologists and engineers." Society of Economic Paleontologists and Mineralogists, Short Course No. 22.

Faul, Henry. 1966. *Ages of Rocks, Planets, and Stars*. New York: McGraw-Hill.

Grim, Ralph. 1968. *Clay Mineralogy*. New York: McGraw-Hill.

Hang, Pham Thi, and G. W. Brindley. 1970. "Methylene blue absorption by clay minerals: Determination of surface areas and cation exchange capacities (clay-organic studies XVIII)." *Clays and Clay Minerals*. 18: 203–212.

Hart, Bruce S., Joe H. S. Macquaker, and Kevin G. Taylor. 2013. "Mudstone ('shale') depositional and diagenetic processes: Implications for seismic analyses of source-rock reservoirs." *Interpretation*. 1: B7–B26.

ICDD website. www.icdd.com.

Jaboyedoff, M., F. Bussy, B. Kübler, and P. Thelin. 2001. "Illite 'crystallinity' revisited." *Clays and Clay Minerals*. 49: 156–167.

Jeans, C. W. 1994. "Clay diagenesis, overpressure and reservoir quality: An introduction." *Clay Minerals*. 29: 415–424.

Kammeyer, Carl W., and Donald R. Whitman. 1972. "Quantum mechanical calculation of molecular radii. I. Hydrides of elements of periodic groups IV through VII." *Journal of Chemical Physics*. 56: 4419–4421.

Katahara, Keith. 2008. "What is a shale to a petrophysicist?" *Leading Edge*. 27: 738–741

Lasswell, Pat. 2013. "Steady-state permeability analysis in unconventional plays." Paper presented at GeoConvention, Calgary, AB, May 6–10.

Lis, Grzegorz P., Maria Mastalerz, Arndt Schimmelmann, Michael D. Lewan, and B. Artur Stankiewicz. 2005. "FTIR absorption indices for thermal maturity in comparison with vitrinite reflectance Ro in type-II kerogens from Devonian black shales." *Organic Geochemistry*. 36: 1533–1552.

Löhr, S. C., E. T. Baruch, P. A. Hall, and M. J. Kennedy. 2015. "Is organic pore development in gas shales influenced by the primary porosity and structure of thermally immature organic matter?" *Organic Geochemistry*. 87: 119–132.

Loucks, Robert G., Robert M. Reed, Stephen C. Ruppel, and Daniel M. Jarvie. 2009. "Morphology, genesis, and distribution of nanometer-scale pores in siliceous mudstones of the Mississippian Barnett Shale." *Journal of Sedimentary Research*. 79: 848–861.

Mavko, Gary, Tapan Mukerji, and Jack Dvorkin. 2009. *The Rock Physics Handbook: Tools for Seismic Analysis of Porous Media*. Cambridge, UK: Cambridge University Press.

McHardy, W. J., M. J. Wilson, and J. M. Tait. 1982. "Electron microscope and X-ray diffraction studies of filamentous illitic clay from sandstones of the Magnus Field." *Clay Minerals*. 17: 23–40.

McHargue, Timothy R., and Rex C. Price. 1982. "Dolomite from clay in argillaceous or shale-associated marine carbonates." *Journal of Sedimentary Research*. 52: 873–886.

Mineralogical Society of Great Britain and Ireland. "Images of clay." www.minersoc.org/gallery.php?id=2.

Nayak, Preeti Sagar, and B. K. Singh. 2007. "Instrumental characterization of clay by XRF, XRD and FTIR." *Bulletin of Materials Science*. 30: 235–238.

Norrish, K., and R. M. Taylor. 1962. "Quantitative analysis by X-ray diffraction." *Clay Minerals Bulletin*. 5: 98–109.

O'Connor, Stephen, Richard Swarbrick, Jamaal Hoesni, and Richard Lahann. 2011. "Deep pore pressure prediction in challenging areas, Malay Basin, SE Asia." In *Proceedings of the Indonesia Petroleum Association, Thirty-Fifth Annual Convention & Exhibition, IPA11-G-022*.

Parker, Carol J., John C. Hathaway, and Paul D. Blackmon. 1956. "Some curves from a portable differential thermal analysis unit." *US Geological Survey Bulletin*. 1021: 237–251.

Pollastro, Richard M. 1993. "Considerations and applications of the illite/smectite geothermometer in hydrocarbon-bearing rocks of Miocene to Mississippian age." *Clays and Clay Minerals*. 41: 119.

Poppe, L. J., A. H. Eliason, J. J. Fredericks, R. R. Rendigs, D. Blackwood, and C. F. Polloni. 2013."Grain-size analysis of marine sediments: Methodology and data processing." Chapter 1 in *U.S. Geological Survey Open-File Report 00-358*.

Poppe, L. J., V. F. Paskevich, J. C. Hathaway, and D. S. Blackwood. 2001. "A laboratory manual for X-ray powder diffraction." In *U.S. Geological Survey Open-File Report 01-041*, 1–88.

Rask, J. H., L. T. Bryndzia, N. R. Braunsdorf, and T. E. Murray. 1997. "Smectite illitization in Pliocene-age Gulf of Mexico mudrocks." *Clays and Clay Minerals*. 45: 99–109.

Reijers, T. J. A., and P. Bartok. 1985. "Porosity characteristics and evolution in fractured Cretaceous carbonate reservoirs, La Paz field area, Maracaibo basin, Venezuela." In *Carbonate Petroleum Reservoirs*. New York: Springer, 407–423.

Rieke, Herman H., and George V. Chilingarian. 1974. *Compaction of Argillaceous Sediments*, vol. 16. New York: Elsevier.

Riley, Ron, and Mark Baranoski. 2011. "Regional stratigraphic and facies relationships of the Ordovician Utica/Point Pleasant interval in the Appalachian Basin." thepttc.org/workshops/eastern_062111/eastern_062111_Riley.pdf.

Rine, James M., Erin Smart, William Dorsey, Kultaransingh Hooghan, and Michael Dixon. 2014. "Comparison of Porosity Distribution within Selected North American Shale Units by SEM Examination of Argon-Ion Milled Samples." *Electron Microscopy of Shale Hydrocarbon Reservoirs*. AAPG Memoir 102: 17–21.

Roaldset, E., He Wei, and S. Grimstad. 1998. "Smectite to illite conversion by hydrous pyrolysis." *Clay Minerals*. 33: 147–158.

Roberson, Herman E., and Richard W. Lahann. 1981. "Smectite to illite conversion rates: Effects of solution chemistry." *Clays and Clay Minerals.* 29: 129–135.

Rowland, Richards A. 1952. "Differential thermal analysis of clays and carbonates." *Clays and Clay Minerals.* 1: 151–163.

Shale Experts. www.shaleexperts.com/images/Utica-Shale-Log.PNG.

Shi, G. R. 2009. "A simplified dissolution-precipitation model of the smectite to illite transformation and its application." *Journal of Geophysical Research: Solid Earth.* 114 (B10).

Slatt, Roger M., and Neal R. O'Brien. 2011. "Pore types in the Barnett and Woodford gas shales: Contribution to understanding gas storage and migration pathways in fine-grained rocks." *AAPG Bulletin.* 95: 2017–2030.

Sliwinski, Jon, Matt Power, Peta Hughes, and J. Harrington. 2010. "Integrated shale gas evaluation: A study of QEMSCAN, SEM optical petrography, XRD and geochemistry of selected potential and producing North America gas shales." Paper presented at AAPG Annual Convention and Exhibition, New Orleans, April 11–14. *Search and Discovery* article 50290. www.searchanddiscovery.com/documents/2010/50290sliwinski/ndx_sliwinski.pdf.

Smykatz-Kloss, Werner, and Slade S. J. Warne, eds. 1991. *Thermal Analysis in the Geosciences*, vol. 38. Berlin: Springer-Verlag.

Surdam, Ronald C., Steven W. Boese, and Laura J. Crossey. 1984. "The chemistry of secondary porosity. Part 2. Aspects of porosity modification." In *Clastic Diagenesis*, David A. McDonald (ed.). AAPG Memoir 37: 127–149.

Tissot, Bernard P., and Dietrich H. Welte. 1984. *Petroleum Formation and Occurrence.* Berlin: Springer-Verlag.

Visher, Glenn S. 1969. "Grain size distributions and depositional processes." *Journal of Sedimentary Research.* 39: 1074–1106.

Waples, Douglas W. 1980. "Time and temperature in petroleum formation: Application of Lopatin's method to petroleum exploration." *AAPG Bulletin.* 64: 916–926.

Whitney, Gene. 1990. "Role of water in the smectite-to-illite reaction." *Clays and Clay Minerals.* 38: 343–350.

Worasith, N., B. A. Goodman, J. Neampan, N. Jeyachoke, and P. Thiravetyan. 2011. "Characterization of modified kaolin from the Ranong deposit Thailand by XRD, XRF, SEM, FTIR and EPR techniques." *Clay Minerals.* 46: 539–559.

Yariv, Shmuel. 2003. "Differential thermal analysis (DTA) in the study of thermal reactions of organo-clay complexes." In *Natural and Laboratory-Simulated Thermal Geochemical Processes*, Raphael Ikan (ed.). Dordrecht, The Netherlands: Kluwer Academic, 253–296.

3

BIOSTRATIGRAPHY, PALEOCLIMATE, PALEOGEOGRAPHY, AND ANOXIA

TAKEAWAY

- Biostratigraphy—determines the exact age of the sediments, their environment of deposition, paleoclimate, and the presence of condensed sequences; can be applied in the definition of rate of deposition.
- Planktonic fauna have wide distribution and thus provide correlation markers
- Specific organisms, conodonts and pollen, can be used to determine thermal maturity of sediments
- Definition and properties of oceanic anoxic events
- Carbon and oxygen isotopes—applied to paleoclimate and related to anoxic events

BIOSTRATIGRAPHY

Unconventional resources shales, particularly marine deposits, require detailed study of their biostratigraphy. Lacustrine shales are dependent mainly on pollen and other freshwater fauna. Aside from information that is useful in determining the age of the sediments, fossils can provide valuable information on the environment of deposition of the sediments under investigation, as well as on the paleoclimate and paleoecology; establish rates of sedimentation; and assist in defining relative levels of organic maturity of the sediments. Moreover, they play a clear role in identification of the condensed sequences that are often associated with hydrocarbon source beds. By definition, a condensed sequence has a rate of sedimentation of 1–10 mm per thousand years and has abundant and highly diverse species of fauna associated with it.

Are there specific climatic conditions that control how organic-rich shales are deposited? If condensed sequences are common in geologic history, why are there not more source rock intervals widely distributed in time and space?

The basis for biostratigraphy is paleontology. Early naturalists must have wondered how seashells could be found on hilltops thousands of feet above sea level. One of these naturalists was Xenophanes (c. 560 to c. 478 BCE). Although better known as a philosopher and poet, Xenophanes made the early observation that marine fossils on landmasses could be explained only by their having been below sea level at the time of deposition. Thus, according to Xenophanes, there had been two periods in Earth's history: a period when the sea covered the land, based on the existence of fossil seashells onshore; and a period when the land was exposed (Lesher, 2014). This profound observation remained disregarded for centuries.

The first scientist to begin documenting fossils was Leonardo da Vinci, who in 1500 CE drew a trace fossil, or *ichnofossil*, so precisely that it was later identified as *Paleodictyon* (Baucon, 2010). However, it was not until the late 1800s that paleontology began to flourish as a science. By the 1940s, Professor Raymond Moore, at the University of Kansas, began what is currently one of the basic references for invertebrate paleontology. He assembled an encyclopedic treatise, revised continuously and now encompasses over 50 volumes. The University of Kansas continues to be the principal coordinator of Moore's treatise. A similar endeavor, by Brooks Ellis and Angela Messina and sponsored by the American Museum of Natural History, was initiated in 1970s and continues to the present day. The institution publishes and maintains an extensive catalogue of microfossil taxonomy. Because these specimens are microscopic and well preserved in sediments, they are perfectly suited for the analysis of cuttings samples from wells. Renz (1948) wrote a classic book that not only described in detail the age of the foraminifera in East Falcon, Venezuela, but also drew attention to the benthonic foraminifera to establish their environment of deposition and age. Subsequent study in both Trinidad and Venezuela later confirmed these early results, including the work of paleontologists H. Kugler, R. Stainforth, and H. Bolli, who published the first global planktonic zonation timescales. Bolli's timescales have subsequently been updated by Haq et al. (1987) and are currently used as the basis for sequence stratigraphy (popularly known as the *Haq curve* or the *Vail curve*; we will discuss sequence stratigraphy more in chap. 4).

For now, we will focus on the application of microfauna and macrofauna in shale studies. After a discussion on foraminifera, particular attention will be paid to conodonts as they apply to Lower Paleozoic sediments. Even though not discussed in detail here, macrofauna (especially ammonites) and other microorganisms, such as dinoflagellates, nanofossils, pollen, fusulinids, and graptolites, are also commonly used in shale studies.

All of the sedimentary sequences on Earth are divided into two eons: the Phanerozoic, or time of significant speciation of fauna (Cambrian to recent); and the Precambrian, or time before speciation. The only living organisms in the Precambrian were bacteria and algae. They were instrumental in changing the early atmosphere on Earth from a CO_2 atmosphere (that would be very toxic to humans) to an oxygen-rich atmosphere. A best estimate is that over 90% of the hydrocarbons in the world were formed during and are stored in sediments from the Phanerozoic. This will be the focus of this chapter.

SOURCE ROCK DISTRIBUTION

Within the Phanerozoic, there are three distinct eras (Cohen et al., 2013): the Paleozoic (541 million years ago [Ma] to 252 Ma, lasting 289 million years); the Mesozoic (252 to 66 Ma, lasting 186 million years); and the Cenozoic (66 Ma to now, lasting 66 million years to date). In other words, 50% of Earth's Phanerozoic living history was during the Paleozoic, 35% was during the Mesozoic, and merely 15% was during the Cenozoic. On a statistical basis, 50% of the hydrocarbons would be expected to be found in Paleozoic sediments, yet this is not the case; in fact, the bulk of world's oil and gas was generated from Mesozoic sediments, during the Late Jurassic (Kimmeridgian to Tithonian, 154 to 145 Ma) and Early to Late Cretaceous (Aptian to Turonian, 125 to 90 Ma) (fig. 3–1). The Miocene (23 to 5 Ma) is followed in importance by the Late Devonian to Early Carboniferous (383 to 323 Ma). Several others are also known (Silurian, Ordovician, Permian, Triassic, etc.). Because of their importance, the main Mesozoic, Tertiary, and Middle Paleozoic will be reviewed in more detail as to their stratigraphy, biostratigraphy, and geochemistry (in chap. 7).

Aside from deterministic faunal speciation, what else can we do with the foraminiferal assemblages? The first observation made on sedimentary sequences was that specific fauna, especially planktonic fauna, were restricted to unique strata and correlatable on a global scale. As climatic and water-chemistry conditions varied with time, foraminiferal species either evolved to adapt to their new condition or became extinct. The second observation was that volcanic ash deposits could be sampled near or within layers containing specific faunal assemblages. That volcanic ash is essentially ubiquitous in time and distribution during Earth's history makes it an excellent age determinant. Volcanic ash traps radioactive isotopes that were instantly encased in the volcanic debris.

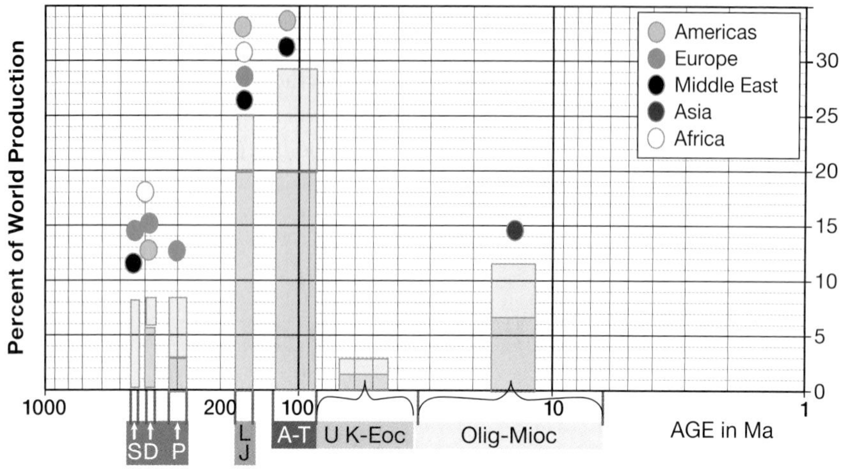

Fig. 3–1. Distribution of oil source rocks worldwide by age. Darker gray denotes oil, and lighter gray denotes gas. Note that this is not representative of the volume of hydrocarbon generated and is not all inclusive. (Based on Klemme and Ulmishek, 1991)

Because radioactive isotopes have been studied, documented, and correlated worldwide since the early 1900s, ample data exist to corroborate their accuracy. Bertram Boltwood, in 1907, observed that lead-to-uranium ratios were higher in older rocks than in younger rocks. Ernest Rutherford, in 1908, documented the process of nuclear decay used in age dating. Since then, geochronology laboratories have become common, and currently many specialized laboratories and universities also have mass spectrometers. Henry Faul, at the University of Pennsylvania, further revolutionized the field by studying the details of the geochronology using potassium-argon dating beginning in the 1960s (Faul, 1966). Over the past 50 years, in combination with other isotopes, potassium-argon dating has allowed age dating to a significant precision level of most volcanic layers worldwide. For Cretaceous sediments, the current precision is within 100,000 years (given the age range 145 to 66 Ma), and for Cenozoic sediments (66 Ma to present), within 50,000 years.

The New Mexico Institute of Mining and Technology (McIntosh and Heizler, 2016) has provided a description of the techniques used in age dating and the limitations thereof. The principle is simple: When the half-life of the isotope is known, the ratio of the daughter product to the parent is an exponential function of time. For example, this technique can be applied to compare the volcanic flows of the Deccan Plateau of central India with the fauna in the interbedded fossil-rich marine shales. The flood basalt flows on the Deccan spread during a critical time in geological history—namely, the Cretaceous-Tertiary (KT) boundary (also known as the Cretaceous-Paleogene [KPg] boundary), and best known for its mass extinction. The Deccan Plateau has three sets of megaflows that have been dated in detail: phase 1, 67.6 to 67.2 Ma, comprising 6% of the lava volume; phase 2, 65.5 to 65.2 Ma, comprising 80% of the flood basalts and corresponding to the KT boundary extinction; and phase 3, 65.0 to 64.8 Ma, comprising 14% of the flows. Additional flows associated with the Bombay High have been dated as young as 60 Ma (Sheth, et al., 2001).

The interbedded marine shales and their corresponding foraminiferal assemblages have been analyzed in detail. A team of geoscientists sponsored by The Geological Society of London have compiled the data on the radiometric dating and foraminiferal assemblage of the Deccan flows (Punekar et al., 2014). Prior to phase 2 volcanism, there was a global cooling, documented by the $\delta^{18}O$ concentration in the tests (shells) of foraminifera. The application of oxygen, carbon, strontium, and sulfur isotopes to stratigraphy and paleoclimatology will be reviewed below. There was a rapid warming phase during the time of the phase 2 volcanism and a cooling phase again at the KT boundary. Because this time period was short, a catastrophic extinction event took place.

Extensive literature describes the faunal assemblage of the KT boundary, and the associated extinctions. Some species adapted, while a large number became extinct. Interestingly, the massive Chicxulub impact in Mexico has been dated between 66.032 and 66.038 Ma (Renne et al., 2013). Note that the precise dating available today allows for an improved definition of the effects of the Deccan volcanics and the Chicxulub impact on the KT boundary. Most publications have used the age of the asteroid impact as the KT boundary. However, recent paleontological studies reveal that the mass extinction was gradual and covered the period from 65.5 to 64.8 Ma, or slightly later than the date of the asteroid impact. A few species survived, evolved, and flourished into the Tertiary Period. Contrary to popular belief, it appears that the large volcanic flows of India and associated rapid global warming were the principal cause of the extinction, exacerbated by the rapid cooling associated with the Chicxulub impact. The region of the Deccan Plateau is in close proximity to the Bombay High offshore field, and it is well documented that the principal source rocks for the Bombay High oils extended from the Lower Paleocene into the Eocene (Wandrey, 2004). What were the marine conditions in the region during the Early Paleocene? As mentioned above, submarine volcanism continued during the Early Paleocene.

Contemporaneous sediments in North America contain only a few locally significant source rock intervals. Mass extinctions in other time periods appear to have had longer durations, and that may be the trigger mechanism for major source rocks. Certainly all major source rock shales were deposited during periods when fauna was stressed. We will discuss this in greater detail later in the text. We now understand that the biostratigraphy can be dated via ash flows. What other faunal assemblages can we use?

If foraminifera had been abundant and well preserved throughout geological history, then the analysis of stressed environments and paleoecology would have been simplified. Foraminifera have a distinct disadvantage: They became very abundant and widespread from the Cretaceous onward, developing in the Jurassic. Prior to the Jurassic microfauna, the world was affected by several great mass extinctions punctuating the end of the Devonian, the end of Permian, and the end of the Triassic. Ammonite zonations are most often referenced for the Cretaceous and Jurassic sediments. Van Hinte (1976) attempted to unify the age dates of the Cretaceous stages, and other investigators are continuously updating the limits.

For the pre–Upper Jurassic, fusulinids (extinct in the Upper Permian) and conodonts (extinct in the Triassic) are used to establish correlations and bracket sequences. Conodonts have two advantages: Speciation and wide diversity allow for excellent correlations; and possibly the most important aspect is their composition (Hautmann, 2012). Because of their calcium carbonate phosphate content, conodonts are insoluble in acid and sensitive to heat, resulting in color changes. Being less soluble in mild acids, they can readily be separated from limestone beds. In shales, they often require considerable sample size to provide a good assemblage. After careful documentation of the conodonts, it was observed that a coloration index could be established and readily related to equivalency with other maturation indices such as Fourier transform infrared (FTIR) spectroscopy (see chap. 2). Results from FTIR spectroscopy can be translated into an equivalent vitrinite reflectance value and other maturity indices. Vitrinite is available only in sediments younger than the Carboniferous. In chapter 7, maturation criteria will be discussed in greater detail. For now, conodont coloration variations, called *conodont alteration index* (CAI), have been tabulated and cross-referenced to other maturation tools. The temperature ranges for the alterations can be carefully calibrated by heating samples of immature conodonts (fig. 3–2). Because a significant portion of the world's resource shales lie in the Paleozoic, having an independent means of establishing the maximum temperature attained during burial allows for an accurate prediction of expected reservoir fluids.

Conodonts are unique fossils. Even though they have been described for over a century, their origin was not known until the 1980s, when an elongate soft-bodied fossil was discovered in Scotland with conodonts in the mouth region (Barrick, n.d.). For example, the conodont biozone ages were used to correlate the Devonian to the Lower Carboniferous of the western United States and, by extension, Western Canada (fig. 3–3).

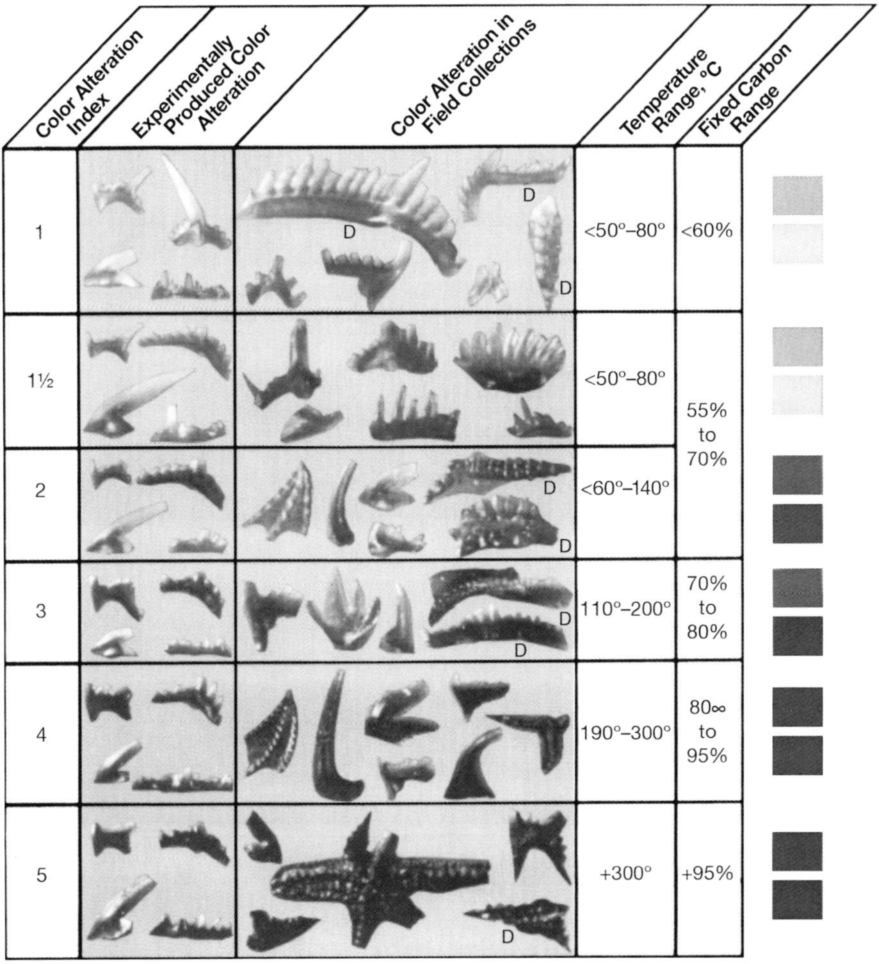

Fig. 3–2. CAI used to establish the level of maturity of the Lower Paleozoic sediments (Reproduced with permission from the U.S. Geological Survey as a public domain document; Epstein et al., 1977)

Johnston et al. (2010) have correlated the Upper Paleozoic (Devonian and Carboniferous) of the Western Canada Basin with the conodont zonation of North Dakota as well (fig. 3–3). Other fossils often used include graptolites for Lower Paleozoic and fusulinids for Upper Paleozoic. Several other organisms can be used to establish the age for sediments of Cambrian to Devonian Periods, including *Acritarchs* and *Chitinozoa*. They appear to have coloration with maturity as well. Because Ordovician shales (Utica shale) have significant reserves, thermal history indicators other than conodonts may be useful to evaluate (Bertrand, 1990).

Period	Stage	Age Ma	Conodont Zonation	N. Dakota Stratigraphy
MISSISSIPPIAN	Tournaisian	351	U Crenulata	Lodgepole
		355.5	L Crenulata	
			Quadruplicata	U. Bakken
		356	Sandbergi	
		357	U Dulicata	M. Bakken (C,D,E,F)
		358	L Dulicata	
			Sulcata	
DEVONIAN	Famennian	361	Praesulcata	M. Bakken (A,B)
		362	Expansa	L. Bakken
		365	Postera	Three Forks
		366.5		

Fig. 3–3. North Dakota stratigraphy based on conodont ages (Adapted from Johnston et al., 2010; Sarg, 2012)

Spores and pollen can also be used, but their utility in Lower Paleozoic rocks is limited. From the Carboniferous and younger, pollen grains can be studied for age determination and provide some clues to understanding paleoclimate conditions. One of their most important contributions is as a maturity indicator. As was the case for conodonts, spores and pollen undergo a color change from pale yellow to dark brown when subjected to higher temperatures. The index is known as the *thermal alteration index* (TAI). For further reading on TAI, refer to the work by Hackley and Cardott (2016).

One of the most notable mass extinctions occurred immediately preceding the Famennian, during the deposition of the Three Forks Formation of North Dakota. On a global scale, all tetracorals became extinct. Of particular note was the *Kellwasser event*, during which it is estimated that 60% to 80% of all animal species of the time became extinct. Extinction events continued up to the end of the

Devonian. Katz et al. (2007) have described the period as the *time of perturbation to the carbon pool.* Sim et al. (2015) have speculated that the significant changes in sulfate levels in the oceans, noted from sulfur isotope data, may be a possible cause for the Kellwasser event. Because this is the time of deposition of the Barnett, Bakken, Exshaw, and Woodford shales, as well as several others, further work is needed on the conditions that led to the observed concentrations of kerogen. Can any technique establish where conditions were most favorable to source rock development?

Because there are multiple geologic time scale tables from various authors, care must be taken to note the reference used. In 2016, the International Commission on Stratigraphy (ICS) published the most widely accepted geological time scale (Cohen et al., 2013). The importance of consistency will become evident in the burial history chart described later (see chap. 7).

OCEANIC ANOXIC EVENTS

> *The anoxic aquatic environment is a mass of water so depleted in oxygen that virtually all aerobic biological activity has ceased. Anoxic conditions occur where the demand for oxygen in the water column exceeds the supply. Oxygen demand relates to near surface biological productivity, while oxygen supply in intermediate and deep water depends on water circulation, which is governed by global climatic patterns and the Coriolis force.*
>
> —*Demaison* and Moore (1980)

For several decades, it has been known that source rocks are deposited under anoxic conditions. Stratigraphic analyses in multiple basins have fine-tuned the description of the *oceanic anoxic events* (OAEs) and correlated their ages. The Cretaceous seas serve an excellent example to better understand these phenomena. It is assumed that most OAEs are associated with maximum flooding events, but why are they so widespread? And why do they occur at very specific times? The most likely cause is the flood basalt.

The age dating of the principal source rock interval for the Lower Eagle Ford, in South Texas, is 97.5 to 91 Ma, which corresponds to OAE 1d (fig. 3–4). The dating of the La Luna source rock, in the Maracaibo Basin, is the same age (Bartok et al., 1981), and so is the Nahr Umr of the Arabian Gulf (Stoneley, 1990). In fact, the entire time period from Aptian to Turonian (125 to 90 Ma) contains some of the world's most significant source rocks. Is the global warming associated with these events related to the development of organic-rich shales? The short answer is a qualified yes. However, studying faunal assemblages under environmental stress may provide clues for the potential exceptions on the conditions for resource shales to deposit. In general, sediments under stress contain fossils dominated by juvenile forms and stunted growth but not necessarily mass extinctions.

One of the prerequisites for formation of an organic-rich shale is to be associated with an OAE. Consequently, interest is currently focused on Pearsall shale, as figure 3–4 shows. Courtillot and Renne (2003) have indicated that several of the flood basalt events are related or in close proximity to OAEs—but maybe not all. Why the exception? The Deccan volcanic event was very short lived. In western India, the Lower Paleocene contains excellent source rocks. Is this OAE related to the Deccan Plateau basalt? Is there a minimum duration required for widespread anoxic conditions to prevail? No other OAEs are associated with the KT boundary.

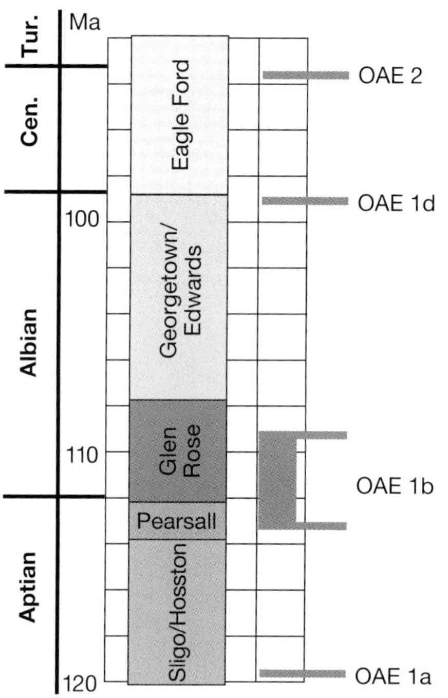

Fig. 3–4. Cretaceous OAEs affecting the northern Gulf of Mexico and extending over the entire Caribbean Basin (Based on Leckie et al., 2002)

It is possible that in some cases the evidence of volcanism is no longer available. In fact, the Late Jurassic (Kimmeridgian to Tithonian) accounts for over a quarter of the oil in the world; yet it is not evidently associated with a major flood basalt event, because they are more difficult to establish. According to BouDagher-Fadel (2012), many species became extinct by the end of the Tithonian, including nearly 30% of all foraminifera. BouDagher-Fadel has proposed several causes: the Shatsky Rise flood basalts, located in the western Pacific near Japan (fig. 3–5); the final opening of the North Atlantic and related volcanism (fig. 3–6); and changes in topographic relief resulting in global climate changes. Given the high extinction rate in Kimmeridgian and Tithonian, there is little doubt as to the stress on the existing ecosystems. The takeaway from this discussion is that the key for organic-rich shales to form is the development of stress in the ecosystem.

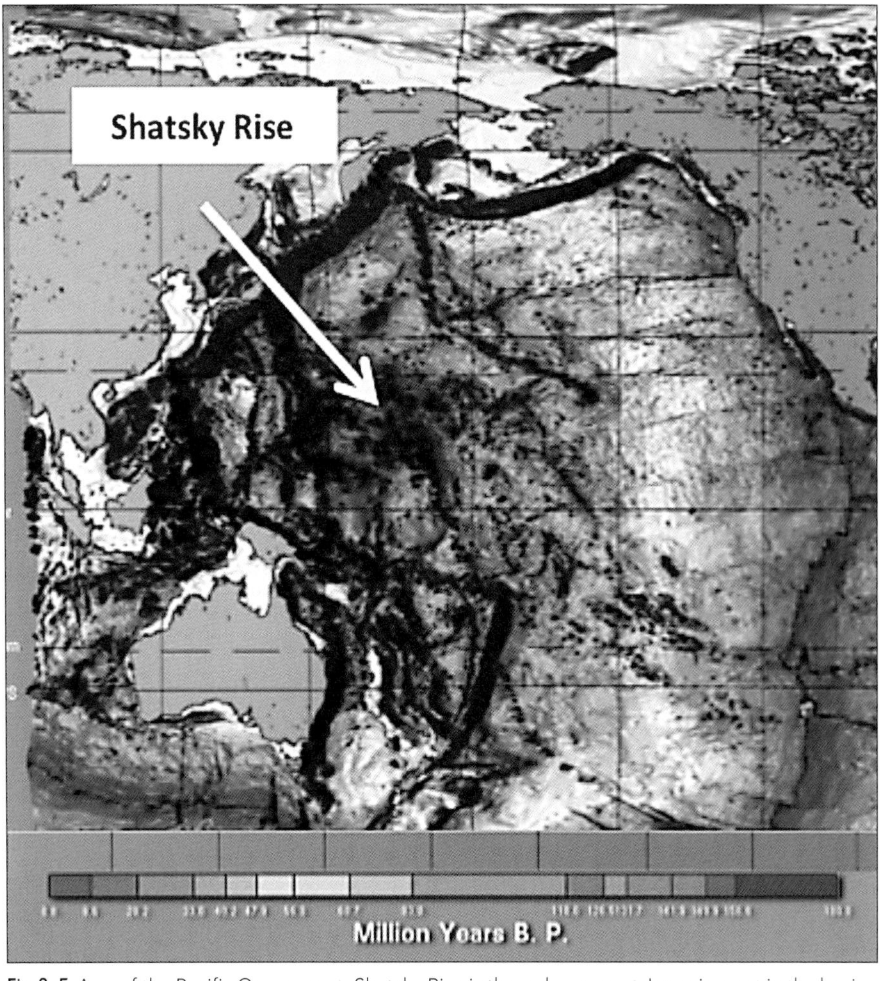

Fig. 3–5. Age of the Pacific Ocean crust. Shatsky Rise is the only remnant Jurassic crust in the basin. (Reproduced from NOAA open-file map in the public domain; Müller et al., 2008)

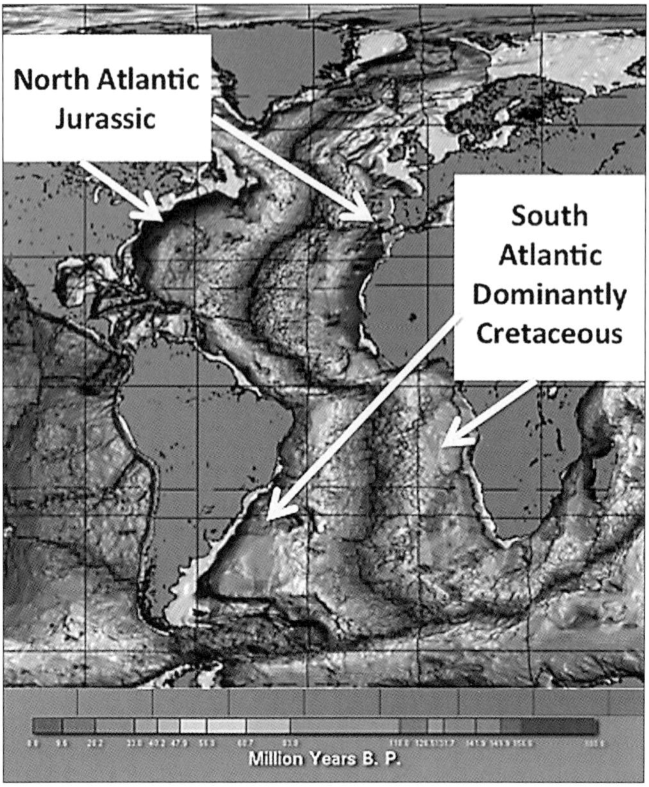

Fig. 3-6. Age of the ocean crust along the Atlantic margin. In the North Atlantic, the Jurassic oceanic crust is present only near the coasts. (Reproduced from NOAA open-file map in the public domain; Müller et al., 2008)

One problem is that, of the vast oceanic regions, only a small portion contain Jurassic ocean crust, and no pre-Jurassic ocean crust is available for study. A possible clue to the presence of extensive pre-Jurassic volcanism is the occurrence of bedded chert in the sediments. Chert is known to be deposited in two forms: nodular and bedded. We will briefly focus on bedded chert, which can form in pelagic environments by radiolarian ooze or associated with volcanism. The latter may result in bedded chert even in neritic environments. One problems with chert deposition is that silica is soluble under pH conditions greater than 7.5–8 (slightly alkaline), which happens to be the average alkalinity of seawater. When marine volcanic activity takes place in the vicinity of the shale deposit, volcanism reduces the pH and silica precipitates (Resing and Sansone, 1999). Therefore, one clue to consider possible source rock intervals in pre-Jurassic sediments is the presence of bedded cherts not associated with radiolarian ooze. The Devonian Marcellus and the Ordovician Utica Formation have bedded chert. Moreover, the Lower Marcellus contains volcanoclasts (Tioga ash beds) near its base (Soeder, 2017). The

Mesozoic source rock shales also have bedded cherts—namely, La Luna (Maracaibo Basin), Niobrara (Colorado) and Eagle Ford (South Texas).

The largest section of Jurassic ocean floor lies near the Shatsky Rise (fig. 3–5). It envelops a small portion of the present-day ocean floor. In the Atlantic Ocean, only a fraction of the ocean floor consists of Jurassic age volcanism, along the East Coast of North America and northwestern Africa and Europe (fig. 3–6). However, the implication of oceanic plateau basalts on the presence of OAEs can be clearly defined for the latest Jurassic to the present. Three OAEs are of particular interest:

- Caribbean phase I, beginning in the Late Jurassic (Mattinson et al., 2008) and extending to the Early Cretaceous, during which there was spreading and arc magmatism
- Phase II, Late Cretaceous, 110 to 63 Ma
- Columbia Plateau, northwestern United States, 17 to 15 Ma

The Caribbean Plateau phase I volcanics are more closely related to the Tithonian Haynesville in the southern United States and the Mexican coeval source rocks. They are poorly represented in northern South America, with the exception of the Guajira Peninsula (Bartok et al., 2015). The Caribbean Plateau basalt in phase II is closely related to the age of the Late Cretaceous OAE—and, hence, resource plays associated with the Eagle Ford (South Texas, USA), Agua Nueva (Mexico), Niobrara shale (Colorado, Kansas, USA) and La Luna/Querecual (Venezuela, Colombia). The biostratigraphy of the La Luna Formation has been detailed by Romero et al. (2003), who have pointed to the Cenomanian-Turonian boundary event as the principal OAE within the sequence. In 1951, Robert Perkins, one of the great Cretaceous macropaleontologists, carried out a detailed analysis of coral species in the Eagle Ford shale. He observed that the corals had stunted growths at several levels, a clear indication of a highly stressed environment (Perkins, 1951). De Baets et al. (2016) have also demonstrated the ammonite extinctions during the Mesozoic, and Parra et al. (2003) have noted that significant chert and phosphate are also present during the late phase of the Turonian OAE. In addition, during the deposition of the Eagle Ford, significant ash beds have been detected, and proximal pyroclastic flows are suspected owing to the presence of titanium-rich kaolinite (Calvin et al., 2017).

Neither the extensive Brazilian Paraná Basin volcanics, 134.7 ± 0.7 to 134.2 ± 0.8 Ma (Thiede and Vasconcelos, 2010), nor the slightly older Karoo volcanics of South Africa, 184.7 ± 1.0 to 180.3 ± 3.6 Ma (Jourdan et al., 2005; Renne et al., 1996), correlate with the source rock interval in the southern Atlantic. This may be due to lack of access to marine sequences since the South Atlantic had not initiated its rifting phase, or the volcanism may simply have been too short lived (fig. 3–6).

Further investigation is required in order to demonstrate why some volcanic events do not permit source rocks to develop.

As discussed already, the younger the formation being studied, the more accurate the age dates will be—and, consequently, the stronger the relationship to associated sediments will be. Plateau basalts are less frequent in the Cenozoic. Two stand out as significant events with potential impact on global climate and, as a result, on potential OAEs: the Columbia Plateau of Oregon, specifically the Grand Ronde basalt; and the Ethiopia Plateau basalts. The Columbia volcanics erupted from 17 to 14.5 Ma, during the Miocene (Reidel, 1998). The Ethiopian basalts erupted at mean age of 30 Ma (range, 40 to 24 Ma, from latest Eocene to early Oligocene) (Hofmann et al., 1997). Eocene source rocks have been described in Ethiopia but may not attain maturity except in offshore Sudan, where high heat flows are present (as noted in chap. 7) .

Because of their large abundance and variety in the Miocene, corals were very well suited to establish water conditions and paleoclimate during that period. A unique event in the sediments of the late Miocene, known as the *Miocene climatic optimum* (MCO), was associated with a significant global warming event followed by a rapid cooling. The MCO ranged in age from 17 Ma to 15 Ma (Foster et al., 2012). Note that this is the same age as the Columbia Plateau. On the basis of coral data, the CO_2 concentration is estimated to have been as high or higher than 280 ppm during that period. By contrast, the average during the Miocene was 200 ppm (Zhang et al., 2013). The significant increase in CO_2 may have been brought about by impact of large-scale flood basalt eruptions on the paleoclimate. Throughout the Pacific Rim, several major source rock intervals are associated with the MCO including the Monterey shale (15 to 8 Ma), in California (Obradovich and Naeser, 1981).

Associated shales in the Sparta Formation, a deltaic deposit in central Louisiana, have been studied by McDade et al. (1993). This formation was deposited during the Late Eocene, in what was then the Gulf of Mexico. McDade et al. (1993) also investigated the organic content of the Sparta Formation and noted that it was a significant source rock. The Late Eocene is also noted for significant faunal extinctions and a relatively short lived but intense global icehouse effect (Hofmann et al., 1997). The Late Eocene OAE, icehouse effect, and extinction may have resulted from the Late Eocene Ethiopian flood basalt or similar event in the northern Gulf of Mexico (Yancey and Guillemette, 1998). The Sparta shales need to be evaluated as a potential resource play.

From the preceding discussions, it is evident that the development of extensive plateau basalts and consequent atmospheric changes are a major driver creating stress in the paleoclimate and faunal distribution. Courtillot and Renne (2003) have provided a good summary for the ages of the plateau basalts. Anoxic conditions are closely related to these events. The next question to ask is how long do the stress periods last? In other words, if a sedimentary section is 300 m thick and ranges in age from 157 to 145 Ma, given that the greatest ecosystem stress occurred at 157 Ma, how far do the effects extend into the upper portions of the column? The logical answer is probably not more than 20%. What has been observed repeatedly is that the lower portions of the resource shales tend to be the richest interval.

As discussed previously, we can only surmise that similar volcanic events were associated with pre-Jurassic OAEs. Isotope data for all the documented OAEs from the Late Jurassic to the Miocene suggest a close relationship between each OAE and major volcanic eruptions. By analogy, if the OAEs from Jurassic to Miocene consistently show a relationship between abrupt climate changes and OAEs and can be demonstrated by various isotope ratios, what can we say about the pre-Jurassic events? On a global scale, the pre-Jurassic source rock intervals can be divided into two families: those with global prominence; and source rocks that are more local. By analogy, as stated already, abrupt cooling and warming are related to major stress and or extinctions. The Antler orogeny is one example that underscores how detailed studies are required in order to establish the relationship between major volcanism and source rock development. The Antler orogeny extended along the present-day U.S. West Coast and occurred during the latest Devonian and Early Carboniferous (Saltzman et al., 2000). This age corresponds to the Exshaw (Western Alberta), Bakken (North Dakota), Barnett (Central Texas), and Woodford (Oklahoma) shales. Over a century, detailed paleontological studies worldwide have shown that at specific times, including the pre-Jurassic, major extinctions took place. However, the discussion thus far also points to high-stress periods at various key geological stages that constitute the principal hydrocarbon source intervals.

A different approach to biostratigraphy and paleobotany uses global plate tectonic reconstruction maps. Professor Chris Scotese, at the University of Texas at Arlington, has combined this approach with information on specific periods of geology with a global reconstruction model (Scotese and McKerrow, 1990), and selections of these maps are also available online for reference. Previously, Habicht (1979) set the stage for establishing paleoclimates by assuming that reefs were present in similar latitudes as in the present day. (Today reefs are bound between the 30th parallels.) What is the distribution of the ancient reefs? Once plants were present on land—and especially swamps in peat bogs—they were commonly associated with more temperate climates. The pollen in the fossil record is indicative of the prevalent vegetation in the region. Where were the ancient coal deposits? For the most part, source rocks tend to develop under more tropical paleoclimates. Professor Blakey, at Northern Arizona University, has provided a series of maps showing plate tectonic reconstructions, with attention to local topographic relief (Blakey, 2003). Topography also plays a vital role in defining conditions at the depositional sites. More detailed mapping of the paleoclimatic zones focusing only on the interval bounding of the source rock may provide clues as to the limiting conditions for each OAE and major volcanic events.

The Devonian is a particularly interesting period because of the significant concentration of source rocks. It is suspected that overall climatic conditions then were warmer than in other periods in the Early Paleozoic. Carbonate reefs extended beyond the 30th parallels. In fact, in South America, the Tequeje and Retame formations (in present-day Peru and Bolivia) are significant shale source rocks (Baby et al., 1995), yet their paleolatitudes were approximately 40°S latitude (Torsvik and Cocks, 2011). Both of these formations have yet to be tested for their resource potential. In addition, the nature of fossiliferous and bioturbated zones interbedded with the source rock shales is cyclical, and the black anoxic shales overlying them are usually in very abrupt contacts. Do volcanic ash deposits trigger these abrupt changes? The interbedded oxic zones are often a few centimeters thick (fig. 2–2).

Thus far, we have discussed how organisms can be deposited onto the sea bottom and preserved in sediments. If the organisms are planktonic (i.e., floaters), then they serve only to yield the age of the deposits, although they may provide information on the ambient temperature via isotopic composition. Benthonic fauna (bottom dwellers) provide information on water depth and age but may be transported. The *ichnofacies* are an assemblage of trace fossils in which the animal itself has not been preserved, but the fossil shows instead where it lived or browsed. There are five important groups, or *ichnogenera*, of ichnofacies: *Glossifungites*, *Skolithos*, *Cruziana*, *Zoophycos*, and *Nereites* (fig. 3–7). The University of Alberta (Ichnofaunal Research Group; see Frey and Pemberton, 1985), the University of Kansas (Seilacher, 1964), and other universities have extensively researched this topic and serve as additional information sources.

Glossifungites are the simplest trace fossils to understand. They correspond to the family of boring creatures capable of penetrating through hard sediments or seashells to establish their dwellings. This can result in the pinholes observed on oyster shell surfaces. Boring organisms are also found on *carbonate hard grounds*, which form when carbonate-rich sediments are bored while they laid bare on the sea bottom for an extended period of time and were later covered by shale beds. Interestingly, in the Ordovician Utica/Point Pleasant shale play, brachiopod packstone hard ground without any boring has been observed (fig. 2–2), suggesting that anoxic conditions prevailed soon after deposition of the shell beds.

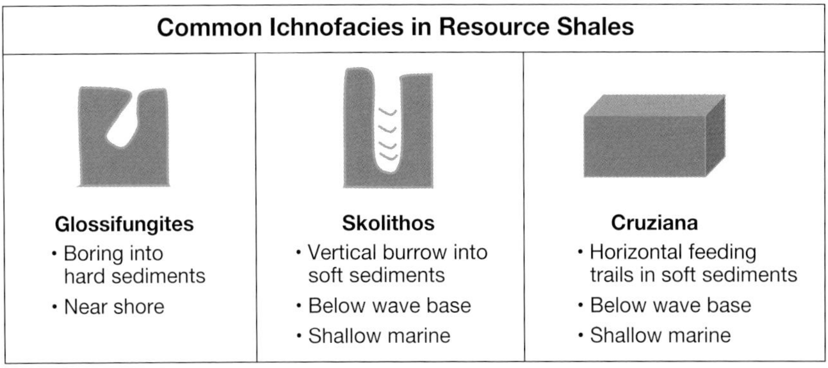

Fig. 3-7. Three classes of ichnofacies commonly found in sediments. The latter two are often observed in resource shales.

Burrowing organisms left trace fossils grouped as *Skolithos*. These include crabs, shrimp, and gastropods. Fauna that are burrowers can be distinguished simply from bottom dwellers: Burrowers, such as high-spired gastropods, usually have pristine shells and are never attacked by boring organisms like scaphopods. Modern-day organisms with pristine shells serve as good analogues for their geological counterparts. Sediments overlying burrowed surface always fill the burrows. Organic fluids from various organisms can result in chemical reactions that harden the walls and the burrowed chamber itself. It is important to document the presence of burrowing animals. Burrowers tend to homogenize the sediment instead of preserving discrete laminae. They indicate that oxic conditions were necessary for the animal to subsist. The conclusion from the fossil record is that periodic oxic conditions can exist even in a dominantly anoxic sequence. The Bakken also has such oxic intervals and often contain *Skolithos* burrows.

Cruziana are dominantly tracks and therefore are always bed parallel. Often they have a chevron shape resulting from displacement of the animal. The resulting grooves can be filled by sediment and thus preserved. Therefore, they can be preserved only in low-energy environments, definitely below effective wave base.

Zoophycos have the appearance of feeding burrows. Often they show internal layering as the animal passes through the sediment. They can also form complex dendritic patterns and U-shaped burrows.

Nereites are feeding trails in deepwater environments. Their unique characteristic is that trails, whether helicoidal or in another geometric shape, never cross a previous feeding path. They are indicative of abyssal deepwater. The final category of bioturbated shales are completely churned sediments with no remaining visible structures.

Why are ichnofossils so important? They represent hard evidence of the water depth at the time of deposition and oxic conditions. In addition, because of the internal chemical reactions taking place in the burrows, they often form resistant pillars within the sediment and thus inhibit compaction. As a banded shale undergoes compaction, anisotropy in the shale is enhanced. A sediment is anisotropic when any physical parameter has a directional component. For example, porosity might improve in a particular direction owing to anisotropy. By contrast, the sediment may be isotropic if there is no preferred direction for physical parameters. Heavily bioturbated shales tend to be more isotropic. As will be discussed in chapter 6, techniques exist to detect the anisotropy in shales on the basis of their geophysical characteristics. For this reason, it is important to compare cores from the same interval and determine the degree of bioturbation and its effect on anisotropy.

Cores from both the Paleozoic Bakken Formation of North Dakota and the Cretaceous Eagle Ford shale of South Texas show clear banding of highly organic-rich beds with no bioturbation and interbedded with intervals showing bioturbation (fig. 3–8). Sarg (2012) has also described the presence of the bioturbated intervals in the Bakken Formation. In general, the bioturbated intervals do not fluoresce, whereas thinly laminated sediments do fluoresce, indicating the presence of hydrocarbons. It is not clear what can cause a temporary reduction in anoxic conditions. In summary, highly laminated sediments are indicative of very calm conditions, showing few effects of bioturbation. All source rock intervals demonstrate this aspect.

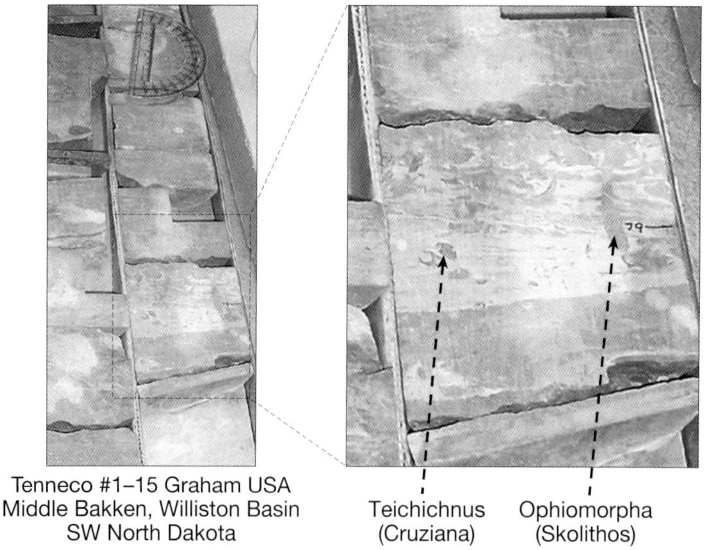

Fig. 3–8. Tenneco well 1-15 (Graham USA), showing bioturbated Bakken Formation, Williston Basin, southwestern North Dakota

ANOXIA AND ITS RELATIONSHIP TO CARBON, OXYGEN, AND OTHER ISOTOPES

The dominant source rocks in the world are associated not only with long periods of sea-level rise, or time of *maximum flooding surface* (MFS), but also with unique global events—namely, OAEs. In fact, the Devonian source rocks are not associated with a significant rise in sea level (fig. 3–9). The most studied OAEs to date are those of the Late Jurassic to Early Cretaceous and the Middle Cretaceous (fig. 3–4). That is, many hydrocarbon source rocks are associated with MFS, but not every MFS is associated with source rocks. In fact, good source rocks have also been associated with early transgressive system tracts (described in chap. 4).

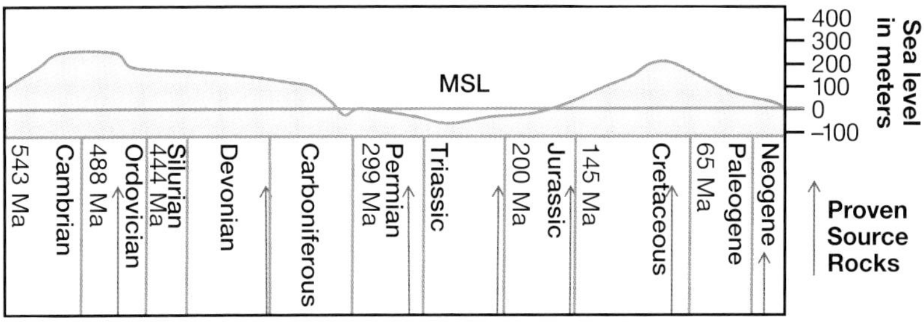

Fig. 3–9. Sea-level changes and principal source rocks do not correlate. "MSL" denotes present-day mean sea level. (Adapted from the British Geological Society website)

ISOTOPE AS PALEOCLIMATE INDICATORS

Carbonates trap carbon in their lattice, and the ratio of ^{13}C to ^{12}C isotopes can be measured. The same can be studied in organic material. Usually both isotopes are plotted, and the difference between the two isotopes is the $\delta\ ^{13}C$ value.

What is the driving force relating specific isotopes to paleoclimate events? Why are there differences in $\delta\ ^{13}C$? Two concepts about carbon isotopes are evident because normal atmospheric CO_2 has a different isotopic composition than CO_2 derived from organic material. First, during periods of global warming, atmospheric CO_2 is released in large quantities owing to photosynthesis; because more atmospheric CO_2 is available to be absorbed by plants, there is a general increase in $\delta\ ^{13}C$ (less negative) during warming trends. Second, during periods of cooling, the opposite effects occur, leading to a decrease in $\delta\ ^{13}C$ (Nordt et al., 2003). Due to the sensitivity of the analyses, further corroboration is recommended.

Note that OAEs are independent of sea-level rise and fall. Rather, they are atmospheric changes. How do we know the validity of this observation? The Pee Dee belemnite (PDB) carbon isotopes were used as a reference standard in studies up to very recently. Currently, other standards are used but always with reference back to the PDB. Values are either above or below this standard.

It is evident from the Cretaceous charts published by Leckie et al. (2002) that increases in $\delta\,^{13}C$ are directly correlated to OAEs. There have been two periods in geological time when sea levels reached a maximum: the Early Ordovician and the Late Cretaceous (fig. 3–9). Of the two Cretaceous OAEs—the Aptian/Albian (OAE 1b) and the Cenomanian (OAE 2)—only the latter was associated with a major sea-level rise (fig. 3–9).

Because $\delta\,^{13}C$ can also be studied in Paleozoic rocks, a similar conclusion can be reached. A quick review of major known Paleozoic source rocks indicates a close relationship. Are there other potential untested source rocks during other periods with high $\delta\,^{13}C$ events that have yet to be explored? The two major source rock intervals are the Mohawkian (456 Ma) and Frasnian/Famennian (375 to 359 Ma), with deposition of the Utica and Trenton shales in the Northeastern United States and Eastern Canada during the first and the Marcellus and Bakken and other Devonian shales during the second. Although the Ordovician OAE is associated with the large sea-level rise, the Devonian OAE is by far the most important oil-generating interval. Note that at the end of the Famennian, a major faunal extinction was recorded.

Because $\delta\,^{13}C$ is also affected by maturity (Schoell and LeFever, 2011) care must be taken in the analysis. In fact, the same authors point to its use in distinguishing migrated higher maturity gas versus in situ gas in shales.

A second set of isotopes studied in carbonate-bearing fossils are the oxygen isotopes. Of these, ^{16}O and ^{18}O are the dominant isotopes; ^{16}O is very abundant whereas ^{18}O is scarcer. Careful studies have shown that the ratio between the two is directly proportional to the ambient temperature. Water containing more ^{18}O condenses faster and thus precipitates in warmer environments and is depleted in colder climates. Therefore, as more ice is forming, less ^{16}O becomes available in the oceans and colder climates prevail. Today's standard, set by the Vienna Atomic Energy Commission, is *standard mean ocean water* (SMOW). The concentration of ^{18}O in a fossil sample is related to the standard as follows:

$$\frac{^{18}O\text{ sample} - {}^{18}O\text{ standard}}{^{18}O\text{ standard}} \times 1{,}000 \qquad (3.1)$$

Note that the Devonian had a particularly warm climate with extensive reefal conditions beyond the 30th parallels, followed by a significant cooling event during the latest Devonian to Early Carboniferous (Saltzman et al., 2000). During the Cenomanian, Earth had no ice caps.

RATES OF DEPOSITION

Biostratigraphy is used in estimating rates of deposition. Care must be taken to understand the impact of sedimentary compaction on these estimates. Given that caveat, an example is used from the Texas-Louisiana Gulf Coast. A well in Calcasieu Parish, Louisiana, shows the first appearance of the benthic fauna *Siphonia davisi* and *Cibicides jeffersonensis*. Importantly, when the first appearances occur, they are part of a group of fauna commonly associated in the same strata. Because their ages are well documented, the rate of sedimentation can be clearly established (fig. 3–10). The ages are provided by Paleo-Data (2017), based on thousands of wells studied in the Gulf of Mexico region. This interval has a rate of deposition of 26 mm per 1,000 years.

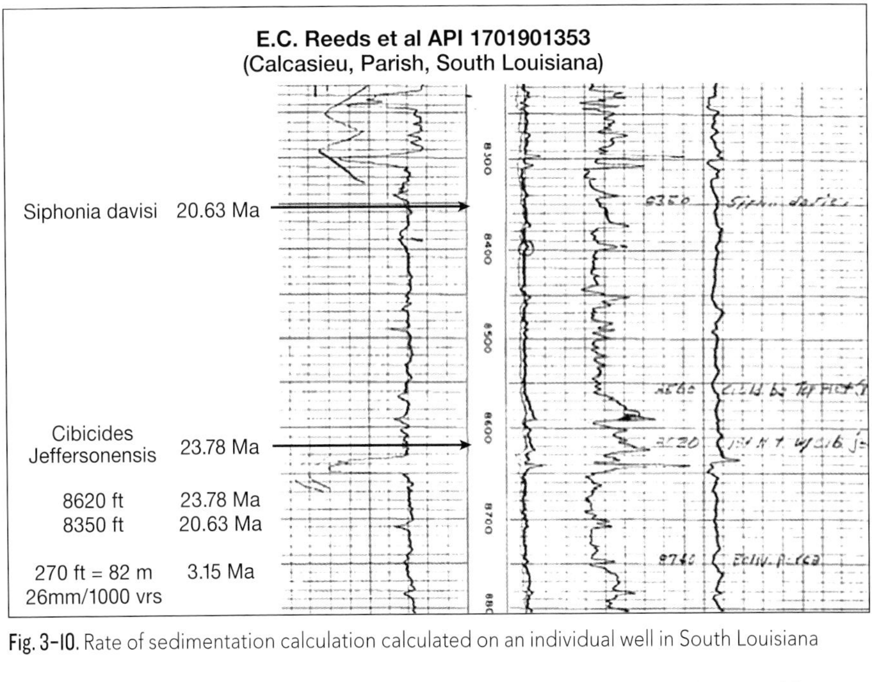

Fig. 3–10. Rate of sedimentation calculation calculated on an individual well in South Louisiana

The calculation for rate of deposition is critical to establish the burial history of the basin for both temperature and pressure. Deltas have sedimentation rates averaging 1,000 mm per 1,000 years. Stable shelf sediments may have rates as low as 50 mm per 1,000 years (Sadler, 1999). It is important to use this information on the burial history plots with care. The following assumptions are needed: first, the interval is not faulted; second, there are no unconformities; and third, there is no compaction or compaction is corrected. Corrections for each of these conditions may have to be integrated.

SUMMARY

Biostratigraphy serves many purposes:

- Age determination of the target horizon
- Detailed correlation between various sedimentary units in the basin
- Definition of the environment of deposition
- Time gap along unconformity horizons is provided
- Determination of the rate of deposition of the sediments
- Definition of MFS on the basis of faunal distribution and abundance
- Deduction of paleoclimatic conditions (distribution of reefs)
- Understanding of anoxic conditions resulting in environmental stress, reflected in the fauna by stunted growth

What controls anoxic conditions and distribution of sore intervals worldwide? Most of the world's oil was generated during the Mesozoic and particularly the Jurassic and Cretaceous. Why are there such few intervals generating oil? The answer appears to be because of prolonged periods of volcanic activity. Flood basalts associated with the opening of the Atlantic were of particular significance. Isotopes are commonly used to demonstrate the presence of anoxic events.

Several key maturation indicators have faunal associations. In particular, conodonts—and, to a lesser degree, acritarchs and graptolites—provide Paleozoic sediments with coloration indices to grade maturity. Pollen is used for the same purpose from the Mesozoic to recent time for estimation of sedimentation rates on the basis of fauna and thickness.

REFERENCES

Baby, Patrice, I. Moretti, Bertrand Guillier, R. Limachi, E. Mendez, J. Oller, and M. Specht. 1995. "Petroleum system of the northern and central Bolivian sub-Andean zone." In *Petroleum Basins of South America*, A. J. Tankard, R. Suárez Soruco, and H. J. Welsink (eds.). AAPG Memoir 62: 445–458.

Barrick, James. n.d. "Microscopic marvels of the Paleozoic: Conodonts." barricklab.org/twiki/pub/Lab/LinkList/Barrick_Conodonts.pdf.

Bartok, Peter, Maria Carolina Mejia-Hernandez, and Murad Ismael. 2015. "Paleogeographic constraints on Middle- to Late-Jurassic tectonic reconstruction of the Maya Block of southern Mexico and equivalent strata of northwestern South America." In *Petroleum Geology and Hydrocarbon Potential of the Colombia Caribbean Margin*, Claudio Bartolini and Paul Mann (eds.). AAPG Memoir 108: 201–216.

Bartok, Peter, Tom J. A. Reijers, and István Juhasz. 1981. "Lower Cretaceous Cogollo Group, Maracaibo Basin, Venezuela: Sedimentology, diagenesis, and petrophysics." *AAPG Bulletin.* 65: 1110–1134.

Baucon, Andrea. 2010. "Leonardo da Vinci, the founding father of ichnology." *Palaios.* 25: 361–367.

Bertrand, Rudolf. 1990. "Correlations among the reflectances of vitrinite, chitinozoans, graptolites and scolecodonts." *Organic Geochemistry.* 15: 565–574.

Blakey, R. C. 2003. "Carboniferous–Permian paleogeography of the assembly of Pangaea." In *Proceedings of the XVth International Congress on Carboniferous and Permian Stratigraphy. Utrecht, August 10–16.* Chicago: University of Chicago Press. Vol. 10, p. 16.

BouDagher-Fadel, Marcelle K. 2012. "Biostratigraphic and geological significance of planktonic foraminifera." *Developments in Paleontology and Stratigraphy.* Vol. 22. Amsterdam: Elsevier. www.sepmstrata.org/page.aspx?pageid=743.

Calvin, Christina, H. Gamero-Diaz, L. Mosse, R. Malpani, C. K. Miller, J. Xu, and K. Fisher. 2017. "Diagenesis of ash beds in mudrocks and their impact on production." Paper presented at Houston Geological Society Applied Geoscience Conference, The Woodlands, TX, March 7–8.

Cohen, K. M., S. C. Finney, P. L. Gibbard, and J.-X. Fan. 2013. "The ICS international chronostratigraphic chart." *Episodes.* 36: 199–204.

Courtillot, Vincent E., and Paul R. Renne. 2003. "Sur l'âge des trapps basaltiques" [On the age of basalt traps]. *Comptes Rendus Geoscience.* 1: 113–140.

De Baets, Kenneth, Renee Hoffmann, Jocelyn A. Sessa, and Christian Klug. 2016. "Fossil focus: Ammonoids," *Palaeontology Online.* 6: 1–15.

Demaison, G. J., and G. T. Moore. 1980. "Anoxic environments and oil source bed genesis." *Organic Geochemistry.* 2: 9–31.

Epstein, Anita G., Jack B. Epstein, and Leonard D. Harris. 1977. "Conodont color alteration: An index to organic metamorphism." USGS Professional Paper 995: 1–27.

Faul, Henry. 1966. *Ages of Rocks, Planets, and Stars.* New York: McGraw-Hill.

Foster, Gavin L., Caroline H. Lear, and James W. B. Rae. 2012. "The evolution of pCO_2, ice volume and climate during the middle Miocene." *Earth and Planetary Science Letters.* 341: 243–254.

Frey, Robert W., and S. George Pemberton. 1985. "Biogenic structures in outcrops and cores. I. Approaches to ichnology." *Bulletin of Canadian Petroleum Geology.* 33: 72–115.

Habicht, Johann Konrad Albert. 1979. "Paleoclimate, paleomagnetism, and continental drift." *AAPG Studies in Geology.* Vol. 9. Tulsa: American Association of Petroleum Geologists.

Hackley, Paul C., and Brian J. Cardott. 2016. "Application of organic petrography in North American shale petroleum systems: A review." *International Journal of Coal Geology.* 163: 8–51.

Haq, Bilal U., Jan Hardenbol, and Peter R. Vail. 1987. "Chronology of fluctuating sea levels since the Triassic." *Science.* 235: 1156–1167.

Hautmann, Michael. 2012. *Extinction: End-Triassic Mass Extinction.* Chichester, UK: John Wiley & Sons.

Hofmann, C., V. Courtillot, G. Feraud, P. Rochette, G. Yirgu, E. Ketefo, and R. Pik. 1997. "Timing of the Ethiopian flood basalt event and implications for plume birth and global change." *Nature*. 389: 838–841.

Johnston, David I., Charles M. Henderson, and Michael J. Schmidt. 2010. "Upper Devonian to Lower Mississippian conodont biostratigraphy of uppermost Wabamun Group and Palliser Formation to lowermost Banff and Lodgepole formations, southern Alberta and southeastern British Columbia, Canada: Implications for correlations and sequence stratigraphy." *Bulletin of Canadian Petroleum Geology*. 58: 295–341.

Jourdan, Fred, Gilbert Féraud, Hervé Bertrand, Ali Bashira Kampunzu, Gomotsang Tshoso, Michael K. Watkeys, and Bernard Le Gall. 2005. "Karoo large igneous province: Brevity, origin, and relation to mass extinction questioned by new $^{40}Ar/^{39}Ar$ age data." *Geology*. 33: 745–748.

Katz, David A., Matthew R. Buoniconti, Isabel P. Montañez, Peter K. Swart, Gregor P. Eberli, and Langhorne B. Smith. 2007. "Timing and local perturbations to the carbon pool in the lower Mississippian Madison Limestone, Montana and Wyoming." *Palaeogeography, Palaeoclimatology, Palaeoecology*. 256: 231–253.

Klemme, H. D., and G. F. Ulmishek. 1991. "Effective petroleum source rocks of the world: stratigraphic distribution and controlling depositional factors." *AAPG Bulletin*. 75: 1809–1851.

Leckie, R. Mark, Timothy J. Bralower, and Richard Cashman. 2002. "Oceanic anoxic events and plankton evolution: Biotic response to tectonic forcing during the mid-Cretaceous." *Paleoceanography*. 17: 13-1–13-29.

Lesher, James. 2014. "Xenophanes." In *The Stanford Encyclopedia of Philosophy*, Edward N. Zalta (ed.).plato.stanford.edu/entries/xenophanes/.

Mattinson, James M., Emile A. Pessagno, Homer Montgomery, and Clifford A. Hopson. 2008. "Late Jurassic age of oceanic basement at La Désirade Island, Lesser Antilles arc." *Geological Society of America Special Papers*. 438: 175–190

McDade, Elizabeth Chinn, Roger Sassen, Lloyd Wenger, and Gary A. Cole. 1993. "Identification of organic-rich lower Tertiary shales as petroleum source rocks, south Louisiana." *Gulf Coast Association of Geological Societies Transactions*. 43: 257–267.

McIntosh, William C., and Matthew T. Heizler. 2016. "New Mexico Geochronology Research Laboratory K/Ar and $^{40}Ar/^{39}Ar$ Methods." *New Mexico Bureau of Geology & Mineral Resources*. geoinfo.nmt.edu/labs/argon/methods/home.html.

Müller, R. Dietmar, Maria Sdrolias, Carmen Gaina, and Walter R. Roest. 2008. "Age, spreading rates, and spreading asymmetry of the world's ocean crust." *Geochemistry, Geophysics, Geosystems*. 9.

Nordt, Lee, Stacy Atlchley, and Steve Dworkin. 2003. "Terrestrial evidence for two greenhouse events in the latest Cretaceous." *GSA Today*, 13, (12): 4-9. December 4–9.

Obradovich, J. D., and C. W. Naeser. 1981. "Geochronology bearing on the age of the Monterey Formation and siliceous rocks in California." In *The Monterey Formation and Related Siliceous Rocks of California*, R. E. Garrison and R. G. Douglas (eds.). Los Angeles: Pacific Section, Society of Economic Paleontologists and Mineralogists, 87–95.

Paleo-Data. 2017. "Biostratigraphic charts." www.paleodata.com/biostratigraphic-charts/.

Parra, Marien, Lorena Moscardelli, and Maria Antonieta Lorente. 2003. "Late Cretaceous anoxia and lateral microfacies changes in the Tres Esquinas member, La Luna Formation, western Venezuela." *Palaios*. 18: 321–333.

Perkins, Bob Frank. 1951. *Hindeastraea Discoidea White from the Eagle Ford Shale, Dallas County, Texas*, no. 2. Dallas: Southern Methodist University Press.

Punekar, Jahnavi, Gerta Keller, Hassan Khozyem, Carl Hamming, Thierry Adatte, Abdel Aziz Tantawy, and Jorge E. Spangenberg. 2014. "Late Maastrichtian–early Danian high-stress environments and delayed recovery linked to Deccan volcanism." *Cretaceous Research*. 49: 63–82.

Reidel, Stephen P. 1998. "Emplacement of Columbia River flood basalt." *Journal of Geophysical Research*. 103: 27393–27410.

Renne, Paul R., Alan L. Deino, Frederik J. Hilgen, Klaudia F. Kuiper, Darren F. Mark, William S. Mitchell, Leah E. Morgan, Roland Mundil, and Jan Smit. 2013. "Time scales of critical events around the Cretaceous-Paleogene boundary." *Science*. 339: 684–687.

Renne, Paul R., Jonathan M. Glen, Simon C. Milner, and Andrew R. Duncan. 1996. "Age of Etendeka flood volcanism and associated intrusions in southwestern Africa." *Geology*. 24: 659–662.

Renz, Hans Hermann. 1948. *Stratigraphy and Fauna of the Agua Salada Group, State of Falcon, Venezuela*. Vol. 32. New York: Geological Society of America.

Resing, Joseph A., and Francis J. Sansone. 1999. "The chemistry of lava–seawater interactions: The generation of acidity." *Geochimica et cosmochimica acta*. 63: 2183–2198.

Romero, Linda M., Irene M. Truskowski, Timothy J. Bralower, James A. Bergen, Oscar Odreman, James C. Zachos, and Francia A. Galea-Alvarez. 2003. "An integrated calcareous microfossil biostratigraphic and carbon-isotope stratigraphic framework for the La Luna Formation, western Venezuela." *Palaios*. 18: 349–366.

Sadler, P. M. 1999. "The influence of hiatuses on sediment accumulation rates." *GeoResearch Forum*. 5: 15–40.

Saltzman, Matthew R., Luis A. Guzman, and Kryger C. Lohmann. 2000. "Earliest Carboniferous cooling step triggered by the Antler orogeny?" *Geology*. 28: 347–350.

Sarg, J. Frederick. 2012. "The Bakken—an unconventional petroleum and reservoir system." *Oil and Natural Gas Technology*. Final scientific/technical report, prepared for U.S. Department of Energy (award no. DE-NT0005672). http://geology-consortia.mines.edu/MUDTOC/.

Scotese, Christopher R., and William Stuart McKerrow. 1990. "Revised world maps and introduction." *Geological Society, London, Memoirs*. 12: 1–21.

Seilacher, Adolf. 1964. "Biogenic sedimentary structures." In *Approaches to Paleoecology*, Imbrie, John (ed.). New York: Wiley, 296–316.

Sheth, Hetu C., Kanchan Pande, and Rajneesh Buthani. 2001. "^{40}Ar-^{39}Ar ages of the Bombay trachytes: evidence for a Paleocene phase of Deccan volcanism." *Geophysical Research Letters*, 28: 3513–3516.

Schoell, Martin, and Julie LeFever. 2011. "Carbon and Hydrogen isotope systematics in gases from horizontal Bakken shale wells." *Search and Discovery, AAPG*, #110149.

Sim, Min Sub, Shuhei Ono, and Matthew T. Hurtgen. 2015. "Sulfur isotope evidence for low and fluctuating sulfate levels in the Late Devonian ocean and the potential link with the mass extinction event." *Earth and Planetary Science Letters.* 419: 52–62.

Soeder, Daniel J. 2017. *Unconventional: Natural gas development from Marcellus Shale.* Vol. 527. New York: Geological Society of America.

Stoneley, R. 1990. "The Middle East Basin: A summary overview." In *Classic Petroleum Provinces*, J. Brooks (ed.). *Geological Society, London, Special Publications.* 50: 293–298.

Thiede, David S., and Paulo M. Vasconcelos. 2010. "Paraná flood basalts: Rapid extrusion hypothesis confirmed by new $^{40}Ar/^{39}Ar$ results." *Geology.* 38: 747–750.

Torsvik, Trond H., and L. Robin M. Cocks. 2011. "The Palaeozoic palaeogeography of central Gondwana." *Geological Society, London, Special Publications.* 357: 137–166.

Van Hinte, J. E. 1976. "A Cretaceous time scale." *AAPG Bulletin.* 60: 498–516.

Wandrey, Craig J. 2004. "Bombay geologic province Eocene to Miocene composite total petroleum system, India." *U.S. Geological Survey Bulletin.* 2208F.

Yancey, Thomas E., and Renald Guillemette. 1998. "Major volcanic ash units in the Eocene of East Texas." *Gulf Coast Association of Geological Societies Transactions.* 48: 511–516.

Zhang, Yi Ge, Mark Pagani, Zhonghui Liu, Steven M. Bohaty, and Robert DeConto. 2013. "A 40-million-year history of atmospheric CO_2." *Philosophical Transactions of the Royal Society A.* 371: 20130096.

4

SEQUENCE STRATIGRAPHY

TAKEAWAY

- Definition of sequence stratigraphy
- System tracts—lowstand (LST), transgressive (TST), and highstand (HST), along with review of maximum flooding surface (MFS)
- Sequence stratigraphy techniques—Vail (seismic), Galloway (logs), and Hubbard (tectonosequence) methods
- Application to stable shelf, listric faults, and foreland basins
- Outcrop and core descriptions
- Interpretation techniques for seismic and logs

GENERAL CONCEPTS

Shale source rocks have been well documented for over 50 years. However, only over the past 30 years has this study been placed in the context of sequence stratigraphy. The first true sequence stratigrapher was James Hutton, who in 1787, while hiking along Siccar Point, east of Edinburgh, observed and documented the first angular unconformity. Up to that point, Earth history was assumed to be several thousand years old, and the processes involved in creating geological features were believed to be instantaneous. Hutton realized that for a significant sequence boundary to exist, defined by a major unconformity, thousands of years (or more) had to have elapsed. As geological phenomena such as faunal extinctions and major tectonic events were recorded, Earth's history was progressively subdivided and fine-tuned into eras, periods, and stages. Modern sequence stratigraphy has simply reinforced these preexisting subdivisions. The nomenclature used today is somewhat different, but the context is similar. The question addressed in this chapter is whether sequence stratigraphy can be used to identify potential source rock intervals within the resource play.

Modern sequence stratigraphy is based on changes in eustatic sea levels overprinted by local tectonic events. There are three methods in common use to describe sequence stratigraphy. For all three, as sea level rises into newly created accommodation spaces, the sediment units transgress gradually onto preexisting unconformities. The process continues until marine flooding reaches a maximum, so that the accommodation space is filled and the rate of sedimentation is reduced to a minimum. This interval is known by two names: a *condensed sequence*, or *maximum flooding surface* (MFS). As the sea regresses, a *highstand system tract* (HST) sets in, and sedimentation spills over the extended shelf. Once the sea level drops to its new baseline, the sequence repeats itself.

However, the actual geological phenomena may not be as clear-cut as this theory suggests. In seismic, based on closely tied regional lines, the interpreter must determine the significance of the missing intervals and the consequences of that history. Thus, by analogy, a seismic line is to a stratigrapher what a novel is to a reader when chapters have been lost and the reader must determine what is the most likely plot and sequence of events. In addition, contacts between formations may be onlapping, downlapping, or conformable.

Chronostratigraphic charts known as *Wheeler diagrams* are used to unravel the geological history of the sequences (Wheeler, 1958). For detailed examples of sequences, refer to the work by Van Wagoner et al. (1992). The basic principle of the Wheeler diagram is to focus on the age of the units, using biostratigraphy, rather than the thickness. In the diagram, distance is given along the *x*-axis and age or time along the *y*-axis. Modules are currently available to generate Wheeler diagrams on computer workstations (Brouwer et al., 2008) (fig. 4–1).

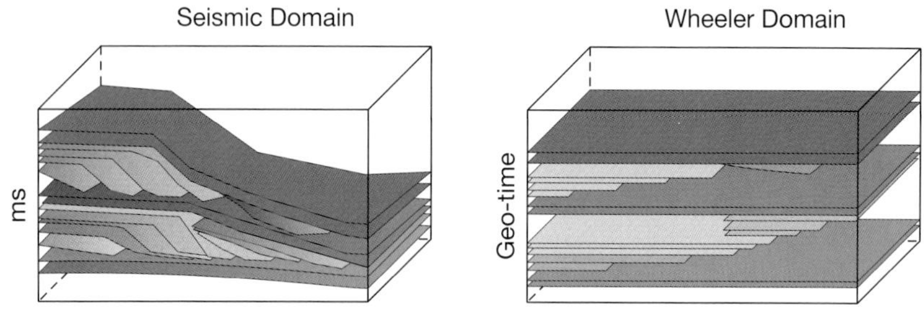

Fig. 4–1. Seismic to Wheeler domain (Reproduced with kind permission of F. Brouwer and the AAPG; Brouwer et al., 2008. Any subsequent reproduction requires permission from the AAPG.)

An additional element of sequence stratigraphy is to define the size and range of the sequence. The set of sequence tracts have been grouped as *orders*. A first-order cycle corresponds to the interval encompassing entire geological periods (~50–100 million years). It is of negligible use in exploration. Three examples are described using the U.S. midcontinent region as an example, based on the work of Sloss (1964): The Tejas cycle covers the entire Tertiary, the Zuni cycle covers the Jurassic and Cretaceous, and the Absaroka covers the Late Carboniferous to the Jurassic (Sloss, 1964) (fig. 4–2). Notably, there is general agreement between the first-order cycles and major tectonic events.

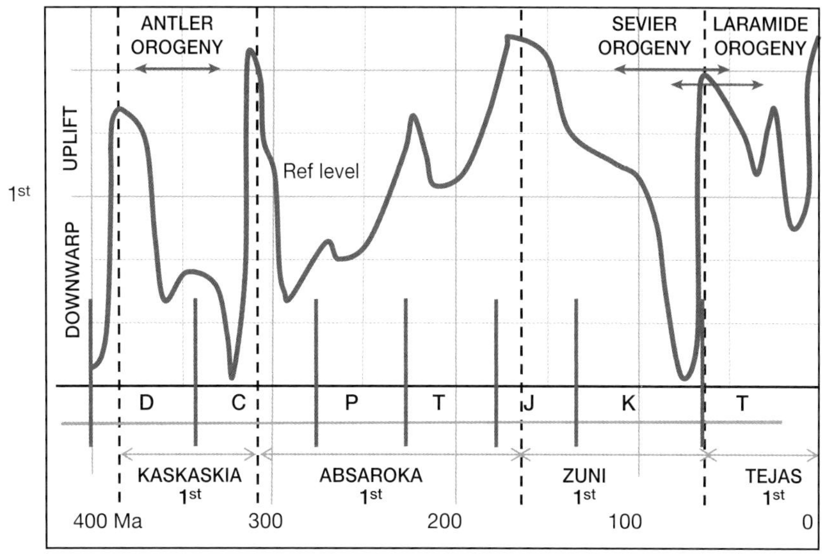

Fig. 4–2. First-order cycles for the U.S. midcontinent. Note that the time span of the cycles is ± 100 million years. Uplift and downwarp are differentiated. (Adapted from Sloss, 1964)

Second-order cycles are still very coarse and may or may not play an important role in exploration. The Lower Cretaceous of South Texas is one example (fig. 4–3). Decades of studies by the Bureau of Economic Geology of Texas are summarized in figure 4–3, and the subdivisions are based on the work of Phelps et al. (2013). Basically, the entire Albian constitutes the second-order supercycle, and it encompasses several formations and significant facies changes.

Figure 4–3 also shows the greater relevance of the third-order cycles. The third-order cycles have a duration of 0.5–5 Ma and thicknesses of 200–500 m. As seen in figure 4–3, they do correspond to major facies changes and are clearly noticeable on seismic lines. Even for the Eagle Ford Formation (at the top of fig. 4–3), the third-order sequence has exploration implications. However, more detailed work has demonstrated that at least two fourth-order sequences can be found in Eagle Ford (fig. 4–4). Further, Breyer et al. (2013) and Donavan et al. (2012) have provided evidence for a possible third subdivision.

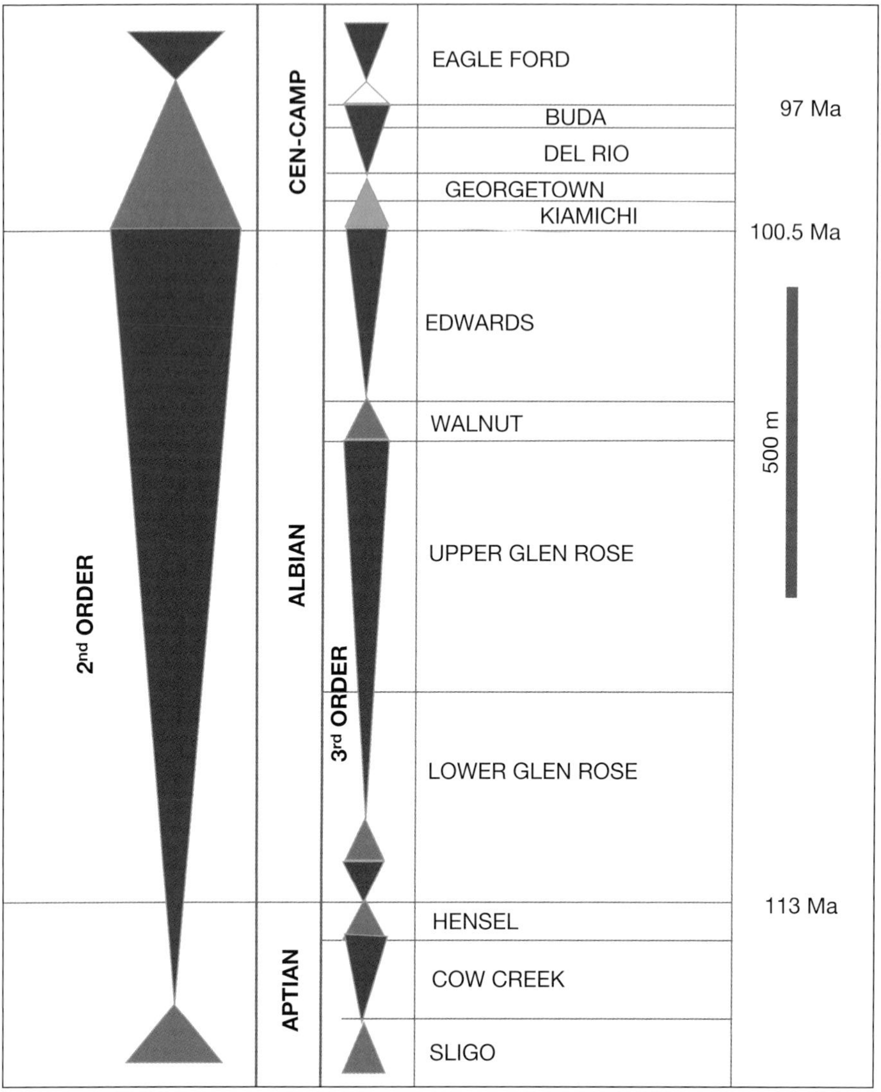

Fig. 4–3. General sequence stratigraphy for Cretaceous of South Texas. Second-order (left track) and third-order sequences (middle track) are shown. (Adapted from Phelps et al., 2013)

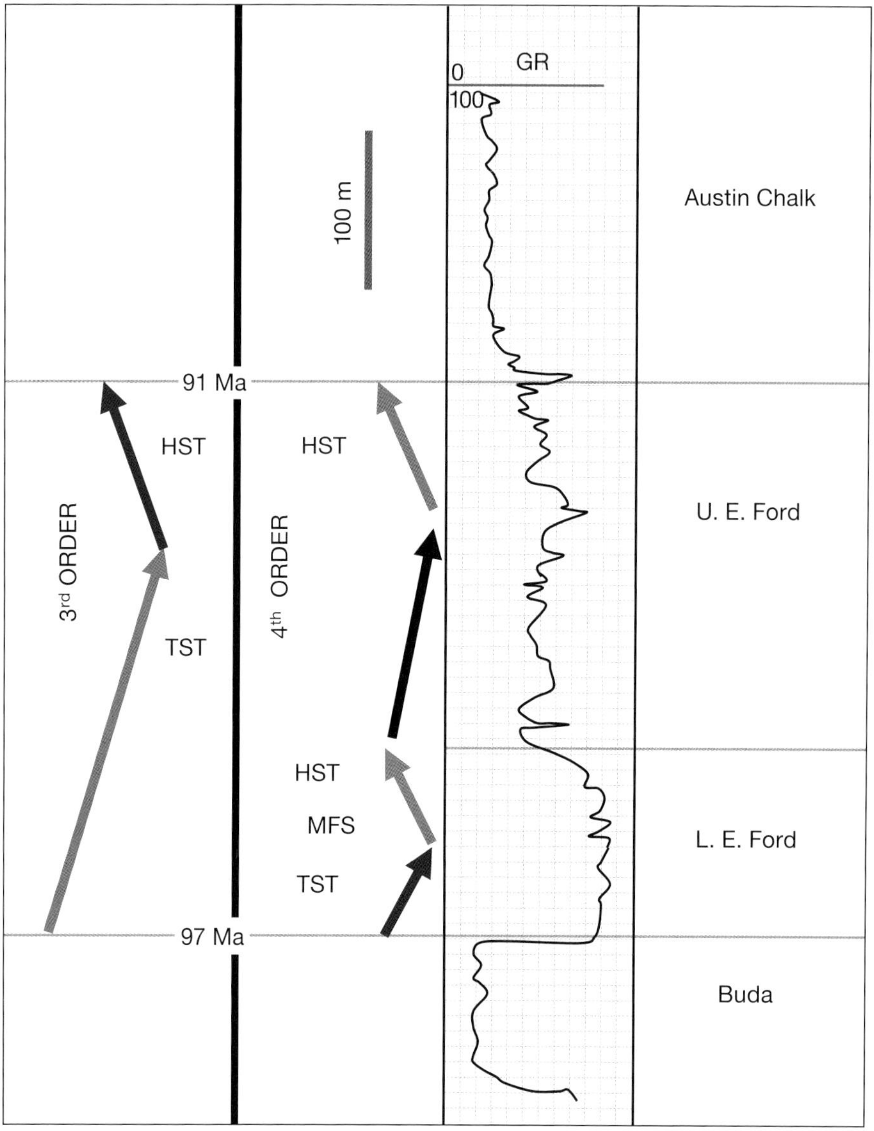

Fig. 4–4. Fourth-order sequence for the Eagle Ford Formation in South Texas, using composite type log. The MFS in the lower Eagle Ford is referred to as the "middle Shale member." The age boundaries are those defined by Lowery et al. (2014).

The optimum shales are in lowermost Eagle Ford Formation. Biostratigraphy and sequence stratigraphy are closely linked. Lowery et al. (2014) have pointed out that few benthic fauna and low diversity were observed in the Lower Eagle Ford, yet they represent a significant population in the Upper Eagle Ford. The Eagle Ford Formation ranges in age from 97 to 91 Ma (Lowery et al., 2014). The thickness of the Lower Eagle Ford is approximately 100 m. Its age is bracketed by the foraminiferal *Rotalipora cushmani* biozone, which ranges from 96 to 93 Ma, and the Buda Formation at its base. Its rate of deposition is 30 mm per 1,000 years, a slightly higher range than most shale resource plays. This is possibly related to their high carbonate content. Estimates vary as to the amount of compaction that may have taken place in the Eagle Ford shale. The upper *Rotalipora cushmani* zone coincides with an increase in $\delta\,^{13}C$ and the presence of the OAE-2 event (Lowery et al., 2014).

An additional example to study sedimentation rates is provided by the Late Paleozoic Bakken shale (Exshaw in Canada). It is dated on the basis of its conodont assemblage (fig. 3–3). The basal contact of Bakken with the Three Forks is the contact between the *Expansa* and *Postera* conodont zones, which dates to 365 Ma. The Middle Bakken spans the boundary between the Famennian and Tournaisian, with the Upper Bakken lying in the *Quadruplicata* conodont zone (356 Ma) (fig. 3–3). The Lower Bakken lies entirely within the *Expansa* zone (365 to 362 Ma) and has an average thickness of 10 m; during these 3 million years, the rate was 3.3 mm per 1,000 years, indicative of a very condensed sequence. Typically, the more condensed the sequence is, the better will be the chances for a high total organic carbon (TOC) content and for the source rock not be diluted by high sediment contamination.

Two questions need to be considered to understand sedimentation that are related to the creation of accommodation space: First, is sedimentation keeping pace with subsidence? Second, is it less than the subsidence rate does it exceed the accommodation space? The answers to both questions yield varying deposition patterns. Therefore, the regional context must be understood to properly estimate the sedimentation rate for a unit.

An example of the procedure and how to integrate the distribution of TOC is as follows:

- Note any significant unconformities and estimate the eroded section
- Establish if the interval is faulted and estimate the thickness of the missing section
- Determine the diagenetic overprint and its timing with respect to faulting and fracturing
- Estimate the compaction effect
- Note the reservoir heterogeneity
- Establish continuity for the reservoir, both lateral and vertical
- Identify and document any unconformity horizons and establish the sequences

- Determine the sedimentation rate
- Using the estimated TOC data for the well, establish its distribution in the context of the sequences (TOC content will be described in detail in chap. 7, and chap. 8 discusses the variability in TOC distribution because of its impact on reserve calculations.)

How can sequence stratigraphy best be applied? To answer this, view an outcrop of a source rock shale and focus on the following two features: first, the shale is not homogeneous; and second, there is *cyclicity* to the depositional pattern, which forms the basis for the theory of sequence stratigraphy. This outcrop study method will be reviewed later in this chapter.

The best example of cyclical sedimentation is provided by the Carboniferous cyclothems of the U.S. midcontinent and Illinois Basin. Each cycle is composed of sandstone or shale/limestone/coal or black shale/shale, with variable thicknesses for each cycle. Each cycle was deposited in 400,000–500,000 years. Each unit in the cyclothem may be 5–10 m thick. Moore (1964) suggested that the cyclothems are related to brief cooling and warming periods. The challenge is to blend the seismic scale with the logs, the core, and outcrop scale in the context of sequence stratigraphy.

The role of preexisting structures is also important. Examples include the Appalachian Basin, the Jurassic in East Texas, central Mexico, and the Middle East. In the Appalachian Basin, the Lexington, Galena, and Trenton platforms (fig. 4–5) are the focus of exploration for the Marcellus and Utica shales. The Galena shelf is a developing Utica play in Ontario, Canada. The relationship between the producing trends on the platforms and the Precambrian Rome graben is evident in figure 4–5. The bulk of the Utica production is in the Sebree trough and Galena platform of Ohio.

The Late Jurassic Haynesville shale—combined with its lateral equivalent in Mexico, the Pimienta Formation—serves as another example that demonstrates problems with oversimplifying geological history. Both basins show indications of Triassic to Early Jurassic rifting. Mexico also shows the strong overprint effects of Late Jurassic rifting (Bartok et al., 2015). The maximum thickness for the Haynesville is 100 m, and the age of the unit is 156 to 151 Ma. The rate of sedimentation is also 20 mm per 1,000 years. Interestingly, the Haynesville's thickest intervals are associated with the salt basins of East Texas and North Louisiana (fig. 4–6). They are unusually thick, as a response to the sag basins that formed due to the Triassic grabens and salt withdrawal, thus increasing the accommodation space. The sag basin fill is created by the differential compaction of the syn-rift sequence typical of the early postrift sedimentation. Otherwise, the unit would retain its regional low rate of deposition.

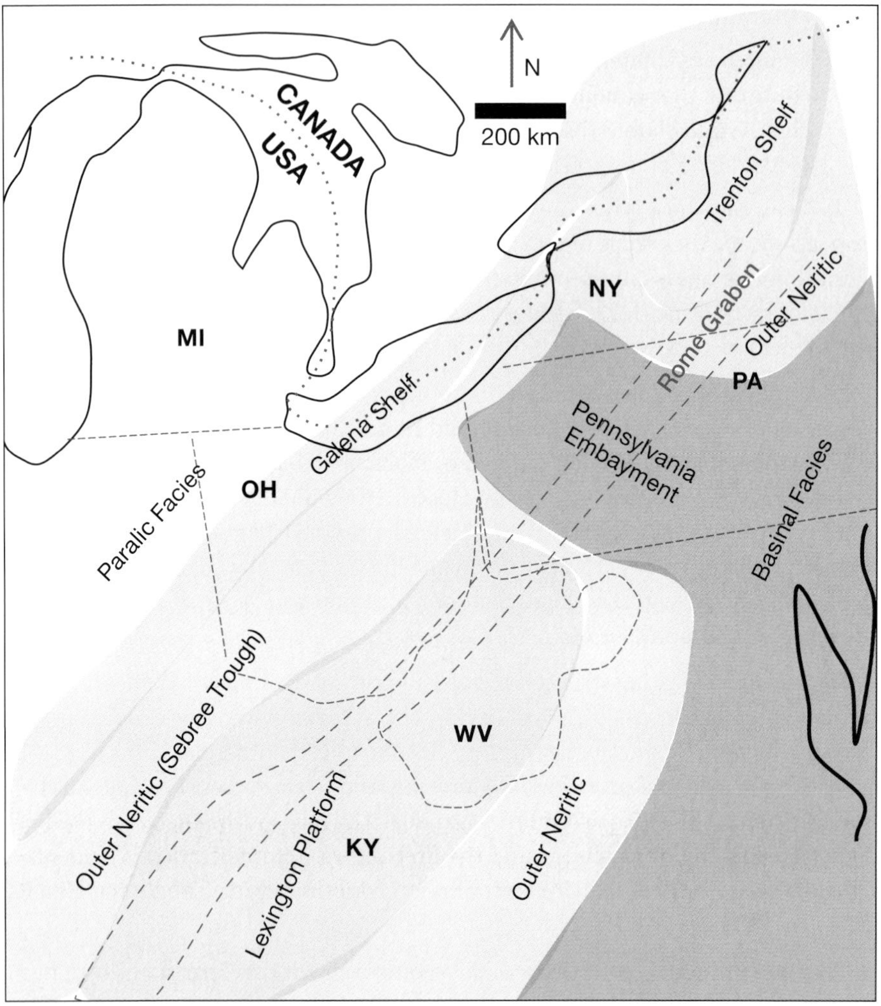

Fig. 4–5. Appalachian Basin paleogeography for the Ordovician (Utica deposition) indicating the Trenton and Lexington platforms. The Rome graben crosses Pennsylvania (PA), West Virginia (WV), and central Kentucky (KY). (Based on Clepper et al., 2011)

The Haynesville Formation is a good example of a developed shale resource play that fits the model of an early postrift in the MFS. The regional cross section (fig. 4–7) clearly demonstrates that the thickness of the Haynesville is directly proportional to the position of the underlying graben, and the thick distribution contains the optimum targets for the play. Note that the overlying unit, the Cotton Valley, is a classic prograding sequence and follows the pattern of an HST overlying the Haynesville.

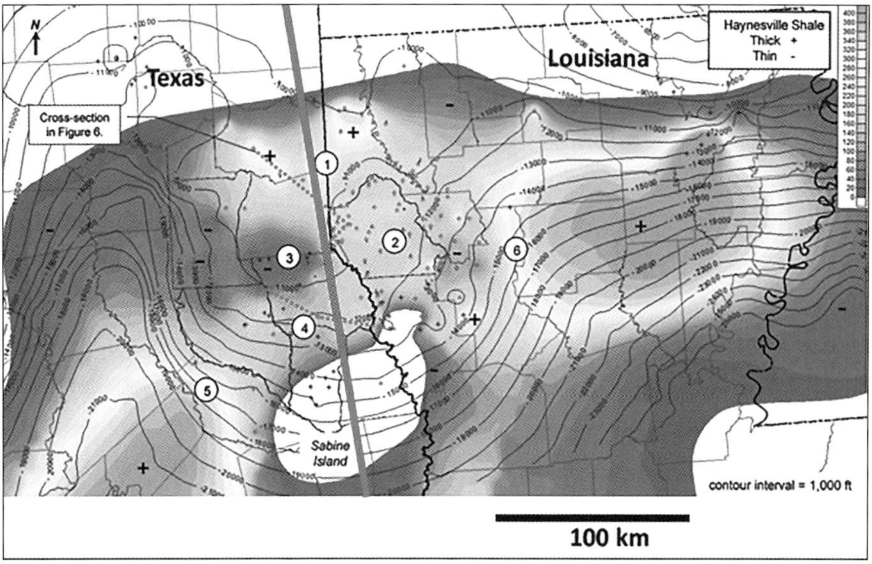

Fig. 4-6. Haynesville (Kimmeridgian) East Texas, West Louisiana, USA, note that the maximum, in general lies in the center of the East Texas and North Louisiana salt basins. Cross section is shown in figure 4–7. (Reproduced with kind permission from the AAPG; Durham, 2010. Any subsequent reproduction requires permission from the AAPG.)

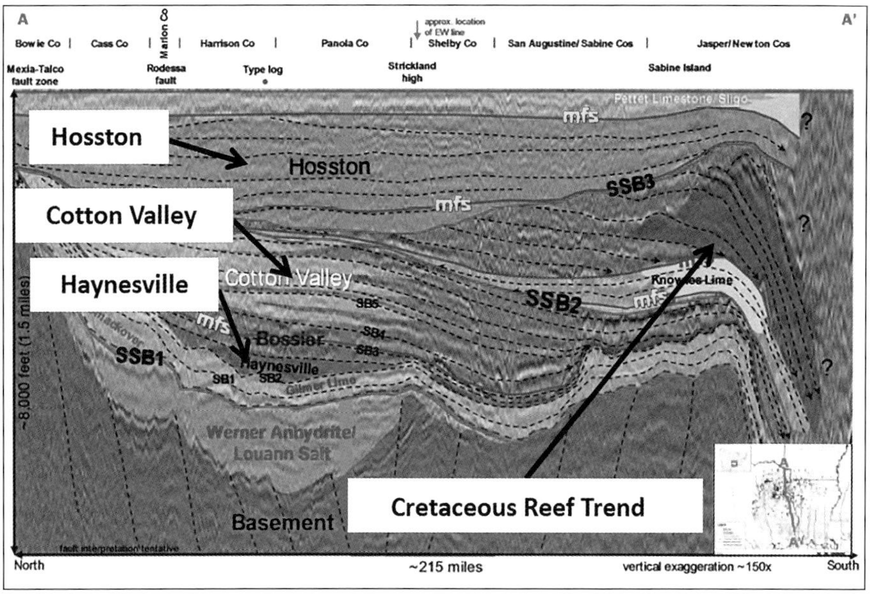

Fig. 4-7. North Louisiana seismic cross section showing Haynesville distribution. (Reproduced with the kind permission of GSCAGS; Steinhoff et al., 2011)

The equivalents to the Haynesville in northern Mexico are the Pimienta, Taman, and San Andres formations (fig. 4–8). These potential resource plays require further evaluation and the application of lessons learned from the Haynesville shale. The San Andres carbonates are confined to the rift shoulders. In contrast, the optimum targets for the Pimienta and Taman would again be near the graben axis. Similar graben conditions apply to the Cuisa shale, located on the Guajira Peninsula of Colombia. They have the same age and were deposited in a similar tectonic setting (Bartok et al., 2015). The difference between the Haynesville and the Pimienta is that the latter suffered a second rift phase. Moreover, the Laramide orogeny of the Sierra Madre Oriental compressed the graben, resulting in the observed central dome, and thus modified the fracture pattern affecting the resource plays. Along the flanks of the graben, thrusting is also evident. As will be discussed in chapter 7, rifting increases heat flow and therefore significantly affected the thermal history of the Pimienta Formation. In addition, the fracture patterns will be controlled by all tectonic events superimposed. Both the Haynesville and Pimienta formations can be recognized on seismic and have a wide extent. Understanding the underlying rift system will greatly improve predicting sweet spots in both formations. In the case of rift systems, the early postrift often extends beyond the boundaries of the original graben; diagnostic strong amplitudes result because of the cleaner shale contact with respect to both overlying and underlying coarser-grained lithologies. As a point of reference, note that the Lower Cretaceous Stuart City reef trend lies on the horst of the rift (Angelina Caldwell High, fig. 4–7). In the Mexican example, the Tamabra Formation is the reef debris apron of the Cretaceous reef trend associated with the family of highs related to the Tamaulipas High (including the Golden Lane, fig. 4–8).

A better understanding of rates of sedimentation can be achieved by considering the Oligocene Frio Formation of East Texas. It is a deltaic deposit dated from 28 Ma at its base to 25 Ma at its top. The thickness of the unit is in excess of 1 km. The rate of sedimentation is 300 mm per 1,000 years.

In general, shales dominate *transgressive system tracts* (TSTs). They are relatively easy to recognize on seismic data because they onlap a preexisting surface or an older shelf. If the basin is stable, recognition of the TST is straightforward. However, if the basin is undergoing either contemporaneous subsidence or uplift, it may be more difficult to identify. Uplift of the hinterland results in upward curvature of the onlap terminations due to the uplift (Allen and Allen, 2013).

Subsidence creates a more difficult geometry. The downwarp results in apparent downlaps of the terminations. Interpreting the sequence requires structural restoration. This is best accomplished by flattening the horizon on the most representative MFS that is in close proximity to the problem area. On petrophysical logs, core data, and outcrops, the TST shows a fining and thinning upward sequence with a characteristic bell-shaped pattern. The sandstone that can develop at the base tends to become thinner and more shaley as the transgression continues. Ultimately, the TST necks into the MFS. Once a TST is defined, it is often straightforward to establish the presence of an MFS on seismic data. TSTs can contain source rocks but are usually associated with more oxygenated conditions.

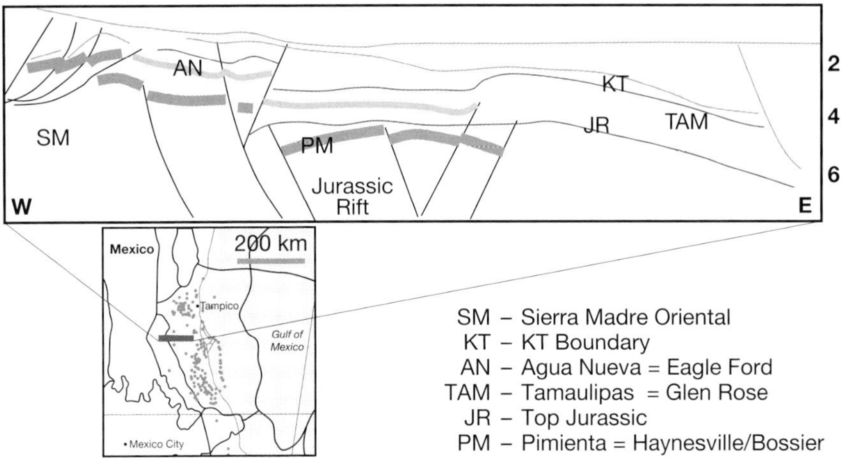

Fig. 4–8. Cross section of Tampico Misantla Basin, North-Central Mexico. Note the similarity to figure 4–7. The Pimienta is Late Jurassic (Tithonian), and the Agua Nueva is Cenomanian, equivalent to the Eagle Ford shale of South Texas. The depth scale is in approximate km. Golden Lane reef trend is at the eastern end of the line. (Modified from U.S. Energy Information Agency, 2015, using author data)

Theoretically, the concept of sequence stratigraphy is sound. In practice, there are many other factors that each play a role. Miall (1984) summarized factors affecting sequence stratigraphy, several of which pertain to the shale resource problem. Techniques of sequence analysis study will be reviewed later in this chapter. Here, suffice it to say that sequence stratigraphy is an excellent descriptive tool that is dominantly controlled by the response of sediments to the accommodation space being created or destroyed in the environment of deposition.

Three methods are commonly applied to study sequence stratigraphy:

- The Vail method, based on sea-level fluctuations (Vail et al., 1977)
- The Galloway method, based on log response (Galloway, 1989)
- The Hubbard method, based on major tectonic events (Hubbard et al., 1985)

The Vail method is the best known and states that sea levels have fluctuated throughout Earth's history and that sedimentation adjusted to the changes (Vail et al., 1977). As seas regress and transgress in reference to the craton, the resulting accommodation space gives rise to distinguishable diagnostic sedimentation patterns that can be classified and mapped. Each segment of the cycle is called a *system tract* and can further be subdivided into *gross depositional environments* (GDEs) such as delta and pro-delta. Vail et al. were keenly aware that local tectonics would complement the eustatic model.

The Vail method should be applied only to marine shelf and shelf margin conditions and is well suited for seismic data. Continental sedimentation is referred to in the Vail method only by the presence of incised valleys. The term *bay-line* is used by Vail et al. (1977) to define the limit of marine encroachment onto the coast and generally defines the upper limit of the Vail cycles. In the deepwater sedimentation realm, the method is less reliable owing to the dominance of shales and the possibility of extreme condensed sequences and diastems (missing intervals in horizontal sequences) that appear as conformable layers. In areas with extensive growth faulting and thrust belts, such as South Texas/South Louisiana and the Rocky Mountain foothills, the Vail method is less reliable because the tectonic overprint overwhelms the global eustatic effect.

The Galloway method is based entirely on log response and is geologically driven (Galloway, 1989). It is also better suited for outcrop studies and simply states that sequences can be identified on the basis of log and outcrop patterns. Petrophysical logs are more difficult to analyze in the context of sequence stratigraphy, with the exception of the dominant shale event described by Vail et al. (1977) as the MFS. It is often referred to as the *shale neck* on gamma-ray and spontaneous-potential curves (both of which are discussed in more detail in chap. 5). One of the best ways to depict the location of the shale neck is to display an inverted gamma-ray log on opposing tracks (fig. 4–9). If gamma ray is not available, an inverted spontaneous-potential log may be used instead.

Galloway (1989) took advantage of the phenomenon associated with the MFS and used it as his reference sequence boundary. Sequences in the Galloway system run from MFS to MFS because most unconformities are difficult to establish on logs. He referred to them as *genetic units*. For shale resource assessment, defining the dominant MFS is crucial. Keep in mind that many resource shale plays along passive shelf margins are located at or near the MFS, but not all MFSs contain resource plays.

The third classification system, championed by Hubbard et al. (1985), focuses on further grouping sequences into *megasequences* defined by major tectonic events. The approach is seismically driven and based on the separation of genetically related tectonosequences. Examples of the megasequences are rifts, listric faults, and foreland basins. All three have distinct units that can be identified, and their internal sequences can be mapped.

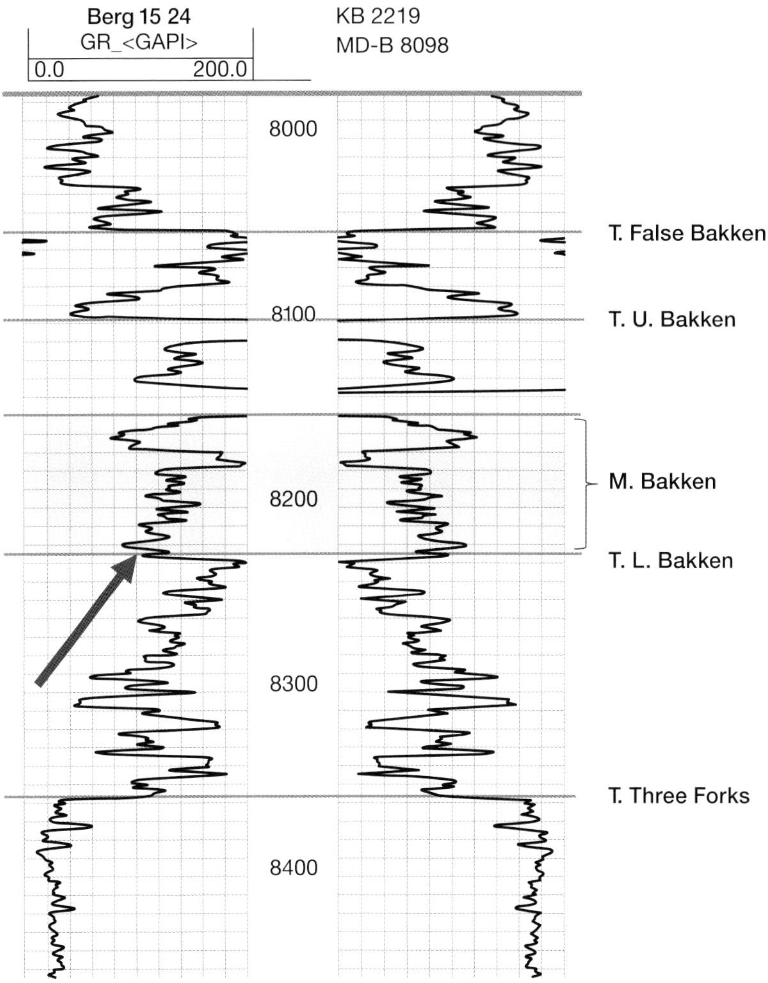

Fig. 4–9. Berg 15-24 well, Bakken Formation, North Dakota, showing tops and inverted gamma-ray track. Lowermost Bakken shows necking, as does the upper portion of the Lower Bakken.

In the case of rifts, the prerift sequences are composed of sediments and other rock types deposited or formed prior to the rifting event. Their internal genetic relationships are inherited from the tectonic events immediately preceding the rift event. They may or may not contain resource plays. The syn-rift consists of sediments rapidly filling a newly created accommodation space and may range from continental to shallow marine sediments. However, if the sediment infill is slow, they may contain resource plays as lacustrine sediments. One of the best modern examples of an accommodation space with associated low rates of sedimentation is Lake Tanganyika, in East Africa, where water depths in excess of

1,000 m and stagnant lake-bottom conditions create a perfect environment for the accumulation of organic-rich shales. Another similar and originally lacustrine system is the Black Sea. Water depths there are in excess of 2,000 m, and organic-rich black shales dominate the depositional fill. After the last glacial event, marine conditions entered a previously lacustrine environment and flooded the Black Sea. Many of the East China basins and Brazil fall into this category of rift-associated lacustrine systems. In the Aptian rift system in Brazil (130 Ma), the Lagoa Feia shales of southeastern Brazil contain rich lacustrine organic shales with resource potential in the billion barrel range. The coeval Bucomazi shale and lateral equivalents of Angola and Congo, in West Africa have similar potential. A significant difference between the Vail and Hubbard methods is that the latter can include continental sequences, such as lacustrine environments whereas the former applies only to marine conditions.

In the case of listric faults, the accommodation space is filled with synkinematic and postkinematic fill (Hossack, 1995). MFSs appear in the postkinematic phase. The Lower Eocene of the Gulf of Mexico is a particular case in point. The sequence was deposited in an overall deltaic setting with corresponding listric faults associated with downslope gravity sliding. The shale of interest in the Eocene expansion is the Sparta shale and was deposited as part of the Canyon River Member of the Wilcox Group. It is a source rock interval restricted to the Louisiana coast (both onshore and offshore). The unit has yet to be tapped as a resource play. McDade et al. (1993) have provided an analysis of the geochemical characteristics of the Sparta shale. It is possible that the coincidence of an MFS and appropriate climatic conditions gave rise to the source rock development. The environment of deposition of the Sparta shale is associated with the early postkinematic deposition described previously (see chap. 2). The lessons learned from these shales and the related MFS is that all shale resource plays lie either in the third-order MFS, at the very base of the HST, at the TST or in the late synkinematic phase. Very specific conditions must be present for a resource play to develop. As stated earlier, the environment must be stressed, and OAE conditions must prevail.

The same rules apply to foreland basins as applied to the rift basins. Dickinson (1974) described a foreland basin as a retroarc and peripheral basin in which the basin fill is the result of plate collision and is deposited on a continental crust. The basic components of a foreland basin are a *moving block*, resulting in uplift and thrusting, and a *flexing hinterland*, responding to the load from overlapping thrust and fold belt (fig. 4–10). The basin principally responds to the load of the overriding fold-thrust belt, influencing the creation of accommodation space for subsequent sedimentation (Bartok, 2003).

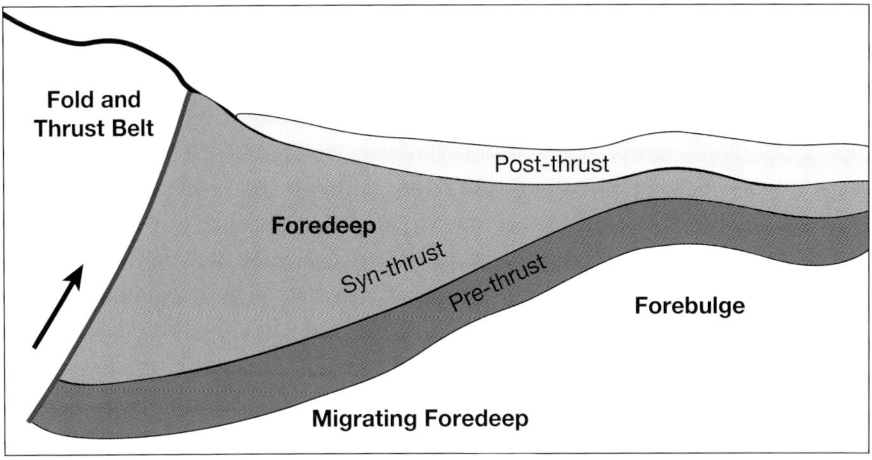

Fig. 4–10. Foreland basin sequence stratigraphy, dominated by the deformation of the overriding block

Foreland basin sequence analyses are important because they contain many unconventional resource plays. The sediments are associated with either a syntectonic phase or the early and late posttectonic sequence (fig. 4–10). The best resource shales tend to prevail in the early posttectonic phase. Often tectonic belts are episodic. In the case of the Rocky Mountains, the initial foreland and orogeny is the Antler orogeny (390 to 340 Ma), followed by the Sevier orogeny (140 to 50 Ma) and then the Laramide orogeny (70 to 35 Ma), with several internal stages within each major event (fig. 1–1). During the episodic uplift of the fold belt, the foreland basin responds with dampened subsidence, resulting in a synkinematic transgressive system tract. The MFS follows during the early postkinematic phase. Figure 4–8 indicates an example of source rock interval deposited in active foreland basin. The Agua Nueva Formation is Cenomanian and was deposited during the Sevier orogeny.

In the northern Rocky Mountain system, the main resource plays are as follows:

- Winnipegosis (lowermost Devonian, 387 Ma), in southern Saskatchewan, Manitoba, and eastern Montana
- Duvernay, Montney, and Horn River Devonian, Lower Frasnian, 383 Ma), in western Alberta, northeastern British Columbia, Canada
- Sappington/Exshaw/Bakken (latest Devonian, 372 to 370 Ma), in western Alberta, southern Saskatchewan, eastern Montana, and North Dakota

All of these plays are essentially coeval with the Antler orogeny. In the Basin and Range Province (Southwestern United States), the Niobrara, Colorado, Pierre, Mancos, and Green River shales are the principal targets for shale resource plays. The first three are associated with the Late Cretaceous OAEs and related to the Laramide orogeny. The Green River was an Eocene saline lacustrine system (fig. 1–1). Similar associations can be made for the Appalachian trend, where the Taconic orogeny (550 to 440 Ma) was followed by the Acadian orogeny (375 to 325 Ma) and then the Alleghanian orogeny (325 to 260 Ma). The facies are a consequence of the eventual fill of the syntectonic basin. The Utica and Marcellus formations are coeval with their respective orogenies. Both the Taconic and Acadian orogenies had significant volcanic activity associated with island arc collision. The same is true for the Barnett shale in the Fort Worth foreland basin associated with the Ouachita volcanic arc-continent collision. Importantly, in foreland basins, tectonics events overrule global eustatic events.

In the Middle East, the displacement of the Arabian plate and ultimate collision with the Eurasian plate extended from nearly 200 Ma to 25 Ma, ultimately resulting in the Zagros Mountains. The age estimate for the Zagros is based on the youngest unit associated with the thrusting, the Oligocene Asmari limestone. Often there is no simple answer to the tectonic evolution of a region, and the Middle East is no exception. In the Triassic, the Eurasia (Turkey/Iran plate) rifted from the Arabian plate, and the old rift acted as a zone weakness where the closure of the rift system resulted in the Zagros Mountains (Pollastro, 2003). In the process of rifting and drifting, the Tuwaiq Mountain, Hanifa, and Arab-D units (Middle Jurassic) were deposited in sedimentation affected by sag basins controlled by the underlying graben system. The Shuaiba and Nahr Umr (Lower Cretaceous) were associated with the drift phase and the associated passive margin. In southern Iraq, their lateral equivalent is the Sulaiy Formation, with a TOC content of 1% (Al-Ameri and Al-Musawi, 2009). So why do some shale intervals have high TOC content and others do not? The answer is that the proper shale facies were deposited at a time coinciding with global OAEs (see chap. 3).

A second form of sequence stratigraphy that is being applied is known as *chemostratigraphy*. Here, unique elemental compositions of the strata provide valuable correlation tools (see chap. 7). Breyer et al. (2013) and Donovan et al. (2012) have described the application of these tools. Aside from potassium, thorium, and uranium, which are important to calibrate spectral gamma-ray logs (spectralogs), the elemental composition includes chromium, copper, and others. Because the Eagle Ford shale contains ash beds, these units have recognizable chemical signature that serve to identify correlation markers and possible hydraulic fracture barriers. A more detailed description is provided in chapter 7.

SEQUENCE STRATIGRAPHIC INTERPRETATION TECHNIQUES

Log-derived sequence stratigraphy generally follows the guidelines outlined by Galloway (1989). As mentioned previously, sequences in the Galloway system run from MFS to MFS. Biostratigraphy is of great assistance, and the biozones are often used to name third- and fourth-order sequences. Not all first and last appearances of fauna are directly correlatable to sequence boundaries. In practice, however, the sequences are linked to their dominant fauna. For example, in figure 4–11, we would refer to the Late Oligocene, Anahuac Formation member overlying the Frio Formation as the *Marginulina howei* third-order sequence, commonly referred to as the *Marg howei* biozone sequence (25.3 Ma). The TST extends beyond to the top of the *Bol perca* biozone. The same section has continuously been studied by the Bureau of Economic Geology of Texas and is updated periodically. This is why it is important for the biostratigraphers to be involved in the evaluation process and subsequent modifications.

Brown et al. (2005) have described the South Texas S5 type log in greater detail. The same can be done in any basin being worked on. One key log with core data should be analyzed repeatedly by different methods and agencies to obtain the best results and tie the project data to the key well. For East Texas and Louisiana, the same well can be tied to sequence analyses carried out in the region following the starting point of Ewing and Reed (1984). It is important to establish a method that allows for flexibility as new information is introduced.

The petrophysical characteristic of the MFS was described previously, but the seismic characteristics of the MFS is also unique and recognizable. Because of its association with maximum transgression (following the TST), the reflector often represents an original flat or near-flat surface on shelf sediments and therefore is recommended as a primary flattening surface. Because it tends to be predominantly shaley, the MFS contrasts with sandier units above and thus often has a high reflection coefficient. The biostratigraphy of the MFS is also very unique. Because of its nature, it represents a stable period of time in the sedimentary history of the basin associated with low sedimentation rates and condensed sequences, during which microfauna rained down on the sea bottom, creating an environment where abundant and diverse fauna were not diluted by high rates of sedimentation. Occasionally, unique minerals such as phosphates and glauconites are also concentrated in this interval. For the same reason, they often concentrate organic matter. Remember that glauconite sands' response to gamma-ray log is the same as shale's. The contaminants must be considered in the petrophysical log response; XRD and the other tools described in chapter 2 can be used to identify their presence.

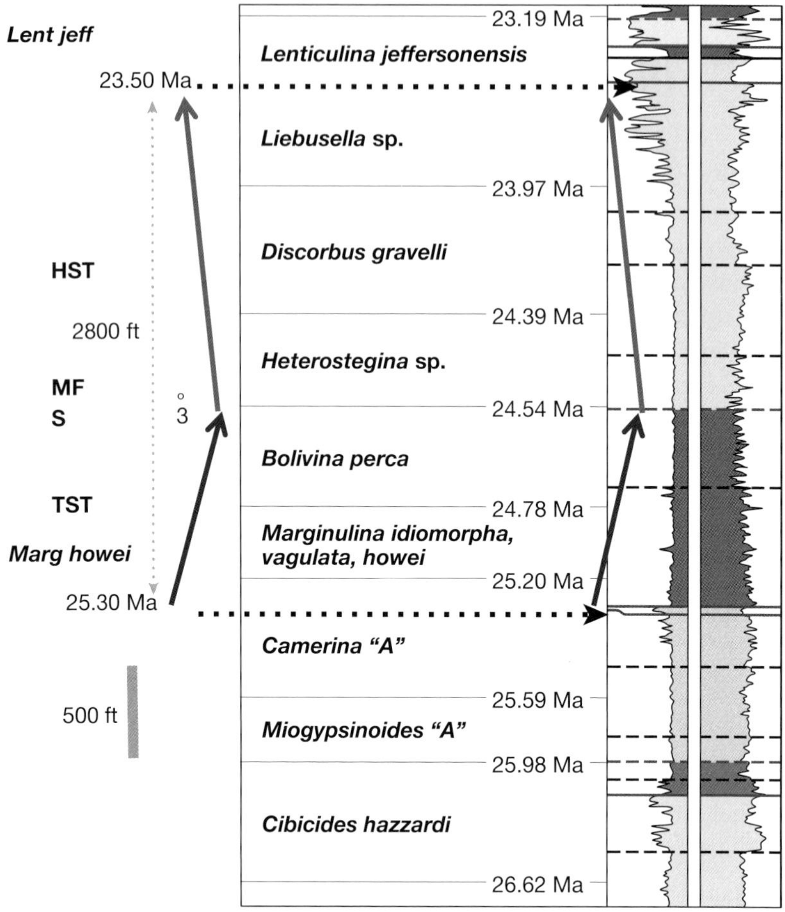

Fig. 4-11. Detailed sequence stratigraphic framework based on foraminiferal biozones. The S5 type log is used by the Bureau of Economic Geology of Texas. Note the sequence designation and that biozones may not coincide directly with sequence system tracts. (Reproduced with kind permission from the Bureau of Economic Geology of Texas. Any subsequent reproduction requires their permission.)

OUTCROP STUDIES

Outcrop studies provide a major source of data for any exploration including resource plays. Here we will use as an example the Cretaceous Eagle Ford Shale outcrop in South Texas because it is a major resource play and has been studied in detail by Donovan et al. (2012).

The outcrop in figure 4–12 represents the Upper Cenomanian to Turonian sequence. It is 180 ft (60 m) thick and is a third-order cycle subdivided into several fourth-order cycles. It is bound by an unconformity at the base of the Austin Chalk and another unconformity at its basal contact on the Buda Formation. Before discussion of the outcrop itself, the process of outcrop analysis should be reviewed.

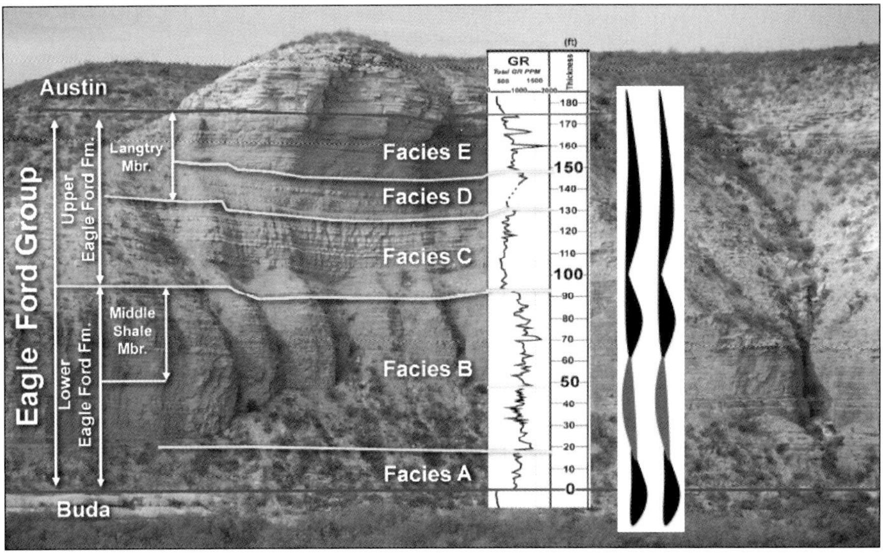

Fig. 4–12. Outcrop of the Eagle Ford, South Texas, and pseudo-seismic trace (Both adapted from Donovan et al., 2012; reproduced with kind permission from Gulf Coast Association of Geological Societies)

Outcrops and core analyses are the only source of accurate rock descriptions and rock property definitions. All other techniques are attempts at deterministic analyses applying models to predict rock properties. Consequently, both outcrop and core descriptions should be treated in a similar manner. They have common elements as well as several differences. The principal difference is that a core, usually 3 in. (10 cm) in diameter, is sliced in half and described. The description therefore is a one-dimensional point in space. The specific point on the core can be described, and its relationship to the overlying and underlying units can be documented. In geology, following Walther's law, within a specific sequence or sedimentary facies that exist in vertical contact, there is a strong likelihood that the lateral contacts will be similar. In cores, there is a notion of lateral continuity but no direct evidence.

For the outcrop, both a vertical and a lateral relationship can be documented. Both sets of descriptions should contain the same amount of detail. In practice, outcrops are rarely described in the same detail as cores. The lithologic description of the outcrop should be as accurate as possible and should be modified once samples and thin sections are studied in a laboratory. For correlation with subsurface wells, a field handheld gamma-ray log can be run (fig. 4–11). In wells where cores are taken, a complete logging suite should be run to catalogue the exact response of the core and log. A whole-core gamma-ray log should be run and compared to the logged interval to ensure a proper tie. Rarely will 100% of the core interval be recovered; therefore, detailed correlation with the well logs is essential.

The constituent particles are described in terms of their grain size, mineralogy or carbonate particle origin, and any observable diagenetic overprint. Once thin sections are taken, the contacts between grains are determined, such as sutured and dissolution contacts. All sedimentary structures, bedding, fractures, or bioturbation structures should described in detail. Wherever present, erosions and abrupt or transitional contacts should be documented. Does the paleontology confirm or contradict the observations of the well logs and outcrop? Do the contacts show subaerial exposure? If so, what is its signature on well logs? Where visible, the porosity amounts and types are described.

Sampling for cores and outcrops should be regular to ensure good ties to logs. In addition, contacts between various units require special sampling. Porosity and permeability must be determined on cores and, although not as significant, may be determined on outcrop data. Specific outcrop and core samples should be thin-sectioned, and XRD analysis is always suggested for shales. Even though surface exposure hinders interpretation of subsurface diagenesis, certain characteristics may still be described.

The end result of the outcrop or core description is a detailed lithology log. For the outcrop, aside from describing the section, it is important to generate a pseudo log based on the core/log correlation from nearby wells. Preferably, it also should be on a 1:200 scale, because this is a common scale for electric well logs. In all cases, the description should reflect the observations and subsequently redrawn to correct for the apparent dip effect, $t = h \cos \alpha$, where t is the true thickness, h is the apparent thickness, and α is the true dip angle. All logs should be loaded onto workstations as conventional log ASCII standard (LAS) files and integrated into the overall interpretation. The AAPG sample examination manual should be used as a reference to ensure consistency in description (Swanson, 1981). Subsequently, other logs can be synthesized by comparing equivalent sequences with logs. Finally, the interpreter should generate synthetic seismic for the outcrop. It will not only clarify the limits of resolution but also show what intervals correspond to the seismic signal (fig. 4–11). Note that the contacts are peaks or troughs and not zero crossing (see chap. 6).

Because core and outcrop studies are costly and very time consuming, it is wise to identify objectives explicitly before embarking on their acquisition and description. In other words, oversample rather than undersample; one can always discard excess samples. Samples should be statistical (e.g., every 10–20 cm). Field samples should be cored for a plug, and the top portion should be retained for a thin section, and another thin section should be taken across the sample. Note that as soon as the core extraction begins, the tins (2.5 ft each) should be filled where the core naturally fractures. Importantly, where the break was forced in order to achieve proper length, mark the break with a small wood or Styrofoam wedge, to note the rock quality designation (RQD) of the interval (see also chap. 8). After clearing the core barrel, wipe the core clean with a dry cloth to remove drilling mud, and immediately paint two stripes on each fragment—black on the left and red on the right—to indicate the orientation of the core. The core in each tin is measured and logged quickly (sand/shale, etc.). In addition, it is important to measure and record the size of each fragment and describe the breaks in the core. Some of the breaks are induced by drilling (usually horizontal), and others are natural breaks (usually diagonal) associated with faults and fractures. The RQD is simply a measure of the length of the individual rock fragments with respect to the length of the cored interval. The RQD index is a percent related to hardness and discontinuities in the rock (e.g., fractures, joints, and bedding planes) (Deere and Deere, 1988).

A small wedge (usually Styrofoam, later replaced by a wooden block) is inserted where the geoscientist collecting the core induced the break or specific samples were removed. The entire core should be photographed to scale, and the length tabulated for each fragment over the length of the core. Changes in length of pieces indicate possible frequency of fractures (drilling may mask the model); determine if there is any fill in the fractures. Fractures in the rock tend to open, especially if there is gas in the system; as the pressure on the gas decreases as it nears the surface, it expands and opens the fracture.

Where special studies (e.g., geochemical analyses) are required, quickly separate the sample, wrap it in foil, and dip it in warm melted wax. The sample should then be placed in an appropriate sealed can and transported in a refrigerated state. The sample gap should be filled with a Styrofoam piece that later will be replaced with a labeled wooden block. Any time a piece of core is removed from the sample bin, it must be recolored to ensure proper orientation. The same should be done with a field sample.

For field samples, it is recommended to place a 4 cm aluminum ring affixed with modeling clay onto the sample. The sample should be labeled with azimuth dip and its orientation noted prior to removal from the outcrop. Then it should be wrapped in tissue and placed in a sample bag. This will allow the sample to be oriented in the laboratory and fracture directions or anisotropies recorded. For cores, some core barrels have a tool to notch the core and hence orient it. The core may also be oriented using paleomagnetic techniques; however, this is time consuming and requires access to a paleomagnetic laboratory.

As will be discussed in later chapters, the problem of defining the appropriate scale is of paramount importance. Priorities have to be set as to what needs to be described first. Usually, the larger scale units are defined first. The detailed description and sampling should focus on the target section. However, samples from the entire section should be collected to determine grain and bulk density and establish their range. Regional formation sampling should not exceed the seismic vertical resolution known as tuning thickness, which is a function of the recorded frequency (see chap. 6). For the time being, let us consider 20 ft (6 m) as the equivalent seismic tuning thickness. However, reservoir-related samples should have a finer sampling, of 60–75 cm intervals, to improve log scale interpretations. Before analysis of a shale resource play, it is highly recommended that outcrop exposures along the margins of the basin be studied to determine mineralogy, organic carbon content, degree of organic maturity, and facies variations and establish depositional patterns, scales, and variability in the parasequences, as well as carry out adequate sampling. The ultimate objective is to tie the outcrops directly with the subsurface information. From a geochemical standpoint, the samples along the margin of the basin tend to be less mature or immature. This is vital information to understand the reserve estimate of the field. (Reserve calculation is discussed in chap. 8.)

The first step in any field effort is to search the literature for descriptions, columns, maps, and field guides. In the United States, Canada, and Europe, it is difficult to find an area for unconventional shale resources that has not been described in detail (except for remote Arctic regions). As stated earlier, because of the intensive harvesting of the shale resource play, the optimum fields tend to be moderately simple. However, by their very nature as outcroppings, there is the possibility of more intense deformation. Therefore, it will be necessary to do detailed structural analyses. All sections should be restored to their true thickness using conventional outcrop techniques (Compton, 1962).

Unfortunately, today it is difficult to spend several days on an outcrop. Previously, labor was less expensive and field assistants were commonplace. Techniques used by our predecessors to measure outcrop sections are rarely used today. For example, a simple transit level can be used to measure the outcrop and any interval within it, to measure a cliff section. The instrument has a level of accuracy of 1 cm at a range of 75 m and a maximum measured angle of 45°. Optimum accuracy can be achieved by a two-person team—one surveying and the other sampling and acting as *rodman* (fig. 4–13). If the sample locations are flagged, an accurate high-resolution photograph can provide an excellent reference for the outcrop.

Fig. 4–13. Rodman at the outcrop. The Jacob staff is colored every 50 cm. In difficult

terrain, a tape measure, level, and compass can be used to measure the section in detail. (Photograph by the author)

In the example shown in figure 4–12, Donovan et al. (2012) have not only described the outcrop but, by projecting and stretching an adjacent well, also generated a pseudo-synthetic log of the outcrop. This demonstrates the influence of tuning thickness, discussed further in chapter 6. The maximum resolution on seismic is one-quarter the distance from peak to peak. Note that this covers several fourth-order sequences. The exercise also provides a better impression of the limits of the seismic signal. In this case, the half wavelength (center black peak, center red trough) corresponds to 40 ft (12 m). The quarter wavelength would then correspond to 20 ft (6 m). That is the minimum resolution of this seismic as displayed. The strong black peak in figure 4–11 corresponds to the lower member of the Lower Eagle Ford shale in contact with the Buda Formation limestone. The overlying hard unit of the upper B facies provides the trace for the trough. Wells commonly drill into lower and upper B members. For engineers, it is very important to visit the outcrop and see firsthand what they will be drilling into.

If a slightly higher frequency were used in the synthetic log, the resolution might improve slightly. This is why when a synthetic seismic is displayed, it will usually indicate the dominant frequency used (e.g., 40 Hz). Note that the fourth-order system tracts labeled as facies C through E cannot be defined on seismic. Interestingly, a surface gamma-ray log was run on the outcrop, clearly showing the facies transitions and that clean shale is evident in the lower section, facies B. In fact, the lower middle of facies B is the MFS of the third-order cycle. Because of the gamma-ray log and its correlation with nearby subsurface wells, the log can be tied to the subsurface interval and the velocities, and densities from the well can be adapted to the thickness of the outcrop and used to generate a seismic model across the outcrop. The ultimate goal is to correlate the outcrop with the seismic data (fig. 4–12).

Two important considerations have to be made prior to visiting an outcrop: First, have multiple hypotheses been proposed that you would like to test? Basically, if you do not know what you are looking for, you will be hard-pressed to find it. Second, how do you wish to gather the data? One goal should be to measure a complete outcrop and integrate it into your project database as a control point. The best way to achieve that is to generate a series of pseudo logs based on regional knowledge. Another goal should be to measure a second section on the same outcrop and note the variations between the two sections. This is of great importance in determining the lateral continuity of the facies. A quick assessment of sequence analysis on an outcrop is to determine upward thinning or thickening. The former is a TST, and the latter an HST pattern.

More sophisticated, laser-based range finders can also be used. For improved documentation of the outcrop, a three-dimensional light detection and ranging (3D LIDAR)–based instrument should be used in conjunction with the Global Positioning System (GPS). As with any instrument, the tool should be calibrated to determine accuracy. Vsemirnova et al. (2012) have published an example of an outcrop pattern using 3D LIDAR. In addition to the LIDAR section, the outcrop should first be sampled, and bright colored flags should be set at each outcrop sample location. As discussed above, a two-person team is ideal to efficiently measure, flag, and sample the section. Samples should be evenly spaced, at a minimum interval of 1 m, with a few additional samples taken at major points of interest. At least two relatively fine vertical sets of samples should be taken along the outcrop face. Following the image and the sample description, a columnar section can be constructed, and a true vertical thickness pseudo log can be generated. This last step allows the data to be loaded onto a workstation and correlated with existing subsurface wells. If possible, a portable spectral gamma-ray tool should also be run because it allows cross-calibration with adjacent wells and provides valuable mineralogical information (see chap. 5).

INTERPRETING SEQUENCE STRATIGRAPHY ON SEISMIC AND ON WELL LOGS

Establishing the starting point for interpretation is often the most difficult task in sequence stratigraphy. Onlap sequences are often self-evident and thus represent an excellent starting point (fig. 4–14). The surface being onlapped is an unconformity and, by definition, a sequence boundary (SB). The terminations to the unconformity should be clearly marked, allowing the unconformity to be interpreted. Note that an unconformity is neither a trough nor a peak; it is simply where the terminations end. At an early stage in the interpretation, the SB should be labeled as sequential, beginning with a relatively large round number like 500 so that other sequences can be named in order as older or younger. The terminations correspond to the TST. Often, shales intercalated with either sandstones or carbonates dominate the TST. The interval has a banded appearance. As a word of caution, in interpretations, never flatten on an unconformity; they are rarely flat surfaces. It is common to observe a high-amplitude event near the top of the TST. The reason for the anomalous amplitude is that the shale interval is often followed by sands that provide the acoustic impedance contrast. This high-amplitude event usually corresponds to the MFS. If the overlying interval is near a shelf break, the overlying sediments may show progradations, as shown in figures 4–14 and 4–7. If none are evident, it may be necessary to follow the correlation to its shelf edge to mark the prograde. This is important because the overlying sequence is the

TST of the following sequence, and its termination to the HST defines the new sequence boundary.

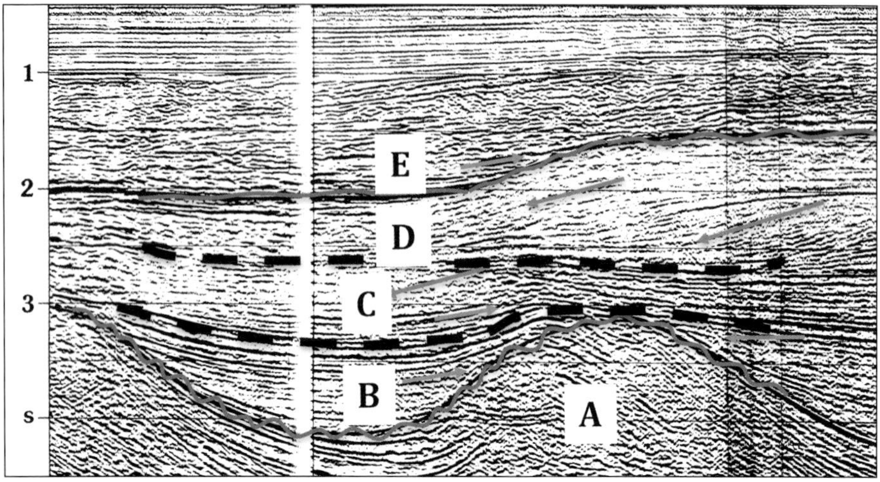

Fig. 4–14. Seismic example of how to label the sequences. TST is labeled "B," black dashed lines indicate the MFS, and "D" denotes the typical HST signature.

Again, wherever possible, you should flatten your interpretation on the MFS because it was likely a horizontal surface. Also note that sequences are labeled from the base. Once a specific age is obtained for the sequence, the assigned age at the base of the sequence is applied to the sequences above. In other words, if an SB 200 is determined to have an age of 82 Ma, its name is changed to SB 82, and its overlying MFS is labeled MFS 82. Shale resource plays tend to concentrate in the MFS but often include the TST and the basal HST.

For petrophysical logs, the convention is to place arrows pointing upward for fining and upward thinning sequences for the TST (fig. 4–15). Arrows pointing up and coarsening upward reflect the HST. Where the two meet in the neck is the MFS of the sequence. Core and outcrop studies work on the same principle. Usually, these sequences are fourth-order parasequences and should be tied with synthetic seismograms to the seismic sequence stratigraphy to establish the third-order sequence. The third-order sequences behave in similar manner but on a broader scale. They thin upward from the TST and thicken upward for the HST. Core studies can supply additional information because unconformities often generate hard grounds bored by diagnostic organisms.

Sequences can further be subdivided by the extent of the unconformity. Major unconformities can significantly erode the HST and often have sands over the underlying unconformity (type I). These unique turbidite sands are known as *lowstand system tracts* (LSTs). If the erosion is deep into the HST, it is possible to break the seal on the MFS source interval and deteriorate its prospectivity. Type II unconformities are more transitional and rarely have an impact on the underlying source interval.

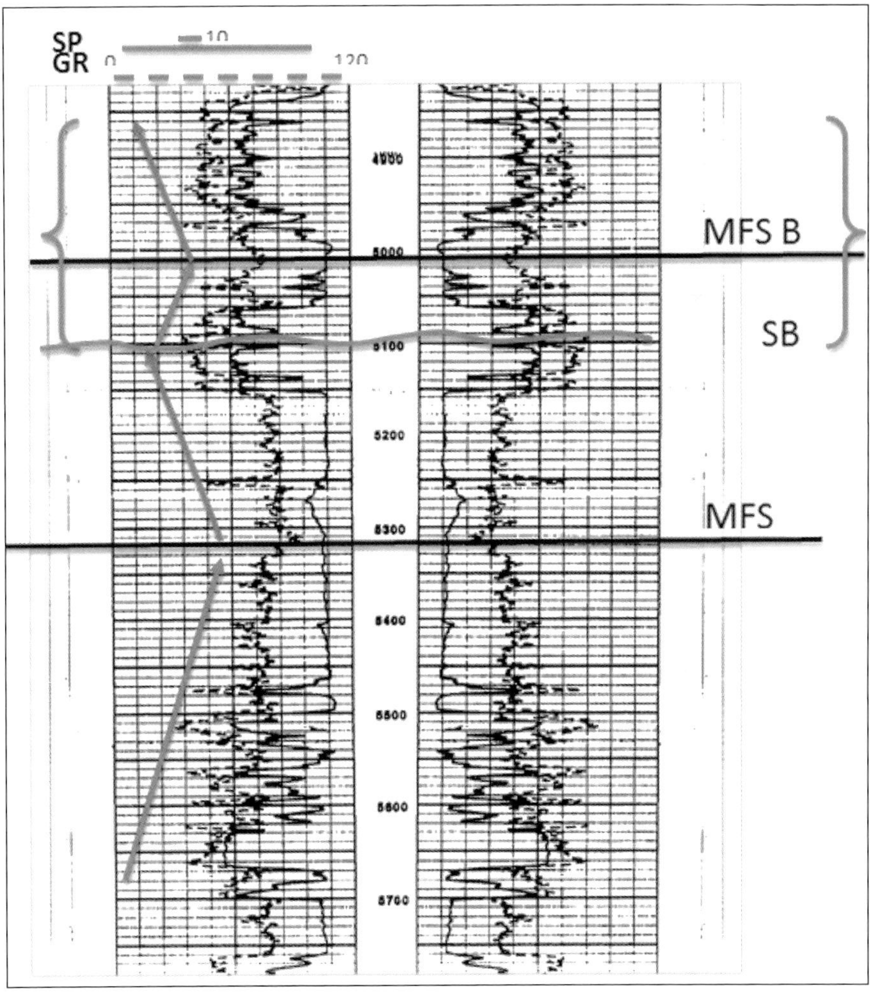

Fig. 4–15. Typical log labeling in sequence stratigraphy. Lower bell-shaped pattern is associated with the TST. The shale neck is the MFS, and the inverted bell shape is the HST. The right-pointing arrows are the TST with fining and thinning upward sequences. The left-pointing arrows are the HST with coarsening and thickening upward sequences.

A typical passive-margin seismic line is used as a reference for establishing general sequence stratigraphy principles (two-way time [TWT] in seconds). Sequence B shows the characteristic onlap sequences (arrows in fig. 4–15) of the TST sequence.

Overlying the TST, a well-behaved reflector with a strong amplitude is usually diagnostic of an MFS (dashed black line, in fig. 4–15). In this case, the SB between B and C conforms to the MFS of the B sequence. Between sequences D and C, the MFS also conforms to the SB. Note the ease with which the MFS can be determined on the log in figure 4–15, as well as the difficulty in determining the SB. Unraveling the history of the sequences is important in determining the quality of any source rock associated with an MFS. Sequence A corresponds to an entirely separate tectonosequence unrelated to the basin fill associated with sequences B through E.

The standard set of sequences called system tracts may include an LST, often defined by incised valleys in the nonmarine environment and *basin-floor fans* (BFFs) in deep marine facies. The BFFs can be either sand prone or mud prone as end members. Mud-prone BFFs tend to be very large, measuring in excess of several hundred kilometers in width, whereas sand-prone BFFs are often tens of kilometers in width. The latter can be detected on seismic data as mounded bodies with significant differential compaction of the shales surrounding the mound. However, LSTs are not always detected.

Following an LST is a TST dominated by shales. TSTs are relatively easy to recognize in seismic data because they generally onlap a preexisting surface or older shelf. If the basin is stable, recognition of the TST is straightforward (fig. 4–15). However, if the basin undergoes either subsidence or uplift, it may be more difficult to unravel. Postdepositional uplift results in upward curvature of the onlap terminations. Subsidence creates a more difficult geometry. The downwarp results in apparent downlaps terminations. Structural restoration is required in order to interpret these sequences.

SUMMARY

Sequence stratigraphy has been defined in terms of unconformity and response to accommodation space. System tracts offer a genetic approach to patterns observed in sedimentation. LSTs are not always present and not significant as a shale resource. TSTs are often the easiest to determine.

MFSs are shale dominant and may be rich in organic material. The MFS usually has low rates of deposition. At the time of deposition and at a local scale, they tend to be horizontal or nearly horizontal.

The Vail method (seismic), Galloway method (logs), or Hubbard method (tectonosequence) should be applied where appropriate. Tectonosequences are better suited for listric faults and foreland basins.

Outcrop and core descriptions should be systematic and detailed. Interpretation techniques for seismic and logs are provided.

REFERENCES

Al-Ameri, Thamer Khazaal, and Furat Ata'a Al-Musawi. 2011. "Hydrocarbon generation potential of the uppermost Jurassic–basal Cretaceous Sulaiy formation, South Iraq." *Arabian Journal of Geosciences*. 4: 53–58.

Allen, Philip A., and John R. Allen. 2013. *Basin Analysis: Principles and Application to Petroleum Play Assessment*. 3rd ed. Chichester, UK: Wiley-Blackwell.

Bartok, Peter. 2003. "The peripheral bulge of the Interior Range of the Eastern Venezuela Basin and its impact on oil accumulations." In *The Circum-Gulf of Mexico and the Caribbean: Hydrocarbon Habitats, Basin Formation, and Plate Tectonics*. Bartolini, Claudio, Richard T. Buffler, and Jon Frederic Blickwede (eds.). AAPG Memoir 79. Tulsa: American Association of Petroleum Geologists, 925–936.

Bartok, Peter, Maria Carolina Mejia-Hernandez, and Murad Ismael. 2015. "Paleogeographic constraints on Middle-to Late-Jurassic tectonic reconstruction of the Maya Block of southern Mexico and equivalent strata of northwestern South America." In *Petroleum Geology and Hydrocarbon Potential of the Colombia Caribbean Margin*, Claudio Bartolini and Paul Mann (eds.). AAPG Memoir 108. Tulsa: American Association of Petroleum Geologists, 201–216.

Breyer, John A., Richard Denne, Jonathan Funk, Tobi Kosanke, and Joan Spaw. 2013. "Stratigraphy and sedimentary facies of the Eagle Ford shale (Cretaceous) between the Maverick Basin and the San Marcos Arch, Texas, USA." Paper presented at the AAPG Annual Convention and Exhibition, Pittsburgh, May 19–22. Search and Discovery article 50899. www.searchanddiscovery.com/pdfz/documents/2013/50899breyer/ndx_breyer.pdf.html.

Brouwer, Friso, Geert de Bruin, Paul de Groot, and David Connolly. 2008. "Interpretation of seismic data in the Wheeler domain: Integration with well logs, regional geology and analogs." Paper presented at the AAPG Annual Convention, San Antonio, TX, April 20–23. Search and Discovery article 40312. www.searchanddiscovery.com/documents/2008/08029scotese/images/scotese.pdf.

Brown, L. Frank, Jr., Robert G. Loucks, and Ramon H. Trevino. 2005. "Site-specific sequence-stratigraphic section benchmark charts are key to regional chronostratigraphic systems tract analysis in growth-faulted basins." *AAPG Bulletin*. 89: 715–724. www.beg.utexas.edu/resprog/starr/papers_reports/S5_Chart.pdf.

Clepper, Marta, Frank Ettensohn, Brandon Nuttall, and Richard Smath. 2011. *New Approaches to Understanding the Upper Ordovician Lexington Limestone in the Bluegrass Region of Central Kentucky: Field Conference Guidebook*. Lexington, KY: Kentucky Society of Professional Geologists.

Compton, R. R. 1962. *Manual of Field Geology*. New York: John Wiley and Sons.

Deere, D. U., and D. W. Deere. 1988. "The rock quality designation (RQD) index in practice." In *Rock Classification Systems for Engineering Purposes*, Louis Kirkaldie (ed.). Philadelphia: ASTM International, 91–101.

Dickinson, William R. 1974. "Plate tectonics and sedimentation." *SEPM Special Publications*. 22: 1–27.

Donovan, Arthur D., T. Scott Staerker, Aris Pramudito, Weiguo Li, Matthew J. Corbett, Christopher M. Lowery, Andrea Miceli Romero, and Rand D. Gardner. 2012. "The Eagle Ford outcrops of West Texas: A laboratory for understanding heterogeneities within unconventional mudstone reservoirs." *GCAGS Journal* 1: 162–185.

Durham, Louise. 2010. "Looking for Hayneville sweet spots: Data boom boosts stratigraphy understanding." *AAPG Explorer*. September. archives.aapg.org/explorer/2010/09sep/haynesville0910.cfm.

Ewing, Thomas Edward, and Roneé S. Reed. 1984. *Depositional Systems and Structural Controls of Hackberry Sandstone Reservoirs in Southeast Texas*. Geological Circular. No. DOE/ET/27111-T4. Austin, TX: University of Texas at Austin and Bureau of Economic Geology.

Galloway, William E. 1989. "Genetic stratigraphic sequences in basin analysis II: Application to northwest Gulf of Mexico Cenozoic basin." *AAPG Bulletin* 73: 143–154.

Hossack, Jake. 1995. "Geometric rules of section balancing for salt structures." In *Salt Tectonics: A Global Perspective*, M. P. A. Jackson, D. G. Roberts, S. Snelson, (eds.). AAPG Memoir 65: 29–40.

Hubbard, Richard J., Joe Pape, and David G. Roberts. 1985. "Depositional sequence mapping to illustrate the evolution of a passive continental margin." Chapter 6 in *Seismic Stratigraphy, Sequence Stratigraphy*. O. R. Berg and D. G. Woolverton (eds.). AAPG Memoir 39. 93–115.

Lowery, Christopher M., Matthew J. Corbett, R. Mark Leckie, David Watkins, Andrea Miceli Romero, and Aris Pramudito. 2014. "Foraminiferal and nannofossil paleoecology and paleoceanography of the Cenomanian–Turonian Eagle Ford shale of southern Texas." *Palaeogeography, Palaeoclimatology, Palaeoecology* 413: 49–65.

McDade, E. C., R. Sassen, L. Wenger, and G. A. Cole. 1993. "Identification of organic-rich lower tertiary shales as petroleum source rock, southern Louisiana." *Gulf Coast Association of Geological Societies Transactions*. 43: 257–267.

Miall, Andrew. 1984. *Principles of Sedimentary Basin Analysis*. Berlin: Springer-Verlag.

Moore, Raymond C. 1964. "Paleoecological aspects of Kansas Pennsylvanian and Permian cyclothems." Symposium on Cyclic Sedimentation. *Kansas Geological Survey Bulletin*. 169: 287–380.

Phelps, Ryan M., Charles Kerans, Robert G. Loucks, Rui O. B. P. Da Gama, Jason Jeremiah, and David Hull. 2013. "Oceanographic and eustatic control of carbonate platform evolution and sequence stratigraphy on the Cretaceous (Valanginian–Campanian) passive margin, northern Gulf of Mexico." *Sedimentology*. 61: 461–496.

Pollastro, Richard M. 2003. *Total Petroleum Systems of the Paleozoic and Jurassic, Greater Ghawar Uplift and Adjoining Provinces of Central Saudi Arabia and Northern Arabian-Persian Gulf*. Reston, VA: U.S. Geological Survey.

Sloss, L. L. 1964. "Tectonic cycles of the North American craton." Symposium on Cyclic Sedimentation. *Kansas Geological Survey Bulletin*. 169: 449–460.

Steinhoff, Ingo, Andrea D. Cicero, Kimberly A. Koepke, Jim Dezelle, Tony McClain, and Cyrus Gillett. 2011. "Understanding the regional Haynesville and Bossier shale depositional systems in east Texas and northern Louisiana: an integrated structural/stratigraphic approach." Paper presented at the AAPG International Conference and Exhibition, Calgary, AB, September 12–15. Search and Discovery article 50379. www.searchanddiscovery.com/pdfz/documents/2011/50379steinhoff/op_steinhoff.pdf.html.

Swanson, Roger G. 1981. *Sample Examination.* E-book, available from AAPG.org.

Turgeon, S. C., R. A. Creaser, and T. J. Algeo. 2006. "Re–Os depositional ages and seawater Os estimates for the Frasnian–Famennian Boundary." *Geochimica et Cosmochimica Acta.* 70: A660.

U.S. Energy Information Administration. 2015. "Technically recoverable shale oil and shale gas resources: Mexico." Washington, DC: U.S. Department of Energy.

Vail, Peter R., R. M. Mitchum Jr., and Samuel Thompson III. 1977. "Seismic stratigraphy and global changes of sea level: Part 3. Relative changes of sea level from coastal onlap: Section 2. Application of seismic reflection configuration to stratigrapic interpretation." *AAPG Memoir* 26: 63–81.

Van Wagoner, J. C., D. Nummedal, C. R. Jones, D. R. Taylor, D. C. Jennette, and G. W. Riley. 1992. *Sequence Stratigraphy Applications to Shelf Sandstone Reservoirs: Outcrop to Subsurface Examples.* Field Conference Guidebook. Tulsa: American Association of Petroleum Geologists.

Vsemirnova, E. A., R. R. Jones, and K. J. W. McCaffrey. 2012. "Studies of fracture network geometry of reservoir outcrop analogues from terrestrial lidar data: attempts to quantify spatial variations of fracture characteristics." *EGU General Assembly Conference Abstracts.* 14: 8972.

Wheeler, Harry E. 1958. "Time-stratigraphy." *AAPG Bulletin.* 42: 1047–1063.

5

PETROPHYSICS

TAKEAWAY IN

- Review purpose for specific logs
- Understand what the logs actually measure
- Review how to interpret the information from the logs
- Discuss porosity, density, interval velocity, formation factor, and fluid content
- Special attention is given to rock physics parameters

INTRODUCTION TO PETROPHYSICS

Well logs are run to evaluate the formation in terms of reservoir quality (porosity/permeability), determine fluid content, and establish the presence of a seal. In addition, taking a full suite of logs allows one to determine rock physics properties, compare logs to their seismic response, model fluid substitution based on logs and seismic, and understand the complexity of the aquifers. Modern calculations even allow for estimation of the total organic carbon (TOC) content in target shales. So that all these goals may be achieved, the interval velocity should be complemented with both compressional velocity (V_p) and shear velocity (V_s) where possible, in addition to basic log data such as resistivity, gamma ray, porosity, and density. If V_s is not available, it must be estimated. To complete the study, the explorationist also needs to determine volume of adsorbed and free gas in the system, as well as have a notion of the permeability in the reservoir. Thus, lithology, rock physics, rock anisotropy, natural fractures, effective porosity, and fluid type can be established and extrapolated. For these investigations, it is essential to understand the fundamentals of logging principles and be able to demonstrate how they relate to rock properties.

The task of the interpreter is to maximize data collection to determine what is critical and subsequently reduce the requirements to fit the needs of the project. Cost and time reduction are of the utmost importance, but they should never be at the expense of using less than the appropriate solution for the task at hand. This leads to the next goal, which is to identify and project the same properties onto seismic lines to extrapolate sweet-spot attributes to undrilled areas. To achieve all these objectives, petrophysicists require detailed information from project geologists that are based on cores and outcrop studies (see chaps. 2 and 4), using rock physics parameters to predict the attributes of the target shale.

Petrophysical data have been compiled on basins for almost 100 years, and the tools are constantly evolving. The list of available logs is too long to provide here—and is made even longer because each contractor tends to use a different name for the same log owing to patent and copyright laws. For the purpose of this chapter, a few of the principal logs used in the shale resource assessment are highlighted, and their applications are discussed. The purposes of the earliest logs were to differentiate between carbonate and siliciclastic reservoirs and to define the oil and water saturations in the reservoir—hence, the *induction log*.

Prediction of fluid content relies on a clear understanding of the lithology and the porosity. Thus, a porosity-determining tool was also critical to estimate the reserve potential of the reservoir. Initially, sonic logs were used to determine porosity, and later tools that measure reservoir density yielded improved porosity determinations. The difference between the known pure matrix density and the observed effective density is the porosity.

Two lithologies were observed to significantly interfere with the *cleanliness* of the sandstone and carbonate reservoirs: clay content and evaporite content (salt or anhydrite). The new focus on shales requires further adjustments to the conventional logs to understand the shale interval as a reservoir. Gamma ray, spontaneous potential (SP), sonic logs (monopole, dipole, and full-waveform sonic [D_t]), and compensated formation density and neutron logs (FDC/CNL) are used in combination to determine lithologies and describe fluids more accurately. Several other specialty logs will be described, including logging or measurement while drilling (LWD or MWD), spectral gamma-ray logs, nuclear magnetic resonance (NMR) log, image logs (e.g., fullbore formation microimaging), and the photoelectric factor log. Regardless which specific logs are used, it is important to understand the environment in which the logs are taken.

Drilling a well into subsurface sediments can never provide a smooth symmetrical borehole—hence, the *caliper log*. Variations in formation compositions, porosities, and formation pressures result in changes in borehole gauge that often affect the logging tool response. Temperature can also modify log response and thus is always recorded. For horizontal wells, logging problems are particularly significant because the tools lie along the bottom of the borehole and provide an asymmetric log response. Wellbores are always filled with drilling mud to lubricate the bit and provide pressure control. By adjusting the density of the drilling mud, a balance is achieved between the formation fluids and the drilling mud used. The composition of the mud can vary widely, ranging from brine-based to oil-based muds. The drilling mud density should be slightly underbalanced to avoid damaging the formation. Barite or hematite is added to the mud to increase its density.

In general, the principles involved in the old and new technologies are similar. Today's modern tools are digital and designed such that they can readily be analyzed mathematically and parameters can be cross-plotted to better identify basic rock properties. Although editing by a trained petrophysicist is still required, the amount of editing has been greatly reduced.

Human intervention is also necessary to determine cutoffs for various parameters, which are often subjective, and criteria vary by company. Cutoffs may require interpretation from multiple logs. The purpose of these cutoffs is to establish commerciality. It is recommended that the geoscientist spot-check the computer-generated results and determine the limits of the interpretation. Data noise can occlude valuable information. Excessive filtering and editing can remove valuable information. Because all modern logs are digital, in log ASCII standard (LAS) format, they can readily be transferred to Excel spreadsheets and cross-plotted as needed. Spreadsheet computations allow for the generation of all rock physics properties. Petrophysical programs are available on all computer workstations.

CORRELATION LOGS

The basic interwell correlation logs are the SP and gamma-ray logs, discussed later in this chapter. The principle behind the SP log is the cation exchange capacity (CEC) process, described in chapter 2. First and foremost, because fluids must be exchanged between the mud and the formation fluids, some permeability must be present. However, the SP log is not a direct measurement of permeability; rather, permeability must be present for a reading to be registered. For cation exchange to occur, an electric current must be generated. The process is known in chemistry as the direct or spontaneous potentiometric reading of the presence of sodium in an aqueous solution. This is also used in medicine to measure the sodium and potassium content in blood. When applied to the subsurface, the tool can distinguish shale from sandstone. Note that oil-based muds will not yield a response because there is no cation exchange. For carbonates, the permeability is not as clear-cut; hence, the response varies. With reference to the discussion on shales (see chap. 2), smectites and illites have significant differences in CEC; therefore, knowledge of mineralogy is important.

RESISTIVITY AND FORMATION FACTOR

The induction log has been the principal workhorse of the oil industry for most of the history of commercial oil production. Modern induction log resolution has improved, but the basic elements of the log are still the same. The concept is to induce an electrical current into a formation and determine if the fluids in the sediment pore space are *conductive*. If sediments contain no fluid, there is no conductivity. Normally, discussion of conductivity centers on its inverse, the *resistivity* of the rock. For example, brine water is highly conductive, and hydrocarbon is highly resistive. The water salinity and temperature should be noted and proper corrections made before the observed resistivity values are interpreted.

One additional factor to consider is the resistivity of the formation water in the area under investigation. In general, local operators have tabulated the results of formation water resistivities for each interval in a given area. An estimate of the formation water resistivity (R_w) can also be calculated (Ushie, 2001), using computations provided by all service companies. Modern log suites include a calculation of the R_w value and register it as a curve. For example, very fresh water is resistive. Thus, resistive water and resistive hydrocarbons will yield similar results and do not permit proper differentiation. Importantly, clay minerals trap freshwater in their lattice. This creates a dual water in the shale reservoir that must be appropriately accounted for, as discussed below.

Because resistivity was initially used to evaluate porous sandstone and carbonate reservoirs, the problems of porosity reduction due to diagenetic overprint became evident. For example, in medium-grained well-sorted clean (shale-free) sandstone, porosities can range from 0% to 35%. The difference is a function of the cementing agents occluding the pore space. To estimate this effect in sandstones, Archie (1942) studied many core samples from wells and compared the porosity results to the resistivity to determine if a formation resistivity factor F could be specified that is dependent on porosity and resistivity in the wet zone (R_o) for a specific rock type. Later, the same approach was used on carbonates. A similar approach is taken with shales, except that the clay mineral participation must also be accounted for. The diagenetic overprint on shales is not only critical in determining its reservoir potential but also far more complex than for sandstones and carbonates. Moreover, the role of catagenesis, the diagenesis of organic material, must also be considered (for details, see chap. 7). For example, pore-filling illite can have a detrimental effect that must be accounted for (fig. 2–21).

The same procedures used to analyze sandstones can be applied to shales; however, the values need to be more precise. In addition, care must be taken that similar mineralogies are compared. A smectite-rich shale will respond differently than an illite-rich shale even if they are lateral equivalents of each other. As discussed already (see chap. 2), in shales, only the water-wet systems are studied to determine the formation factor F.

As later defined by Wyllie (1954), the formation factor is the resistivity of a saturated rock with a conductive fluid (R_o) divided by resistivity of the saturating conducting fluid (R_w) in a wet zone:

$$F = \frac{R_o}{R_w} \tag{5.1}$$

F is also a function of the size and geometry of interconnected pores (ϕ). As in the IBM-SEM analysis (described in chap. 2), knowing the characteristics of the interconnected pores allows for more precise determination of the F factor. The formation factor F can be simplified as any of the following:

- $F = 0.62/(\phi^{2.15})$, for a very porous sand intervals
- $F = 0.81/(\phi^{2})$, for a partially cemented rock
- $F = 1/(\phi^{2})$, for a very tight rock

A more realistic value for the F factor was established by Archie (1942), who noted that tortuosity of flow (a) is as important as the porosity and provided the following general form for the equation:

$$F = \frac{a}{\phi^m} \tag{5.2}$$

Note that in some carbonates, m can range from 2.2 to 3. The tortuosity a can range from 1.0 to 1.73 in siliciclastics and is a function of porosity; the lower value is the most common (Boudreau, 1996).

Determining the m factor in equation (5.2) is critical during the development phase of a field for traditional sandstone and carbonate reservoirs, and for shales, it is important even from the outset because cementation and pore reduction in shales is so complex. The m value can be obtained from core samples under laboratory conditions and is currently an area of ongoing research (Teh et al., 2011). Note that m is basically a cementation factor. Nearby shales are compared to obtain the values of the plot. Wyllie (1954) provided ranges for m, and regional knowledge can be used to determine ranges for the formation factors. When the F factors are plotted against porosity (fig. 5–1), the slope of the line is the m factor.

A more precise value of m can be computed by using a Picket and Hingle plot along the wet zone, generally referred to as a *Picket plot* for short (Dewan, 1983) to yield the Archie cementation factor (fig. 5–1). The Picket plot is constructed by cross-plotting the observed total resistivity against porosity (fig. 5–2). The slope along $S_w = 1$ gives the m value for the Archie formation factor (F). Both plots should be computed and compared.

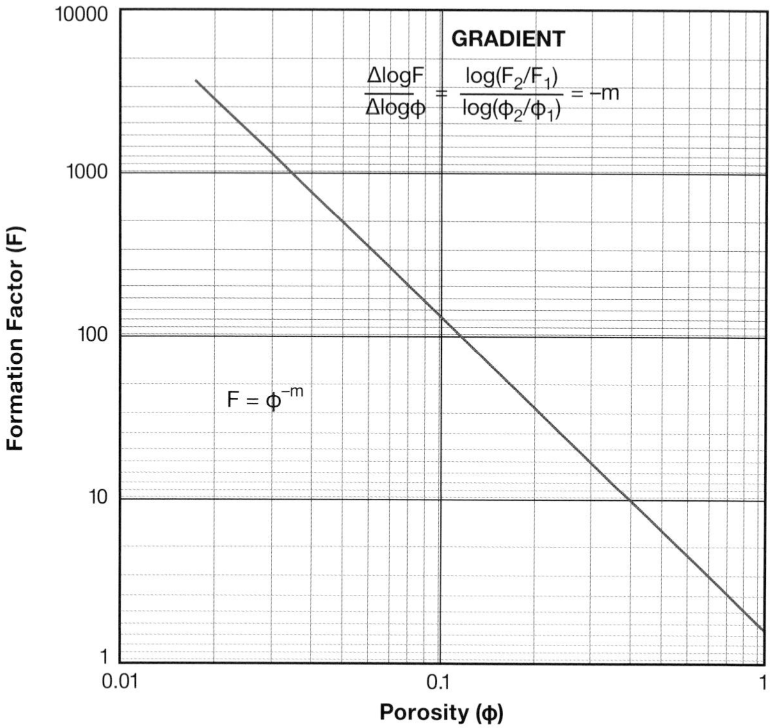

Fig. 5–1. Cross-plot of formation factor F versus porosity to yield value of m

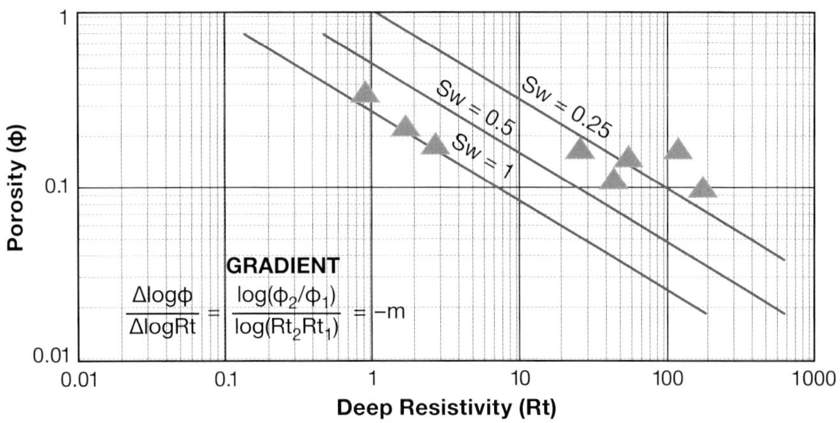

Fig. 5–2. Pickett plot of R_t versus porosity. The diagonal lines are water saturation (S_w), and the slope is the *m* value. The slope is calculated on the basis of the resistivities of the wet zones in the interval of interest.

Plots of the resistivity index (I) versus water saturation (S_w) provide the n values (fig. 5–3), where n is the saturation coefficient. The resistivity index (I) is the ratio of the deep resistivity (R_t) to the wet-zone resistivity (R_o). A basic requirement of both the resistivity index and the Pickett plot is to establish the nearest value to the true resistivity of the formation, R_t. There are no tools available to obtain the actual R_t, but by comparing short, medium, and deep resistivity tools, a reasonable estimate can be obtained. In particular, the deep resistivity tool comes close, with a margin of error of nearly 10% (Dewan, 1983). For a first estimate, the result using the deep resistivity is adequate. Once again, a plot (fig. 5–3) is generated for water saturation, S_w, versus resistivity index (I), to obtain the n factor used in equation (5.3) and in Archie's equation.

$$\log\phi = \frac{(\log aR_w) - \log R_t}{m} + \frac{n\log S_w}{m} \tag{5.3}$$

The Kansas Geological Survey has provided a tutorial for the preparation of the Pickett plot, which is freely available through their website.

For shale resource plays, it will be important to obtain a more accurate reading for n than that obtained from the plot in figure 5–3. The value of n ranges from 1.8 to 2.0. Research is ongoing to improve the basic parameters presented in this summary. For additional information, refer to the work by Lee and Collett (2006).

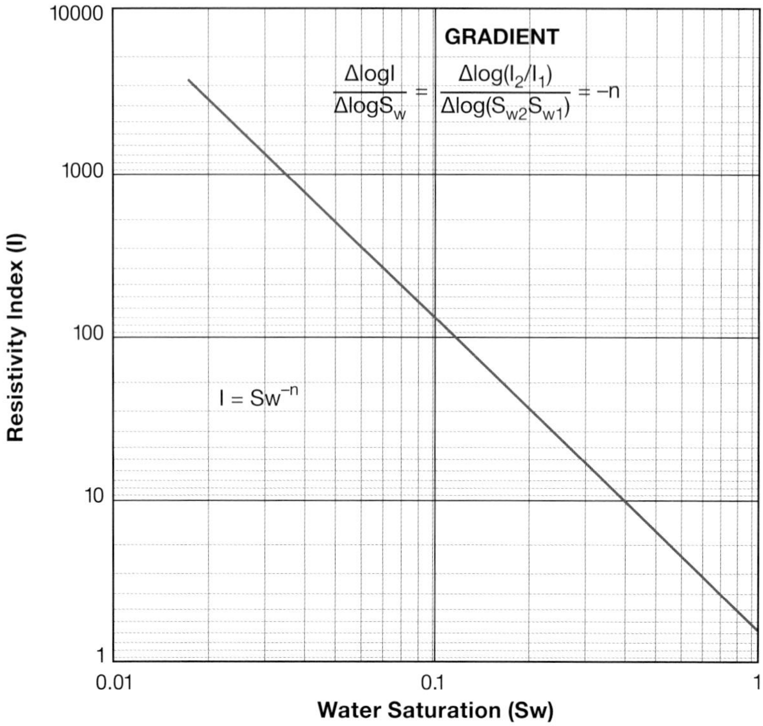

Fig. 5–3. Crossplot of resistivity index (I) vs. water saturation (S_w) to yield value for n. Note that this form provides a simple graphic solution for a linear equation.

Once the F factor is known, the oil saturation can be determined using a relatively simple relationship:

$$S_w = \left(F \times \frac{R_w}{R_t}\right)^{1/n} \tag{5.4a}$$

$$S_w = \left(\frac{1}{\phi^m} \times \frac{R_w}{R_t}\right)^{1/n} \tag{5.4b}$$

where R_t is the deep resistivity of the formation independent of the fluid present, a is the Archie tortuosity of flow, ϕ is the porosity, and n is the saturation coefficient. Estimating the water saturation in the reservoir is fundamental in determining the reserves and is discussed in greater detail later in this chapter.

A relationship based on the Archie equation yields the oil saturation (S_o) in the formation under investigation, and this is related to the water saturation (S_w) as follows: $S_o = 1 - S_w$. For most calculations of equation (5.4), $n = 2$ is adequate; therefore, the square root is sufficient to solve the Archie equation. As can be seen in equation (5.4b), the importance of defining porosity is essential. Determining the effective porosity, which is defined as the pore volume that allows fluid to flow through, will be discussed in detail later in this chapter. The maximum porosity includes space filled with any fluid, including bound water that is immobile, also known as *irreducible water.*

Water systems behave differently in source rock shales than in sandstone reservoirs. Where the hydrocarbons have not been expelled, shales will contain free water and bound water; hence, they yield low resistivities. As hydrocarbons begin to be liberated, free water is displaced, and the resistivity increases while the bound water is retained. Cations released into the formation fluids also impact the resistivity. These factors will be used later to determine TOC content in the shale. As discussed in chapter 2 (see the section "Structure of Clay Minerals"), when smectite is converted to illite, a significant amount of freshwater is released, creating a dual water system.

A better approximation for shale resource analysis is the application of the Simandoux equation to determine S_w. Although several modifications of the Simandoux calculation are copyrighted, the basic elements can be discussed. The Simandoux equation incorporates not only the resistivity but also the neutron log data, thus generating a shale porosity as well. Other, similar corrections are also available and should be discussed with the petrophysicist. The Simandoux approach is not a perfect solution, so others, such as the Poupon-Leveaux model may be applied instead (Doveton, 2001).

The dual water system in the shale—namely, free water and clay-bound water—warrants further discussion. Tools like neutron logs (which will be discussed in further detail later in this chap.) detect water but cannot distinguish water types. The Waxman-Smits equation (Waxman and Smits, 1968) differentiates between free water and bound water by considering the modification in CEC associated with smectite diagenesis. The CEC value is determined in a laboratory as discussed previously (see chap. 2; because the clay mineralogy can be quite complex, the procedures outlined there, such as focused ion beam milling for pore description, XRF for elemental analysis, XRD clay mineral volumes, SEM for organic material pore typing, and Rock-Eval pyrolysis [see also chap. 7], should be carried out prior to interpreting the logs). The basic premise of the Waxman-Smits equation is as follows:

$$R_w C_o = \frac{1 + R_w B Q_v}{F_A} \tag{5.5}$$

where Q_v is the cation concentration per unit volume, B represents the mobility of the ions near the grain surface, F_A is the formation factor, C_o is the core-based conductivity determination, and R_w is the resistivity of the formation water.

Once again, particularly in shales, logs by themselves are insufficient to determine a true resistivity (R_t). Therefore, it is necessary to recalibrate the equations until a best fit can be obtained. The interaction between the core sample data and the petrophysics must be continuous. In the original Waxman-Smits equation, C_o is calculated under laboratory conditions, as are the factors B and Q_v; the latter two parameters are calculated on the basis of CEC (Waxman and Smits, 1968).

Recently, Devarajan et al. (2006) have presented a variation on the Waxman-Smits equation by relating the free water (ω_w) in the pore space and the water adsorbed in the clay or organic matter (ω_o), as follows:

$$\omega_o = \frac{1}{F(\omega_w + BQ_v)} \tag{5.6}$$

where Q_v is the cation concentration per unit volume, B represents the mobility of the ions near the grain surface, and F is the formation factor. The mobility (B) in equation (5.6) results in measurable conductivity associated with the exchange of cations. Note that this applies only to a pure shale. The B factor increases exponentially for low values of ω_w and stabilizes in high-salinity water (Devarajan et al., 2006). Equation (5.6) establishes a proportional relationship between ω_o and ω_w.

Often, clay particles, such as chlorite, can coat the coarser-grained siliciclastic grains. This is one of the main sources of bound water in a reservoir and has implications on the calculation of water saturation and effective porosity. In other words, if a sample is totally dried, as is routinely done in the laboratory, then the porosity and permeability determinations remove both bound water and free water, and a false effective porosity is obtained. Similarly, if a reservoir has bound water, free water, and oil, then the calculation of water saturation must distinguish between bound and free water. An additional problem is to distinguish between matrix that has a higher or lower wettability. In a pore space containing two liquid phases, the dominant mineral may be oil wet or water wet. This is a function of the difference in surface tension between the two fluids and will be discussed in detail in chapter 8.

One way of describing the effect of wettability is by comparing a drop of water that beads on a waxy surface with water that spreads on a smooth, clean surface. The lower the surface energy is, the higher will be the tendency for that fluid to wet the surface. Drelich et al. (2002) have discussed the various methods to calculate a value that describes this tendency. Between three grains of quartz with a clay coating, if the bound water impinges on all pore throats, then no hydrocarbon will flow between them. Only the free water and the free oil will migrate out of the reservoir given a pressure differential.

Not all clay coatings yield a negative result. Chlorite coating enhances the presence of free or capillary water and permits greater oil flow (Ajdukiewicz et al., 1991). The technique to measure the adsorbed water is similar to the detection of adsorbed gas (see chap. 2). NMR logging and electron spin resonance (ESR) techniques can detect the amount of bound water in the sample and are discussed later in this chapter. Once again, there are three types of water trapped in the reservoir: bound water, capillary water (trapped in the small pores between the clay layer by capillary forces), and free water. The bound water is usually a monolayer. Therefore, with knowledge of its molecular size and by comparison with the IBM-SEM cube, an effective pore throat size can be established.

A key to the Archie equation is to determine accurate porosity and permeability values; however, for shale resources, that is difficult, and as the name implies, unconventional approaches must also be applied. The best estimate for permeability is the *minifrac* (leak-off) test, reviewed in chapter 8, and the second-best estimate is obtained from the core analysis.

Bust et al. (2013) have elaborated further on logging problems associated with shales. Traditionally, permeability is measured in crushed rock from which the hydrocarbon is withdrawn, and the sample is injected with mercury in an imbibition study (described in chaps. 2 and 8). The problem is that shales have micropores and microfractures. If a sample is crushed, the permeability measurement will consider only matrix permeability and, as a result, will underestimate permeability. Considering that problem, porosity/permeability crossplots derived from imbibition studies can still be used and provide a minimum curve. As with traditional well logs, all well logs must be calibrated to the core data.

WELLBORE ENVIRONMENT

At this point, it is important to review the appearance and the characteristics of the borehole being logged. There are four basic environments in the borehole during drilling and logging (fig. 5–4). The open hole is filled with drilling mud, forming a mudcake when drilling stops. Clays used in drilling have a property known as *thixotropy*. The purpose of the mud is to lubricate the bit and ensure that the fluids from the formation remain in the formation and avoid formation damage by deep invasion. If a mud has thixotropy, it is fluid when stirred, and it solidifies into a hard gel when unstirred. This is important to maintain stability in the wellbore and form the mudcake.

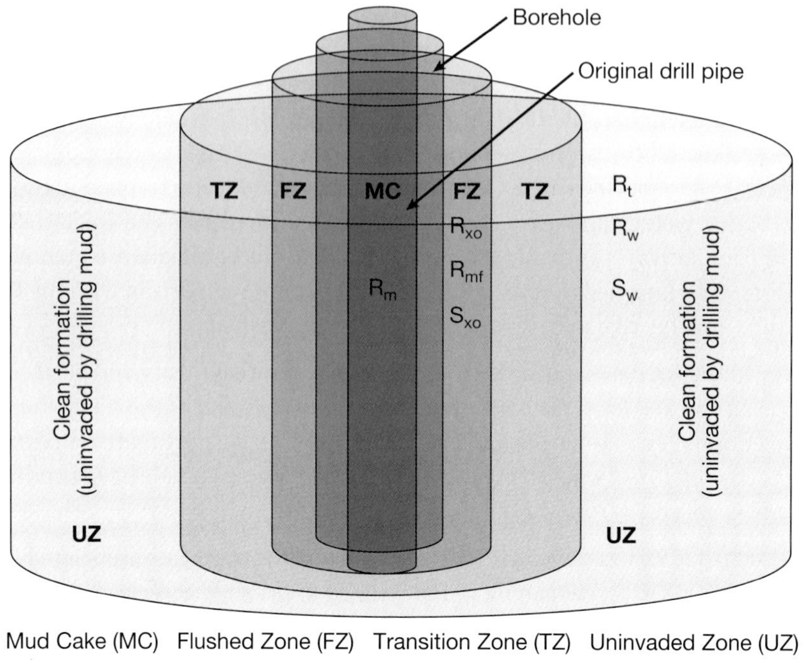

Fig. 5-4. Borehole environment showing the relationship between the various zones affected by drilling fluids

The density of the mud is always kept slightly underbalanced with respect to the formation to ensure balance with the formation fluids. The mudcake at the specific interval under investigation has a resistivity, under surface conditions, of R_m, which is recorded on the log header. In the borehole, the mudcake resistivity is R_{mc}. Tables are available from service providers to calculate R_{mc} values. The second zone is the immediate area affected by drilling fluids penetrating the formation. There, the combined resistivity of rocks and fluids yield an invaded-zone resistivity (R_{xo}), and this is affected by the invading mud (R_{mf}), also reported on the log header. There is often a transition zone.

The ultimate goal is to log the deep formation not invaded by drilling fluids. These results yield the true formation resistivity (R_t), the formation water resistivity (R_w), and finally, the water saturation (S_w). For a given formation, the S_w value is calculated as follows (Asquith and Krygowski, 2004):

$$S_w = \left(\frac{R_{xo}/R_t}{R_{mf}/R_w}\right)^{0.625} \tag{5.7}$$

Modern induction logs include *microspherically focused logs*, which, like proximity logs and short normal logs, measure the value of R_{xo} directly. *Laterologs* are most often used in shale plays. They are able to detect thinner beds. Carbonates are often associated with many shale source rocks and require comments on the significance of the resistivity log. These sediments are often fractured. The intensity of open fractures in a rock can also be related to the ratio R_t/R_{xo} when compared to gamma-ray logs (Bartok et al., 1981).

A second problem with carbonate arises from their complex diagenetic overprint. The same facies can undergo dissolution of $CaCO_3$ in the more exposed section and precipitation in the lower section of the same interval. This results in an increase in resistivity toward the base. The bell-shaped curve can be confused with fining upward sequences when it is simply a diagenetic overprint.

In general, fractures have very limited storage space. As a result, if the reservoir is purely a fractured reservoir, it can produce hydrocarbons because of good communication, but the life of the well will be very limited. The calculation of water saturation is critical in establishing the volume of hydrocarbons in place. Remember that either sandy intervals or carbonate-rich intervals are used to land horizontal wells and hydraulically fracture from the resistant bed into the shale. As shown in chapter 7 (see the "Geochemical Petrophysics" section, the resistivity tool plays a vital role in the establishing estimates of TOC content in the shale under investigation. For more on logging tools, the reader should consult the work of Foster and Beaumont (1990).

An additional application of the resistivity log is to convert resistivity to a pseudo sonic log. The relationship is described by the Faust equation. As a cautionary note, the Faust equation should be applied only in wet zones. The basic form of the Faust equation is as follows:

$$K(ZR_t)^{1/6} = \frac{1,000,000}{D_t} \tag{5.8}$$

where K is a function of the maximum burial depth and age of the rock (and varies between 1,000 and 3,000), Z is the maximum depth of burial of the sediment and should be corrected for major unconformities, R_t is the deep resistivity reading, and D_t is the sonic velocity in microseconds per foot. However, in nearby wells where both resistivity and sonic logs are available, the Faust conversion should be tested and the results cross-plotted to determine the optimum indices and K constant in equation (5.8) (Smith, 2007). Many older wells have resistivity and density logs but lack sonic logs. The Faust equation provides a means to estimate the critical interval's velocity value.

CHALLENGE OF DEFINING POROSITY IN SHALES

Before we can define *effective porosity*, we must define porosity and describe what fills the porosity. By definition, all sediments are comprised of a matrix consisting of allochem and autochem constituents. Any porosity can be reduced or enhanced by diagenetic events. Carbonates are the most dramatic, for which porosities greater than 30% have been recorded by dissolution in reservoirs in the Middle East and Western Canada. It is more common for carbonates to have low porosities, from 5% to 20%, owing to the effect of grain interpenetration associated with dissolution, carbonate cements, and dolomitization partially filling the original pore space (Bartok et al., 1981). In sandstones and shales, other diagenetic processes are evident. Sandstones can have silica or carbonate cements, as well as intergrain penetration. Under different conditions, authigenic clays can further occlude pore space. Shales are affected by compaction, illitization, and authigenic clays. Compaction affects the capillary pressure in the reservoir owing to the reduction in pore throat. Another term often used to describe porosity is *solidity*, which is defined as the complement of porosity:

$$\text{Solidity} = 1 - \text{porosity} \tag{5.9}$$

Solidity is very similar to rigidity, which is in turn related to brittleness. All three lithologies mentioned in the previous paragraph need to be understood by the unconventional resource player, as sediment solidity varies. However, in shales the relationship of solidity and depth must be carefully calibrated. The shales provide the resource. The sandy and calcareous intervals provide the drilling and fracking targets. The importance of solidity will become more evident in chapters 6 and 7, with relation to thermal conductivity of sediments, as well as to fractures and fracking.

Note that organic material may fill the pore space. There is no differentiation between organic matter and oil in the pore space. Hence, it is read as porosity, and as a result, a correction for TOC needs to be added to the porosity determination from the density log. The determination of TOC content from logs is described in chapter 7.

NUCLEAR LOGS

There are several nuclear logs, of which the gamma-ray log is the most common. It is well known that radioactive isotopes will emit alpha and beta particles and gamma rays. Because the alpha and beta emissions are particles, they lose their energy quickly. The gamma rays are emitted as high-energy photons and therefore are more easily detected.

There are three common sources for gamma rays in sediments: uranium, thorium, and potassium. The concentration of each of these elements is related to the mineralogy in the host sediment. The simplest count is based on the absorption of gamma rays by a scintillation counter. For a more in-depth description of the physics of gamma-ray emission and detection, refer to the work of Dewan (1983) and Ellis (1987).

Because of their crystallography, clay minerals can trap all three aforementioned radioactive isotopes better than most other minerals, with the exception of glauconitic and orthoclase-rich sandstones. The difficulty with the latter two sands is their high concentration of potassium, up to 5% by weight for glauconite and 14% by weight in orthoclase associated with orthoclase-rich arkosic sandstones. Gamma-ray logs are read in API units, which represent a weighted average of the gamma rays emitted by the three source elements. In wells, the maximum deflection of the gamma-ray log is considered the baseline for clean shales. The lower the deflection is, the smaller will be the volume of clay in the sample. This relationship is very important when considering the porosity and elastic parameters of the sediment (the latter of which will be discussed in chap. 6.). As already analyzed in detail in chapter 2 and 4, sections with maximum gamma-ray deflection are commonly associated with maximum flooding surfaces and often with shale resource plays; when they are observed, they should certainly be studied in detail. In the previous discussion on resistivity, the formation factor was introduced. It should be calculated for clean shales and then modified as a function of the lithology mixture under investigation.

The conventional gamma-ray tool contains a sodium iodide crystal laced with thallium that acts as photomultiplier (Ellis, 1987). The sodium iodide crystal emits a small flash of light when bombarded by a gamma ray. The light pulses are then recorded. Only the gamma quanta are able to reach the detector. Each atom producing a gamma ray varies with respect to the energy spectrum of the emitted gamma ray, which is measured in million electron volts (MeV). The scattering effect of the gamma ray in the sodium iodide crystal, known as the *Compton effect*, results in an energy that allows the detector to separate the contribution from the various sources of gamma rays in the sediments (National University of Singapore, 2005). The *Compton effect* records the reflecting quanta of gamma ray. After a gamma ray has bombarded an atom and a portion of the energy has been absorbed by the atom, the remainder is reflected.

The traditional gamma-ray tool will provide a total count and serves to distinguish shale and nonshale lithologies. As discussed previously, there are no pure shales in nature. They can contain silica, carbonates, and other contaminants. Usually there are several shale intervals in the section of interest. As a first approach, a gamma-ray index (I_{GR}) can be estimated by comparing the gamma-ray log (GR_L) in the section of interest with the minimum and maximum (GR_{min} and GR_{max}, respectively) within the general zone of interest and associated with similar lithologies:

$$I_{GR} = \frac{GR_{log} - GR_{min}}{GR_{max} - GR_{min}} \quad (5.10)$$

The gamma-ray index can help differentiate the shales identified with the source rock potential by the Passey equation (discussed in chap. 7).

On a given log, the point of interest on a shale is compared with the gamma-ray response of the nearby shales. Their minimum and maximum values are used to estimate the value of V_{sh}. A similar estimate from the SP can approximate the value for a shale, as follows:

$$\frac{SP_{log} - SP_{min}}{SP_{max} - SP_{min}} \quad (5.11)$$

As a word of caution, though, if the level of interest contains source rocks, then the gamma-ray method should not be used.

A second type of gamma-ray log is the spectral gamma-ray log (fig. 5–5) that, instead of summing the gamma-ray sources, separates them into three components. It has been documented that the dominant energy level for potassium-40 (^{40}K) emits an energy peak at 1.46 MeV, thorium-232 (^{232}Th) at 2.164 MeV, and uranium-238 (^{238}U) at 1.764 MeV (Dewan, 1983). Of particular importance are the thorium/potassium ratios; when compared to XRD and XRF data, they can consistently be grouped by mineral type.

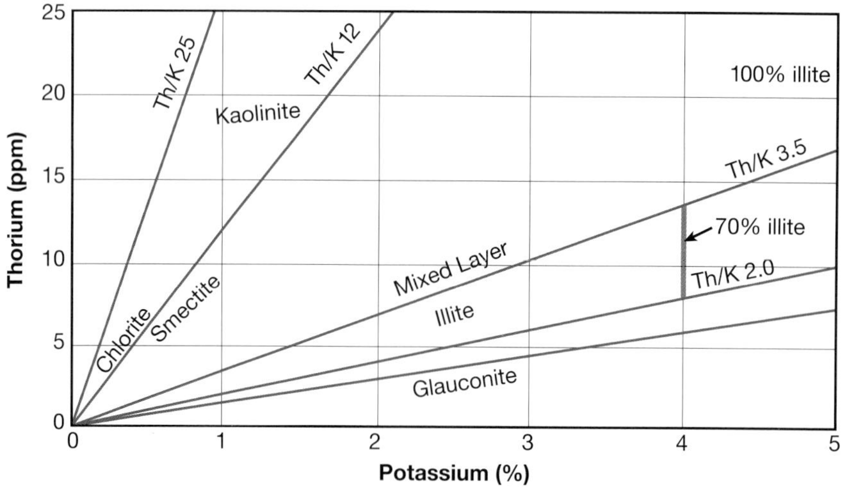

Fig. 5–5. Typical spectral log results used to differentiate clay minerals. For a specific region, the values should be calibrated at least with XRD and XRF.

Equation (5.10) can be applied to each component of the gamma-ray signal, and a relationship can be established in a formation to estimate the contribution of each as follows:

$$I_{Th} = \frac{Th_{log} - Th_{min}}{Th_{max} - Th_{min}} \tag{5.12}$$

$$I_{U} = \frac{U_{log} - U_{min}}{U_{max} - U_{min}} \tag{5.13}$$

$$I_{K} = \frac{K_{log} - K_{min}}{K_{max} - K_{min}} \tag{5.14}$$

Today, all cores analyzed in a laboratory are logged with a spectral gamma-ray tool, and the samples are calibrated to the ratios observed in the subsurface logs. Care must be taken when it is suspected that soil profiles are present. Soils, called *paleosols*, can yield anomalously high radioactivity readings. Halliburton currently uses an enhanced gamma-ray tool called GEM for elemental analysis that differentiates between several mineralogies; the results using GEM match well with XRD data. It is always recommended that the tools be calibrated with core data in the area of study.

COMPENSATED FORMATION DENSITY AND NEUTRON LOGS

The density log is usually run in combination with a neutron porosity log. As the full name implies, compensated formation density and neutron logs (CFD/CNL) are based on nuclear response. A small radioactive source, when open to the formation, excites the atoms in the fluids, and the returning signal can be related to the pore space available in the sediment. The original application of the tool was to identify sandy intervals with gas. For very-low-porosity shales, its usefulness is somewhat diminished. However, its saving grace is the density log, which plays a pivotal role in obtaining the proper calculation of petrophysics and generation of seismic synthetics.

The density log uses cobalt-60 (^{60}Co) or cesium-137 (^{137}Ce) isotopes as the source of gamma rays. The radiation bombards the formation and results in a loss of energy. The Compton effect of the reflected gamma ray is recorded. The main signal is recorded as a function of the amount of material present per unit area to emit a response—hence, a density value for the rock, which in turn is related to porosity. The standard porosity measurement for the density tool is similar in form to the porosity calculation from the sonic log. Three factors control the porosity read by the logging tool:

- Bulk density (ρ_b) of rock,
- Pure matrix density (ρ_{ma})
- Density of the fluids (ρ_{fl}) in the pore space (ϕ)

Thus,

$$\phi_d = \frac{\rho_{ma} - \rho_b}{\rho_{ma} - \rho_t} \tag{5.15}$$

Note that although the term ϕ_d refers to the porosity derived from the density log, this is not an effective porosity.

For the matrix densities of principal lithologies encountered in the subsurface, reference values have been published (e.g., see Asquith and Krygowski, 2004). Because different service companies use slightly different tools, the interpreter should use the set of tables provided by the contractor and not a generic table.

The neutron log was designed to determine the presence of hydrogen in non–shale-bearing porous rocks. The tool uses radioactive isotopes that emit neutron particles, usually from an americium-beryllium source. The neutrons collide with the hydrogen-bearing fluids and can be absorbed by the molecules. Because both water and oil contain similar quantities of hydrogen, the tool is sensitive to the liquid filling the pore space and not the liquid composition. Once absorbed, they emit gamma quanta that are read as described above. Because gas has a significantly lower concentration of hydrogen, its presence in the pore space yields an abnormally low value for the porosity when compared to the porosity from the density log; hence, the crossover in porosity for the overlay of both curves is used as a gas indicator.

Both neutron and density tools are contact logs. That is, they must be in direct contact with the formation by expanding pads that touch the walls of the borehole; the more rugose or irregular the borehole is, the poorer the results will be. Therefore, data must always be checked and edited with a caliper log. Because the assembled FDC/CNL is often slightly tilted, the log is mathematically corrected (i.e., compensated) to read in its vertical or horizontal position.

The neutron log always reads less than density log in the presence of gas. The average porosity favors the density log 2:1 with respect to neutron density; therefore, the effective porosity is (ϕn +2ϕd)/3, in a non–gas-bearing zone. The older density logs reported only the porosity based on a specific reference densities. Care must be taken to ensure that the correct reference value is applied. The common reference used is limestone (ρ =2.7 g/cm^3).

In shale resource assessment, we are interested in the actual matrix density as well as the porosity. This is particularly true where smectites dominate. Modern logs report the observed matrix density values as a separate curve. Density curves are provided with a correction factor. If the correction is too large, the data point should be discarded. As discussed previously, the presence of authigenic illite clays can significantly affect the density of the clay and, hence, the porosity reading.

PHOTOELECTRIC LOG

The photoelectric log comprises a third set of radioactive tools important to shale assessment. It is an advancement of the gamma-ray tool that allows identifying specific atomic numbers from their energy spectra. The atomic number (Z) can be converted to the photoelectric index I_{Pe} as follows:

$$I_{Pe} = \left(\frac{Z}{10}\right)^{3.6} \tag{5.16}$$

Not only does the photoelectric log indicate the clay type, but it also provides an approximate porosity scale (fig. 5–6). It is most often used where carbonates are present in the section of interest. The separation of clays appears to be a function of the iron content in the clay.

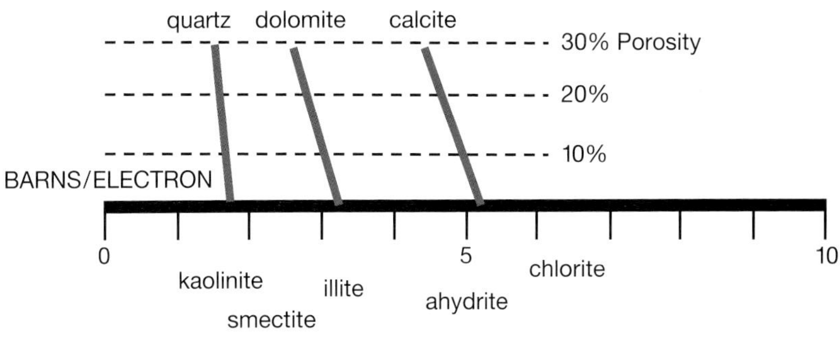

Fig. 5–6. Photoelectric interpretation chart. Note sensitivity to porosity and mineralogy.

SONIC LOGS

The fourth group of logs includes the various types of sonic logs that can be run in the wells. These include borehole compensated sonic (BCS) logs, which record compressional waves only and can be used to determine the compressional velocity, V_p. The dipole sonic contains a simple multicomponent detector. However, as previously noted, sound traveling through sediments can provide V_p and also the shear velocity, V_s. The shear wave can be in the direction of the compressional wave or orthogonal to the P wave; however, only one of these shear waves is recorded by the dipole sonic log, and may not always be that with the lowest value. The full-waveform sonic contains all shear velocities, as well as the Stoneley wave, and would be the preferred log to run on at least a few wells in the region. Reducing costs in drilling and logging are very important considerations. However, at least a few wells with complete log suites are necessary to calibrate the optimum relationship between V_p and V_s both in a well and in its immediate vicinity.

A common error in sonic logs is *cycle skip*. This is caused when there is a discontinuity along the borehole, such as a fracture or washout zone. As a result of such discontinuities, only one detector receives a signal, leading to an abnormally high value. Cycle-skip errors should be edited.

If a dipole sonic is not available, the compressional velocity V_p should be transformed into a shear velocity for similar lithologies at similar depths. The method commonly applied is the Castagna equation (Castagna et al., 1985). Where dipole sonics are available, the dipole sonic log can be edited, either in Excel format (as described above) or in a petrophysical package, and V_p can be converted to V_s to test the Castagna equation:

$$V_s = 0.862\ V_p - 1.172 \tag{5.17}$$

By application of an appropriate cutoff line, all nonshale intervals are assigned a null value, and the shales are analyzed. The Castagna equation is applied, and the new curve is compared with the actual V_s curve. By adjustment of the indices, a reasonable approximation can be made and applied to adjacent wells. The Castagna equation is also referred to as the *mudrock line*.

Several other equations are also available and in general follow the same basic premise as equation (5.17). Cores can also provide valuable information for the relationship. However, differences between dry and wet samples must be considered.

The time difference between two receivers is mathematically reduced to the inverse of interval velocity to provide an optimum reading on a sonic log. Given a particular lithology, the sonic velocity is directly proportional to the porosity. All log contractors provide tables calibrated for their specific tools to provide the sonic porosity. The sonic log works similarly to a seismic source and receiver; the primary difference is that source and receiver are nearly next to each other. As discussed in chapter 6, the primary driver on the sonic log is the interval velocity. Note that velocity is proportional to the density of the medium. It is because of its close relationship to the density of the sediments that the log was used as porosity detection tool. In fact, where density logs are also taken, the sonic and density log are plotted against each other in what is commonly known as an *M-N plot* (Bartok et al., 1981), which provides an indication of lithology and porosity. Moreover, Gardner et al. (1974) established an important relationship that relates density and interval velocity. The equation is based on empirical data:

$$\rho = b\, V_p^c \tag{5.18}$$

where as a first approximation $b = 0.23$ and $c = 0.25$ (Gardner et al., 1974).

By use of the V_p values and adjustment of the coefficients, pseudo-density logs can be generated. Care must be taken where lithology variations are present. The results are coarse but provide a range of potential values in regions where there are no density logs. In fact, in seismic processing, the same technique is applied to generate a density inversion model. However, as a word of caution, the pseudo-density should not be used for synthetic generation except as a last resort. In addition, the density value derived from equation (5.18) should not be used in determining elastic parameters because they are not independent variables. It is strongly suggested that equation (5.18) be calibrated with core data. Note that sonic logs are measured in microseconds per foot, and normal scales range from 40 to 120 and occasionally up to 150 ms/ft. For a reading of 40 ms/ft, this corresponds to (1,000,000 μs/s)/(40 μs/ft) = 25,000 ft/s (7,620 m/s); by similar conversion, a reading of 120 ms/ft corresponds to a velocity of 8,333 ft/s (2,540 m/s). Therefore, the description used for sonic logs is that an increase in *slowness* corresponds to a velocity decrease. If a full-waveform sonic is available in a well and there is a good density log, then all the elastic parameters can be calculated directly.

The porosity is obtained using a similar relationship as with density. The matrix interval velocity and the fluid velocity are used to calculate the porosity as follows:

$$\phi_s = \frac{D_{t\text{-log}} - D_{t\text{-ma}}}{D_{t\text{-fl}} - D_{t\text{-ma}}} \tag{5.19}$$

where ϕ_s is the sonic porosity, $D_{t\text{-log}}$ is the actual log reading, $D_{t\text{-ma}}$ is the time difference for a clean matrix, and $D_{t\text{-fl}}$ is the time difference for a specific fluid. The latter two values are available in all log manuals. Because of the tool design, source, and sets of receivers located away from the source, the minimum resolution of the tool is the distance between source and receiver. Depending on the provider, it may be 1 ft (30.5 cm) to 2 ft (61 cm) apart. It is suggested that a 2 ft resolution be used for log work.

The shear wave data (V_s) combined with compressional wave data (V_p) and the rock bulk density comprise the three basic elements of rock physics and are crucial for the exploitation of the shale resource play. Full-waveform sonic simply adds more receivers to the logging tool. Because there is additional cost with multiwaveform sonic, it is used more often in an early phase of exploration and not commonly used in the appraisal or development phase. There are several drawbacks to dipole sonic and full-waveform sonic. Any log tool that needs to be in direct contact with the formation wall is subject to borehole irregularities and the risk of sticking the tool in the hole.

The difference between the shear wave velocities provides vital information as to the anisotropy in the sediment. In essence, it is a function of the preferred orientation of the clay minerals. Detailed XRD studies have been carried out on shales to determine the characteristics of the anisotropy (Lonardelli et al., 2007). To date, hundreds of publications by numerous investigators (notably, Fred Hilterman, John Castagna, Rob Steward, Dan Hampson, Brian Russell, Gary Mavko, Jack Dvorkin, Bill Goodway, and Marco Perez, among others) have established the basic principles on which rock physics was founded. Rock physics is strongly dependent on petrophysics and therefore will be discussed more in chapter 6.

NUCLEAR MAGNETIC RESONANCE

Nuclear magnetic resonance (NMR) is a relatively new logging tool to geoscientists. The basic principles have been outlined by Coates et al. (1999). NMR is employed extensively in medicine as magnetic resonance imaging (MRI). The same principles apply to imaging from the subsurface, except that the tool is introduced into the borehole. NMR looks only at fluid filling the pore space; it is independent of the composition of the sediments.

As the name implies, NMR generates a magnetic field that causes certain protons to orient themselves and resonate in the magnetic field—particularly hydrogen. When the magnetic field is removed, the hydrogen atom relaxes to its previous state and generates a signal that is a function of amount of fluid present, or the porosity. It can also be used to estimate the permeability (Asquith and Krygowski, 2004). Because hydrogen is present in both water and oil, NMR cannot easily differentiate between the two but can distinguish bound water from free-flowing fluids; hence, it is an important tool in determining the effective porosity in a reservoir. As mentioned previously, the effective porosity represents the pore space available for hydrocarbon flow and therefore is of paramount importance in determining the volume of hydrocarbon that can be produced. In a water-wet system, water tends to adhere to the grain surface, and oil is thus more mobile.

Not all isotopes have a magnetic moment. A magnetic moment is a consequence of fast-spinning atoms. The isotopes ^{1}H and ^{13}C have magnetic moments, and depending on the compounds the ^{1}H and ^{13}C isotopes are associated with, adjoining atoms may interfere with the magnetic moment of the two isotopes. When the results are filtered, the carbon can be shielded, so that the tool can be focused on detecting hydrogen in the gases, light hydrocarbons, and water. Because the main focus is hydrogen, which is present in both liquids (i.e., oil and water), the pore space in shale can be accounted for independently of the adjoining mineralogy. The downside to the NMR tool is that it penetrates only a few inches into the formation (within the invaded zone).

The measurement made is known as either *transverse relaxation* (T2) or *longitudinal relaxation* (T1) of the magnetic interaction between the protons resulting from the precessing atom transferring energy to its surroundings (fig. 5–7). Usually only the T2 curve is reviewed (fig. 5–8). The donor proton relaxes to its lower-energy state as it precesses in a particular direction given the induced magnetic field. For a more detailed discussion, refer to Halliburton's online NMR logging manual (Coates et al., 1999).

A properly calibrated NMR tool can also determine the pore size distribution. In shales, NMR can be used, but corrections have to be made for the clay-bound water. They usually requires knowing the clay type a priori; therefore, NMR should be used in combination with spectral gamma-ray log, XRD, XRF, and EDAX studies to increase its accuracy, rather than as a stand-alone tool. Usually, it is displayed in combination with spectral gamma-ray logs or other lithology indicators (fig. 5–8). Therefore, cores or sidewall cores should be taken at the interval of interest, to measure the total rock porosity regardless of the pore fluid type (Dunn et al., 1999). The two T2 distributions, one from the fully polarized echo train and the other from the partially polarized echo train, are merged together to build a T2 distribution from 0.3 ms to over 3,000 ms (fig. 5–7) to estimate the total porosity.

A specific range of values have been shown to correspond to either clay-bound water or movable fluids (fig. 5–8). The area under the curve is an indication of the pore volume and may in turn be related to pore size (Coates et al., 1999). This is particularly effective when used in combination with IBM-SEM images. Greater amounts of fragments of nonclay minerals increase the uncertainty of obtaining an accurate T2 cutoff value for clay-bound water (Chitale et al., 1999). Although estimation of the T2 cutoff value has been discussed by many authors, there is still uncertainty as to the optimum choice. With reference to the work of Matteson et al. (1998), the standard relaxation time of 3.3 ms is used, but this is appropriate only where the clay type is not kaolinite.

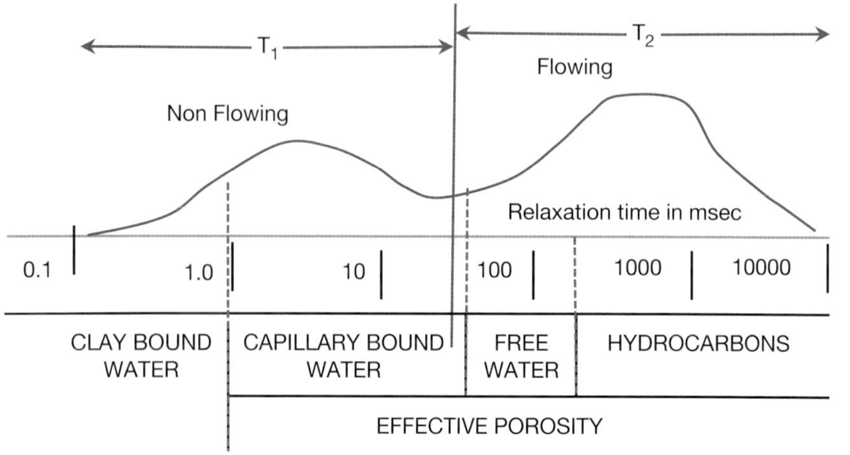

Fig. 5–7. Range of values for NMR relaxation times, indicating mobility of the fluids. Area under the curve is indicative of bulk porosity. The NMR tool measures hydrogen content and therefore is not a direct fluid predictor.

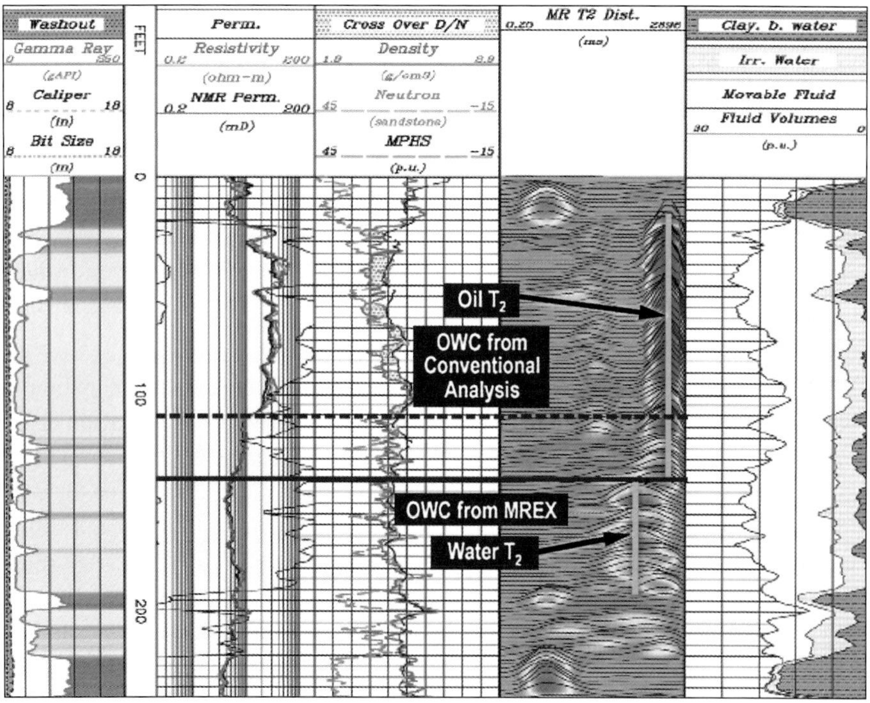

Fig. 5–8. Example NMR log showing T2 curve in the right track. Note that NMR is combined with lithology predicting tools (FDC/CNL) and porosity/permeability data. In the shale sections (upper and lower portions of the diagram), the presence of bound water is shown. (Based on Allen et al., 2000; reproduced with kind permission of Baker Hughes Corp.)

FULLBORE FORMATION MICROIMAGE LOG

There are two approaches to fullbore formation microimaging (FMI). One is based on sonic waves and the other on microresistivity. As will be discussed in chapter 6, all sound-generating waves produce a series of waves. In surface geophysics, there are surface waves and waves that penetrate through the formation (compressional and shear waves). In wellbores, a specific wave, known as a Stoneley wave, is generated that is analogous to the surface wave. It travels along the surface of the borehole and is generated either by a sonic log or from the vertical seismic profile.

The Stoneley wave is useful in determining the presence of fractures and formation boundaries (Saito et al., 2004). As it travels across the surface of the wellbore, any interference—such as fractures, bedding, and, unfortunately, borehole diameter changes—will result in a deflection of the wave. The propagation of the Stoneley wave is recorded by FMI and produces a pseudo image of the borehole. Because the image is recorded 360° around the borehole as a cylinder, it must be peeled back inside out for analysis (fig. 5–9).

The FMI method uses an overlay sine curve graph to decipher the signal information (fig. 5–9). Once again, wherever possible, an image log should be run in key wells within the field of interest, and a selection of cores should be taken to calibrate the image log. The observations can then be translated into geological features. Because there is a clear orientation to bedding, the cores can also be rotated to their original position in the subsurface on the basis of their FMI response. Where this has been confirmed, the cores should also be analyzed using *paleomagnetic* (remanence magnetization) techniques to reorient the core if verification is necessary. Magnetic remanence can also be applied to older cores for which FMI logs are not available. The orientation of cores is critical in determining the orientations of fractures in the subsurface. If the core barrel is notched, then the incoming core can be oriented from its marking. The sonic log has less resolution than the microresistivity log but can be used in oil-based muds. However, the mud should be as free as possible from solids.

The microresistivity log uses alternating current from various electrodes. Because it is an electric current that needs to be recorded, it can be used only for brine-based drilling muds. The results are better, and the method is the same.

The display of the FMI tool (see fig. 5–9) is conceptually the peel of the borehole—hence, the directional indication on the log bottom. Its geometry defines the angle of the fracture or dip of bedding.

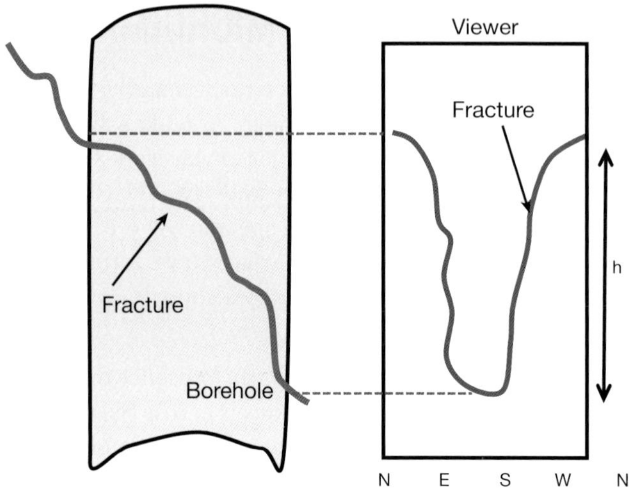

Fig. 5–9. Typical FMI and procedure to evaluate the log. The image on the right is essentially a peel of the wellbore, as noted by the directions at the base of the diagram.

LOGGING HORIZONTAL WELLS

Before discussing logging horizontal wells, it is important to review what information is required in order to appropriately drill the well and *land* the well on target. The initial steps involve traditional exploration techniques. Wells in the vicinity must be correlated in detail. Cross sections in several directions within the area of interest have to be prepared using different log suites. It is essential to determine how basic rock properties vary across the field being penetrated by the borehole. The cross sections are commonly based on the static geocellular model described later, in chapter 8. The fewer variations there are along the trace of the well, the better will be the prediction. Synthetic logs must be generated for all key wells and tied to the three-dimensional seismic over the block. Variations in amplitude and other rock physics properties assist in predicting continuity.

Maps must be generated tying together the seismic and well data. Each projected layer should be modeled in detail, and the expected log response should be documented to ensure it stays on target. Measurement while drilling (MWD) focuses attention on the mechanical properties, directional and inclination data, and resistivity and gamma-ray logs to ensure proper placement. Logging while drilling (LWD) is included in the downhole assembly and includes sonic, density, neutron, and image logs. The most common tracking logs are resistivity and gamma ray. Because resistivity penetrates the formation, it is possible to use it as a tool to see ahead of the bit. Once a deviation from the model is observed, the drill bit is adjusted.

The velocity and the image log can be combined to show the presence of existing fractures. Selectively perforating the formation at optimum sections is a better approach than using ball-and-socket fracturing. The latter has equally spaced windows that can systematically frac the formation. Because significant vertical fractures are often observed in shale resource plays, it is best to avoid larger fractures and to preserve the integrity of the seals.

A FINAL WORD ON PETROPHYSICS

Ensure that the units are in the correct system and format. Manuals from the various log providers should be used in interpretations for the specific logs used. For more sophisticated analyses, including rock physics, it may be necessary to run specialized logging programs. It is highly recommended that estimated values be modeled prior to logging to ensure that they can be properly sampled and adequately populate the seismic data to provide valuable information. Histograms of values provide averages and ranges, and these improve log editing. The investigator should test the data on a spreadsheet to ensure full control of the steps involved. All rock physics values can be computed on spreadsheets. Furthermore, it is recommended that all values be converted to compatible SI units. In other words, densities should be measured in kilograms per cubic meter, velocities should be in kilometers per second, and so forth. In all cases, statistics should be run on the log response to ensure that the data for the target member of the formation under investigation is within the range of the observed value.

There are at least three approaches to the study of the data. The first is to evaluate the mean and the range in values; the second is a more geostatistical approach known as *covariance*; and the third is to apply geostatistics as a predictive tool using programs such as Markov chain and Fourier series (Davis, 1970). (A more detailed discussion on geostatistics is provided in chap. 9.)

My personal preference is to work log interpretations on Excel spreadsheets. This improves the visualization of the input parameters. The speed and accuracy of packages like Hampson and Russell or Jason Geophysical is not questioned. However, when the objective is to apply logging tools to rock physics, the spreadsheet is the best tool.

The first objective of using petrophysics in wells is to evaluate the conditions in situ. In other words, by use of SP, gamma-ray, FDC/CNL, sonic, and image logs, among others, clear estimates on lithology, porosity, and water saturation can be computed. Therefore, the reserves associated with the well can be estimated. As more wells are integrated into the database, maps can be generated to show the distribution of the parameters described above and quantify the reserves of the field. The second objective of petrophysics is to transform the well data into a seismically compatible synthetic wavelet and tie specific rock physics parameters to the seismic. In addition, the reflection coefficients (chap. 6) for the principal formation boundaries can also be calculated and modeled to define potential seismic anomalies. When generating a synthetic wavelet (as with all data), the input data must be edited to eliminate any spurious noise. Two data sets are used: sonic logs and density logs. The primary goal is to generate a wavelet that closely approximates the seismic response.

In an initial tie well, a first data set is generally selected with check-shot data–derived interval velocity or a vertical seismic profile (VSP)–generated interval velocity. The well should also be located in an area where there is only minor deformation. In general, use the simple form $d = vt$, where d is depth, v is average interval velocity, and t is time (one-way time). However, velocity studies worldwide have demonstrated that the velocity function is a hyperbolic curve with a very dramatic increase in velocity within the first 1,000 ft of burial and then have a fairly linear increase with depth beyond that point. The early increase is caused by sediment compaction, resulting in a density increase and a consequent porosity decrease. If check-shot or VSP velocity data are not available in the area under investigation, either the general average velocity curve from a nearby well or, better still, accurate interval velocities should be used.

It is important to avoid areas of overpressure because of its effect on both velocity and density. Caution will ensure that the seismic reflector is properly correlated. In more complex areas, the reflector may be inadequately migrated and thus shifted in space. For correlations with synthetics, always go from simple problems to complex. If a complete sonic log is available, it may be used, but this is uncommon. Governments require only an induction log in the shallow portion of the well to document the presence of aquifers and surface as well as intermediate casing point.

The second data set that must be provided is a set of seismic traces on either side of the well, with the approximate formation tops indicated both in depth and in time. Note that the seismic is in two-way transit time (described in more detail in chap. 6), and the well log is in depth. Once a synthetic wavelet is generated and correlated with seismic, it is useful to compare equivalent velocity and density data across the area of study. Geostatistics provides the necessary tools to ensure that all data used in the model are within the range of expected values.

SUMMARY

Correlation tools include gamma ray and spontaneous potential (SP) logs. For gamma-ray logs, V_{sh} is evaluated by comparing clean and mixed lithologies. Spectral gamma-ray log (spectralog) differentiates clay minerals. Photoelectric log measures porosity and identifies clays.

The Archie equation requires adjustment for shales. In addition, the investigator must correct for the m and n.

Density neutron log were also presented. Note that the density log needs corrections (through use of histograms); this is important for bulk rock density.

Sonic logs are an original porosity indicator. Dipole sonic is important for V_s determination; V_s can be estimated, but calibration is required. Full-waveform sonic is preferred to study both V_s components.

NMR is a good porosity indicator and determines if water is movable or bound. However, NMR does not distinguish between water and oil.

Fullbore formation microimaging (FMI) was presented as a further log type. FMI provides details on bedding and fractures.

For horizontal wells, MWD or LWD is used. FMI are often run to detect fracture systems along the borehole.

The vertical pilot wells determine landing parameters based on pilot and are combined with static models for correlation.

REFERENCES

Ajdukiewicz, J. M., S. T. Paxton, and J. O. Szabvo. 1991. "Deep porosity preservation in the Norphlet Formation, Mobile Bay, Alabama." Paper presented at AAPG Annual Convention, Dallas, Texas, April 7–10. Search and Discovery article 91004.

Allen, David, Charles Flaum, T. S. Ramakrishnan, Jonathan Bedford, Kees Castelijns, David Fairhurst, Greg Gubelin, et al. 2000. "Trends in NMR logging." *Oilfield Review.* 12: 2–19. www.slb.com/~/media/Files/resources/oilfield_review/ors00/aut00/p2_19.pdf.

Archie, Gustave E. 1942. "The electrical resistivity log as an aid in determining some reservoir characteristics." *Transactions of the AIME.* 146: 54–62.

Asquith, George B., and Daniel Krygowski. 2004. *Basic Well Log Analysis*, vol. 16. Tulsa: American Association of Petroleum Geologists.

Avseth, Per, Tapan Mukerji, and Gary Mavko. 2005. *Quantitative Seismic Interpretation.* Cambridge: Cambridge University Press.

Bartok, P., T. J. A. Reijers, I. and Juhasz. 1981. "Lower Cretaceous Cogollo Group, Maracaibo Basin, Venezuela: Sedimentology, diagenesis, and petrophysics." *AAPG Bulletin.* 65: 1110–1134.

Boudreau, Bernard P. 1996. "The diffusive tortuosity of fine-grained unlithified sediments." *Geochimica et Cosmochimica Acta*. 60: 3139–3142.

Bust, Vivian K., Azlan A. Majid, Joshua U. Oletu, and Paul F. Worthington. 2013. "The petrophysics of shale gas reservoirs: Technical challenges and pragmatic solutions." *Petroleum Geoscience*. 19: 91–103.

Castagna, John P., Michael L. Batzle, and Raymond L. Eastwood. 1985. "Relationships between compressional-wave and shear-wave velocities in clastic silicate rocks." *Geophysics*. 50: 571–581.

Chitale, D. V., Peter I. Day, and George R. Coates. 1999. "Petrophysical implications of laboratory NMR and petrographical investigation on a shaly sand core." Paper presented at SPE Annual Technical Conference and Exhibition, Houston, TX, October 3–6.

Coates, George R., Lizhi Xiao, and Manfred G. Prammer. 1999. *NMR Logging: Principles and Applications*. Houston: Halliburton Energy Services. www.halliburton.com/public/lp/contents/Books_and_Catalogs/web/NMR-Logging-Principles-and-Applications.pdf.

Davis, John C. 1970. *Statistics and Data Analysis in Geology*. New York: Springer.

Devarajan, S., E. Toumelin, C. Torres-Verdín, and E. C. Thomas. 2006. "Pore-scale analysis of the Waxman-Smits shaly sand conductivity model." Paper presented at the SPWLA 47th Annual Logging Symposium, Veracruz, Mexico, June 4–7.

Dewan, John T. 1983. *Essentials of Modern Open-Hole Log Interpretation*. Tulsa: PennWell.

Doveton, J. H. 2001. "All models are wrong, but some models are useful: 'Solving' the Simandoux equation." Paper presented at the 2001 Annual Conference of the International Association for Mathematical Geology, Cancún, Mexico, September 6–12. www.kgs.ku.edu/Conferences/IAMG//Sessions/J/Papers/doveton.pdf.

Drelich, J., C. Fang, and C. L. White. 2002. "Measurement of interfacial tension in fluid-fluid systems." *Encyclopedia of Surface and Colloid Science*, vol. 3. Arthur T. Hubbard (ed.). New York, Marcel Dekker, 3158–3163.

Dunn, Keh-Jim, Gerald A. LaTorraca, and David J. Bergman. 1999. "Permeability relation with other petrophysical parameters for periodic porous media." *Geophysics*. 64: 470–478.

Ellis, Darwin V. 1987. *Well Logging for Earth Scientists*. New York: Elsevier.

Foster, Norman H., and Edward A. Beaumont. 1990. *Formation Evaluation I, Log Evaluation*. AAPG Treatise Reprints, no. 16. Tulsa: American Association of Petroleum Geologists.

Gardner, G. H. F., L. W. Gardner, and A. R. Gregory. 1974. "Formation velocity and density: The diagnostic basics for stratigraphic traps." *Geophysics*. 39: 770–780.

Lee, Myung W., and Timothy S. Collett. 2006. "A method of shaly sand correction for estimating gas hydrate saturations using downhole electrical resistivity log data." U.S. Geological Survey Scientific Investigation Report No. 2006-5121. pubs.usgs.gov/sir/2006/5121/pdf/sir5121_508.pdf.

Lonardelli, Ivan, Hans-Rudolf Wenk, and Y. Ren. 2007. "Preferred orientation and elastic anisotropy in shales." *Geophysics*. 72: D33–D40.

Matteson, A., J. P. Tomanic, M. M. Herron, D. F. Allen, and W. E. Kenyon. 1998. "NMR relaxation of clay-brine mixtures." Paper presented at SPE Annual Technical Conference and Exhibition, New Orleans, September 27–30.

National University of Singapore, Physics Department. 2005. "Compton scattering." www.physics.nus.edu.sg/~L3000/Level3manuals/Compton%20Scattering.pdf.

Saito, Hiroyuki, Kazuo Hayashi, and Yoshikazu Iikura. 2004. "Detection of formation boundaries and permeable fractures based on frequency-domain Stoneley wave logs." *Exploration Geophysics.* 35: 45–50.

Smith, Joe H. 2007. "A method for calculating pseudo sonics from e-logs in a clastic geologic setting." *Gulf Coast Association of Geological Societies Transactions.* 57: 657–678.

Teh, W. J., G. P. Willhite, J. H. Doveton, and J. S. Tsau. 2011. "Improved predictions of porosity from microresistivity logs in a mature field through incorporation of pore typing." Paper SPE 1495061, presented at SPE Eastern Regional Meeting, Columbus, OH, August 17–19.

Ushie, F. A. 2001. "Formation water resistivity (R_w) determination: The SP method." *Journal of Applied Sciences and Environmental Management.* 5: 25–38. www.bioline.org.br/pdf?ja01004.

Waxman, M. H., and L. J. M. Smits. 1968. "Electrical conductivities in oil-bearing shaly sands." *Society of Petroleum Engineers Journal.* 8: 107–122.

Wyllie, Malcolm Robert Jesse. 1957. *The Fundamentals of Electric Log Interpretation.* 2nd ed. New York: Academic Press.

6

GEOPHYSICS

TAKEAWAY

- How is seismic acquired?
- What should an interpreter ask before initiating an interpretation?
- Determine the objectives of seismic data and mapping
- Extension of rock properties from wells including brittleness
- Importance of seismic inversion
- Lamé parameters, Young's modulus, and Poisson's ratio—how do they help define the sweet spot?
- Rock physics combined with amplitude versus offset (AVO) studies to predict fluids

BASIC THEORY

Given the proper calibration tools by the geologist, geochemist, and petrophysicist, a seismic interpretation can provide valuable information on the structural configuration of the shale opportunity and the presence of fractures and faults, predict pressures, and, most important, the internal properties of the shale. First, before embarking on an analysis of applications, we need to review how seismic is acquired and processed.

Seismic theory is based on Hook's law and, in its simplest form, can be related to the motion of a spring. The periodic motion of the spring can be described graphically as a sine curve with a specific amplitude of displacement versus time. As the motion is plotted on a curve, it follows a circular pattern from crest to trough. The seismic pulse can be converted into a similar curve. The *phase* of the data refers to whether the measurement is initiated from a crest or a zero crossing. The common rule in seismic is to process the data such that positive reflectors appear as a crest; this is known as *zero-phase seismic*. There may be occasions when the phase is changed to achieve a particular objective. For simplicity, begin with all data in zero phase. The *period* (T) of the waveform is defined as the time difference from crest to crest on the plot of amplitude versus time. The *frequency* of the waveform is the inverse of period and is measured in cycles per second (1 cycle per second = 1 Hz). The actual terms applied to seismic are derived in part from the spring motion and the physics of sound. Depending on the frequency and the amount of energy applied, the amplitude and rock physics response change across the various lithologies penetrated.

Seismic source and signal characteristics can be described by analogy with musical instruments—in particular, the principle of sound in wind instruments. For example, in a trumpet, a stronger airflow is associated with higher energy, and a louder sound is referred to as having higher amplitude. Once a particular set of valves are depressed on a trumpet, a unique frequency is generated that corresponds to a specific musical note. Importantly, the frequency (pitch) is independent of the amplitude (loudness). In music, middle C is tuned to a frequency of 261.63 Hz. For orchestral tuning, the A above middle C is used, which by modern convention has an exact frequency of 440.00 Hz. The exactness of this pitch allows the associated notes to be tuned by reference. By contrast, the amplitude of the sound is the force or pressure emitted and is measured in decibels (dB). Note that decibels are not an absolute scale, but rather a relative one. Is the sound audible in a quiet auditorium (30 dB)? Or is it louder than heavy traffic (100 dB)? Returning to the music analogy, if music is played in an enclosed room, the sound is loud in the room but is attenuated by a wall. Listening to the sound outside the room, you can still hear a pitch as middle C, for example, but the amplitude will be reduced.

The same phenomena can be observed as a wave passes between strata. In seismic, the term applied is *acoustic impedance* (AI), defined as the acoustic pressure differential resulting from density changes across transmitting material. The effect of AI is measured as the *reflection coefficient* (RC), defined as the difference of products of the transmission velocity, V, and the bulk density, ρ_b, below and above the interface divided by the sum of both products, thus generating the reflectors observed on a seismic line. By definition,

$$\mathrm{AI} = \rho_b V \tag{6.1}$$

and

$$\mathrm{RC} = \frac{\rho_{b_2} V_2 - \rho_{b_1} V_1}{\rho_{b_2} V_2 + \rho_{b_1} V_1} \tag{6.2a}$$

$$\mathrm{RC} = \frac{\mathrm{AI}_2 - \mathrm{AI}_1}{\mathrm{AI}_2 + \mathrm{AI}_1} \tag{6.2b}$$

In seismic geophysics, the term *polarity* is used to describe whether the RC is positive or negative. Unfortunately, the U.S. standard is the exact opposite of the European standard. Therefore, it must always be noted which convention is used in the seismic. In the U.S. standard, the positive RC is filled black (Brown, 2011).

PURPOSE OF THE SEISMIC ACQUISITION

Mapping prospective areas in detail, identifying the internal properties of all exploration targets, and predicting potential drilling hazards prior to drilling are the ultimate goals of the explorationist. Seismic data are best suited for this purpose because they allow a means of "seeing" laterally beyond the well control. Following the initial step of accurately mapping the subsurface, the investigation focuses on the internal properties of the interval of interest, including lithology determination, porosity characteristics, fluid prediction, sealing capacity, and pressure regimes. In addition, the ability of the sediments to fracture can be estimated from seismic; it is even possible to estimate organic richness.

Once an unconventional play is detected, an early question that requires an answer is what brittle drillable interval will be sufficiently thick to allow one to land and stay within the proposed objective. Brittleness, in simple terms, is defined by the mineralogy, as described in the shale depositional system (chap. 2) and by its rock physics parameters—namely, the average of its Young's modulus and Poisson's ratio factors. Can fluid types be differentiated? Can porosity and reservoir pressure be estimated?

Seismic exploration data were initially acquired in the mid-20th century to evaluate the presence and location of contrasting lithologies and generate maps of the subsurface. The seismic process is conceptually straightforward and is based on the principle that a generated high-energy vibration source of a given frequency can be transmitted into the subsurface. The emitted wave is refracted and reflected at discrete strata (fig. 6–1). Recording the arrival times of the reflected wave at areally distributed geophones and progressively moving the energy sources provides information that can be processed to generate arrival times into a detailed image that can illuminate the subsurface. If the interval velocity for the horizons penetrated can be obtained from well data, the transit time can be converted to depth. Note that a time map is not an accurate depth structure map, but an approximation of the true structure; a depth map provides the ultimate structure. Modern interpretations of seismic require integrations of multiple rock parameters to describe the targeted objectives. These include porosity, rock mechanics, density information, and other core log data.

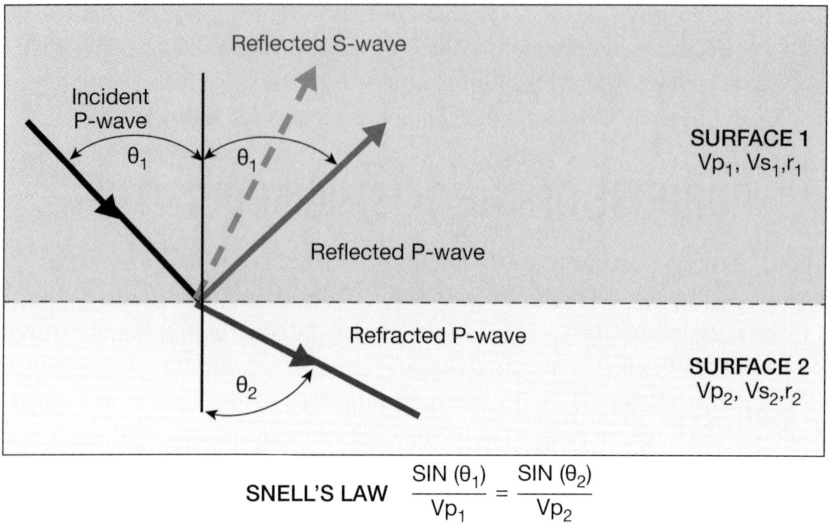

Fig. 6–1. Snell's law: The angle of reflection is equal to the incidence angle if the wave velocity is the same. The reflected S wave is slower. The angle of refraction (θ_2) follows Snell's law.

Seismic dynamite, vibrators, and air-gun sources produce energy that is transmitted through the sediment as a compressional wave (P wave). The most common measure of the geophone is the vertical component of the particle velocity. The energy is reflected and refracted as P and S (shear) waves in the subsurface, as well as various surface waves.

Because most shale resource plays are onshore, concerns about the thickness of the weathering zone are important. If the weathering zone is slower than the underlying bedrock, the upcoming P and S waves are bent toward the normal weathering interface, which is usually flat, such that the P waves are nearly vertically polarized and the S waves are horizontally polarized and not measured by vertical geophones. Therefore, multicomponent geophones are required in order to improve velocity determinations. Because of the mass production of accelerometers used in automobile air bags, micromechanical devices are now affordable for use in multicomponent geophones.

As a survey progresses with additional shots, how many times is the same subsurface point recorded? The sum of the overlap is known as the *fold,* and the higher the fold is, the greater will be the confidence in the combined signal response. Older two-dimensional surveys can have as little as sixfold. Modern two-dimensional surveys have 24-fold or more (e.g., commonly identified in the header as 2,400% fold). Three-dimensional surveys have high folds, up to 400-fold.

These are displayed on maps because, along the edges of the three-dimensional survey, the fold may be as low as six fold, meaning the data are poor.

Importantly, a receiver cannot be on top of the source because the equipment can be damaged. The distance between source and receiver is known as the *offset*. Although zero-offset recordings (i.e., receiver over shot point) are physically impossible, it is a critical measurement that can be estimated once the other offsets have been recorded. Furthermore, as shots are moved down a section, the geophones will read signals from the same *common midpoint* (CMP), but the arrival times will be different. *Common depth point* (CDP) is often used in place of CMP. The preferred term is CMP, given the presence of dipping beds, which will be subject to dip correction; by contrast, CDP applies only to a flat surface.

An initial signal (S4) generated at the surface travels to a CMP, then is reflected and received by a recorder (R4). The ray path travels a longer distance and thus is delayed with respect to a subsequent ray path associated with S2 to the CMP and recorded at R2 (fig. 6–2). As a result, a plot of arrival time of a single shot versus distance to the receivers generates a curved path. The closer shots will receive the signal earlier. That difference is known as the *normal moveout* (NMO) and can be corrected if the velocity of the medium travelled is known. The plot for each ray path plot is known as a *CMP gather*, or *gather* for short. The dip-corrected velocity used for the NMO is listed on the seismic header. After NMO correction, a plot of the offset, showing the amplitude, can be displayed for each CMP.

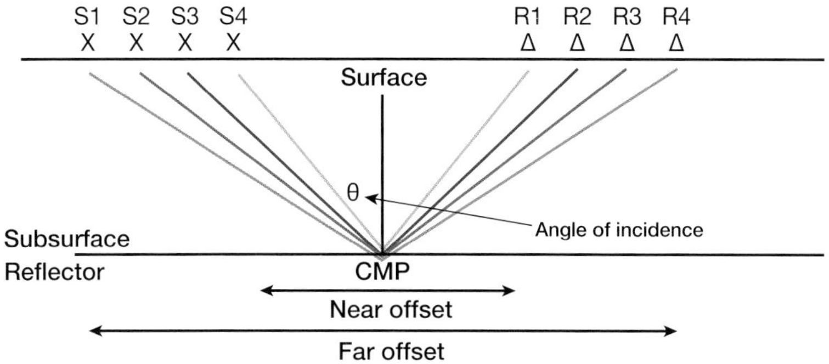

Fig. 6–2. Seismic ray path from source to receiver, showing offset and angle of incidence change versus offset

With respect to the CMP, the S4-R1 path is referred to as the *near offset*; S3-R3 as the *mid offset*; and S1-R4 as the *far offset*. The angle of incidence (θ) increases from far to near. This is the basis for amplitude versus offset (AVO), which will be described later in this chapter (see the section "Internal Properties of the Target Horizon"). What other information should you be aware of when looking at a seismic section for the first time?

TWO-DIMENSIONAL SURVEY INFORMATION

Spread is the distance among the set of geophones recording from each shot. For example, 96 geophones can be spread equally over 21,500 ft (6,553 m). If the number of geophones is reduced and the spread length is high, then the fold at each shot point will be diminished and the impact of the surface waves will be more difficult to suppress. The geophones are commonly set a few meters apart.

The general rule is that the spread length should be equal to the maximum depth to be investigated. Longer spread length can cost more to acquire, can be more difficult to process, but provides information critical to a more detailed velocity and AVO analysis.

What is the fold of your survey (discussed above)?

Ask the geophysicist what filters were applied and to explain why. The goal of the filters is to reduce noise and enhance the signal.

What processing was done and why? Note that some processing can improve the structural interpretation but will diminish the ability to study rock properties. For this reason, several versions are available from the processor and should be compared once the initial interpretation has been completed.

Onshore data also require good surface statics corrections. This is because of the effect of weathering or other near-surface issues, such as glacial till.

THREE-DIMENSIONAL SURVEY INFORMATION

A common three-dimensional acquisition program consists of orthogonal shot and receiver lines. If neighboring shot lines are 220 ft apart and the neighboring receivers are 220 ft apart, then the CMP bin size is 110 ft × 110 ft.

The size of the ideal bin is usually defined by the frequency of the data, the dip of the beds, and the velocity of the interval. It is also a function of the size of the feature being sought. For definition of a depositional channel, the bin size should be no greater than the channel width, and the resulting distance is given as follows:

$$\text{Bin} < \frac{V}{f \sin\theta} \tag{6.3}$$

where V is the interval velocity, f is the dominant frequency ($f = V/\lambda_1$, where λ_1 is wavelength), and θ is the dip angle.

Note that seismic time is two-way time. The complete wavelength from peak to peak should be measured as half the recorded time on the seismic data at the target zone. The period, T, of a wavelength is the time involved for a full wavelength to repeat and is defined as the inverse of the frequency.

Again, what filters were used and what processing was done and why? Ask if the other volumes reveal any significant information. As in two-dimensional surveys, several versions with different processing steps should be available. Three-dimensional seismic records the same signal at all geophones active in the pattern. This provides azimuthal information critical for anisotropy studies (discussed later, in the section "Anisotropy").

The terms prestacked cube and poststacked cube are used to describe whether processing such as migration was carried before or after the traces were stacked. For fluid studies, prestacked data is preferred. To improve imaging, poststacked data is used.

Ultimately, the manager's goal is to obtain the optimum information at the best price in the shortest time. In the case of three-dimensional surveys, it is prudent to model the seismic with pseudo synthetics and pseudo gathers before carrying out the survey, to establish the optimum acquisition parameters and the signal-to-noise ratio. Modern seismic is focused on three-dimensional acquisition. Regional two-dimensional data in the study area should be analyzed to determine the proper three-dimensional acquisition parameters and ensure it is shot in the optimum orientation. It is best to shoot the two-dimensional lines perpendicular to the fault trends, which is the recommended orientation for three-dimensional data as well. As mentioned already, all participants in the shale study should understand the terminology (e.g., gathers, NMO, spread, and migration). Before any detailed interpretation is made on the seismic data, the gathers are *conditioned* or processed to reduce noise and improve signal.

INTERPRETING SEISMIC DATA

Most seismic used in interpretation has frequency ranges of 10–80 Hz. As noted previously, frequency, f, is related to interval velocity, V, and wavelength, λ_1, according to the relation $f = V/\lambda_1$. At normal seismic acquisition ranges (i.e., 10–80 Hz), the velocity is generally independent of the frequency. However, in the lower-frequency ranges, velocity and frequency are proportional. The changes in velocity within this range are known as *velocity dispersion*. The main problem lies in wireline logs and surface waves, because they generate low-frequency signals that are subject to dispersion effects; this should be considered when generating synthetic seismograms (Schwenk et al., 2012).

Dipping beds present an additional problem in two-dimensional seismic. Because the seismic wave is transmitted through the sediment medium as a cone, the precise location of the signal is not clearly defined. The signal must be migrated. However, because two-dimensional seismic is shot in a unique direction, the migration must be carried in the strike direction of the line. It is obvious that the apparent dip on any two-dimensional cross-line may be different than the true dip observed in a three-dimensional survey. After migration, when surveys with different orientations (either two- and three-dimensional surveys or a pair of two-dimensional surveys) are merged, the upper reflectors may tie, but the lower ones will not. The lines cannot be tied directly and may be adjusted for mapping purposes only or smoothed in the mapping package.

Multiple two-dimensional surveys may have different stacking velocities, and the surveys themselves may not match; when this happens, they are called *misties*. One survey is used as the reference level, and the others are shifted to compensate for the difference. Several three-dimensional surveys may have different acquisition parameters. The data are merged and reprocessed together to obtain the best results. Great care must be taken in mistie corrections because the well data may not tie correctly. It is best to adjust the two-dimensional data according to the main three-dimensional data used in the interpretation. The wells can be adjusted by temporarily modifying the well datum. Good bookkeeping of any modification is vital. A journal should be maintained on a workstation available to all interpreters to avoid repetition of processes and note any modifications.

In chapter 5, the synthetic seismogram was introduced. The subject requires further discussion. In principle, the sonic log provides the velocity data, and the density is determined from the FDC/CNL log. Therefore, the RC can be determined using the standard computation for RC discussed previously. However, sonic-derived velocity is an estimate, and additional control may be used to correct the velocity. Interval velocity derived from a check shot or a vertical seismic profile (VSP) provides better information. The check shot is similar to the seismic, except that the geophone is in the well and is moved to the appropriate depth after each surface shot. One-way travel time is recorded. The VSP is similar, except that more shots are taken and a geophone array is lowered into the wellbore. It is processed to provide an actual seismic profile. Having more geophones yields better sampling.

Seismic processing stacking velocity, called V_{rms}, should be used as a last resort, because it is the least accurate of all velocity sources. V_{rms} and can be converted to an approximate interval velocity by applying the Dix equation (Dix, 1981). The phase and the polarity of the synthetic must be clearly indicated and then matched to the seismic line. Because density and velocity values are introduced to the synthetic, a perfect match is sought with the existing seismic directly at the well (fig. 6–3). It should not be projected, if possible. For generation of a synthetic, estimate of formation tops, in both time and depth, should be provided. In addition, several traces should be extracted from the seismic data and used as calibration tie points for the synthetic (fig. 6–4). The synthetic can then be modelled to meet different criteria. Porosity may be changed, as may be the fluid content, the hardness of the rock, and so forth. This is an important step to ensure that the lateral changes observed away from the wellbore and along the seismic line can be properly explained. The procedures for fluid replacement will be reviewed later in this chapter. The synthetic follows the same rules as the seismic signal, and the RC must be converted to a seismic wavelet.

An additional processing step in any seismic interpretation is to analyze frequency and amplitude. As discussed above, the original signal is recorded in a time domain, and a RC can be established. RC can be positive or negative depending on the rock properties traversed.

As interpreters, we are interested in the time domain to determine the depth of the objective, but we require that data be available in the frequency domain to study the rock properties. The conversion from time domain to frequency domain is mathematically intensive, using a process called *Fourier transform*, the details of which are beyond the scope of the current book. In brief, Fourier transform involves generating a series of sine waves that emulate the RC. Introducing a type wavelet allows for a better understanding of the waveform (figs. 6–3 and 6–4). The sample wavelet commonly used is a 25 Hz zero-phase *Ricker wavelet*, also referred to as the *Mexican hat* because of its shape. Other frequencies and other wavelets (e.g., the Butterworth wavelet) may also be used by the processors to better suit the needs of an individual project.

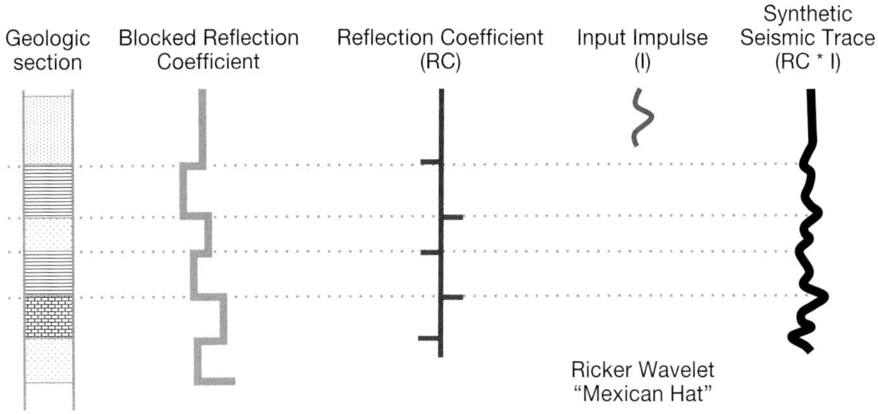

Fig. 6–3. Typical synthetic trace based on petrophysical data. Note the strong lower contrast (shale to limestone). The peak corresponds to the contact and not the zero crossing.

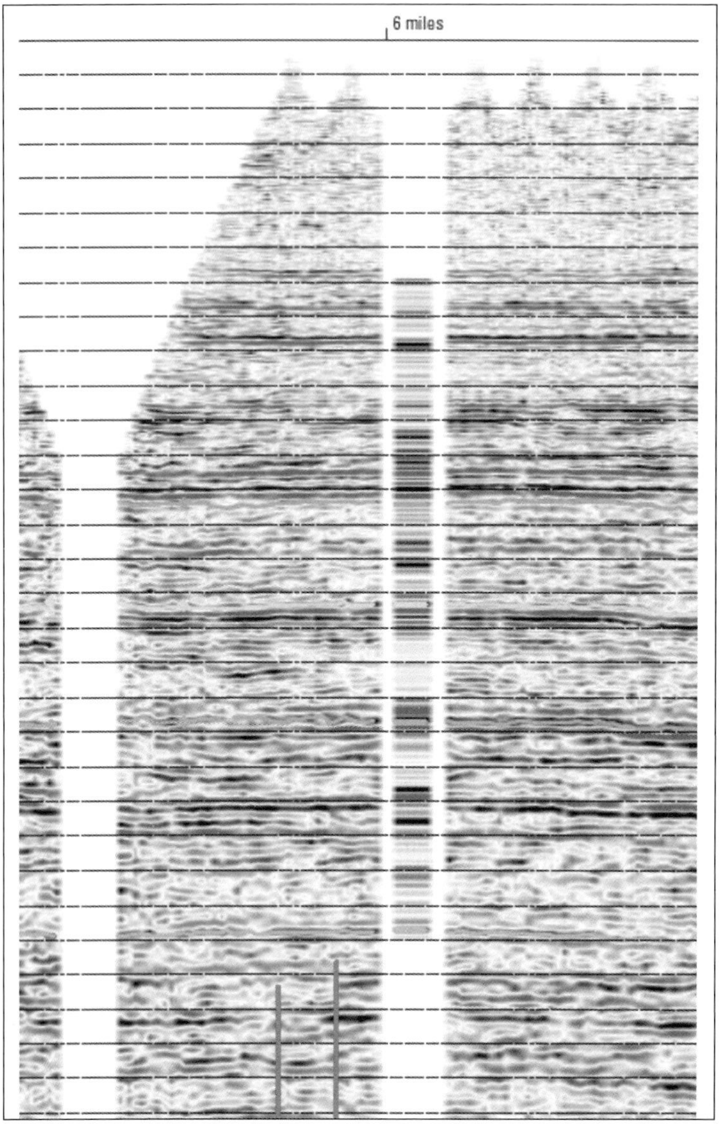

Fig. 6–4. Synthetic for Lear Petroleum Parshall School District 1 well, Williston Basin, North Dakota (Reproduced with kind permission from the U.S. Geological Survey; Anna et al., 2010)

The RC for the P wave is shown as a wavelet, and the resulting seismic line is called an *acoustic impedance display* (AI = I_p). Where the shear velocity is used, the display is known as a *shear impedance profile* (SI = I_s). The higher the available frequency is, the thinner will be the bed resolution. The minimum detectable thickness is known as the *tuning thickness* (fig. 6–5) and is defined as a quarter wavelength ($1/4\ \lambda_1$).

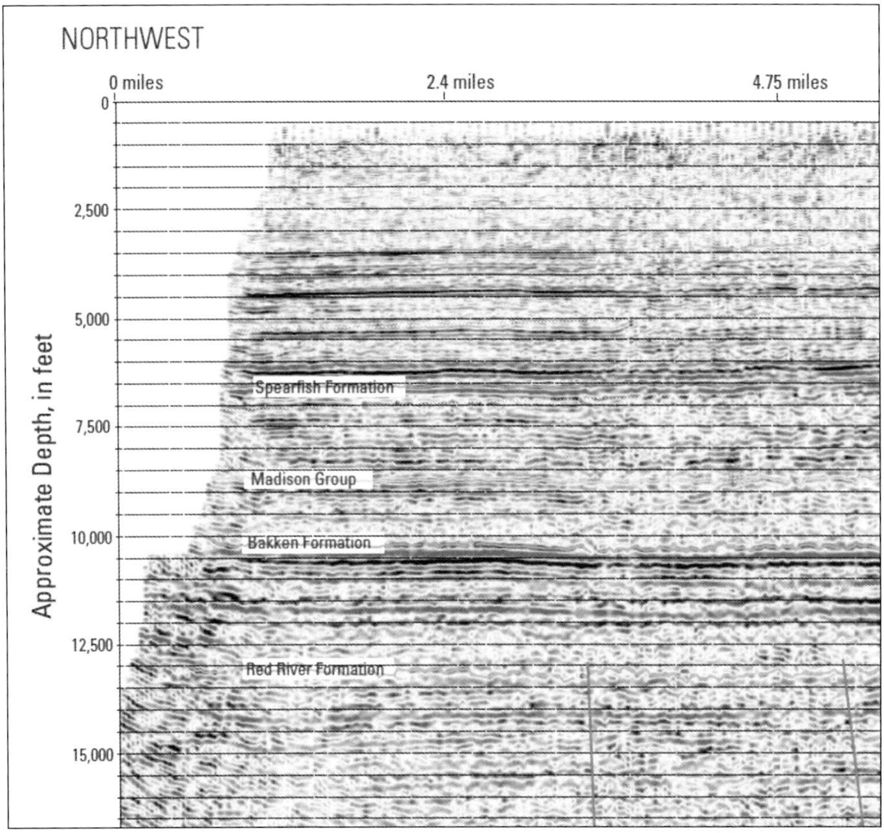

Fig. 6–5. Two-dimensional seismic line, Parshall County, Williston Basin, North Dakota. Note the thickness of the Bakken Formation. Tuning thickness is approximately 94–100 ft (~28.5–30 m). (Reproduced with kind permission from the U.S. Geological Survey; Anna et al., 2010)

It is vital to understand the minimum detectable bed thickness. As mentioned above, the distance in time from peak to trough is a half wavelength ($1/2\,\lambda_1$). However, because the time on a seismic line is measured in two-way time, the true time difference, to determine the tuning thickness, is $1/8\,\lambda_1$ when measured in two-way time. If the one-way time difference for the true $1/4\,\lambda_1$ is 0.01 s (i.e., 10 ms) and the interval velocity is 2,000 m/s, then the tuning thickness is 20 m and is the minimum thickness that can be resolved for an accurate characterization. In figure 6–5, the section is in depth, and peak-to-peak distance is approximately 375 ft ($\frac{1}{4}\,\lambda$ = 94 ft [~30 m]). Tuning thickness is also important when studying the wedge effects of a fluid contact. The maximum extent of a fluid contact can be determined up to the tuning thickness.

The same rules as used for AI (equation [6.1]) apply to the shear velocity; however, instead of an AI, a SI is calculated. SI is a function of shear velocity, V_s, and bulk density, ρ_b:

$$SI = \rho_b V_s \tag{6.4}$$

and

$$RC_s = \frac{\rho_{b2}V_{s2} - \rho_{b1}V_{s1}}{\rho_{b2}V_{s2} + \rho_{b1}V_{s1}} \tag{6.5a}$$

$$RC_s = \frac{SI_2 - SI_1}{SI_2 + SI_1} \tag{6.5b}$$

where RC_s is the reflection coefficient as a function of shear velocity.

The amplitude of the seismic wave is measured in terms of decibels. The dominant frequency can be computed by measuring the time differential, peak to peak, to determine the number of complete cycles per second. Normal ranges are from 5 to 120 Hz. Higher-frequency data improve vertical resolution. In practice, all interpretation packages provide power spectrum analyses that demonstrate the recorded frequency in the data set.

In the case of seismic data, the focus is on generating a similar frequency ("note") that is repeated multiple times. However, as in the orchestra analogy, other notes are played, and a frequency spectrum is generated. Rocks are not perfectly tuned instruments, so noise is also generated, and filters are applied to reduce the noise. In the case of dynamite, a small charge of the same weight is buried at the same specified depth to create a similar source signal. For example, a dynamite charge of 2 kg buried at 15 m will provide a signal similar to a 48,000-pound vibroseis (Suarez and Stewart, 2008). At the same depth of charge burial, the amplitude of the resulting wave from dynamite is proportional to one divided by mass of the charge cubed. The response of a dynamite charge provides a dominant frequency and initial amplitude. Therefore, the charge should be small but able to generate sufficient signal.

As the mass of the charge increases, the low frequencies become detectable. Again, by applying filters, one can better reduce both the high- and low-frequency and focus on the critical signal. This process is known as *band-pass filtering.* Note that the low frequencies might contain valuable fluid information. The maximum high frequency is 125 Hz for onshore data sampled at 4 ms and is known as the *Nyquist frequency.* Sampled at 2 ms, the Nyquist frequency is 250 Hz. Greater frequencies result in *aliasing interference*, yielding a false signal.

Because the laws of physics apply to sound waves, the same rules must apply as sound travels through rock media. The greater the distance is from the power source, the less energy will be available to perform work. As a consequence, with greater depths, the signal-to-noise ratio is reduced. This concept is very important because as the target gets deeper, it becomes more difficult for the energy source to detect and define differences in amplitudes and rock properties.

In all of the preceding discussions, for descriptive purposes, all beds were considered to be horizontal. Any detectable dip in the beds must be corrected by use of the corresponding trigonometric functions.

INTERNAL PROPERTIES OF THE TARGET HORIZON

An occasional change in the RC may be observed on the NMO-corrected gathers as the incidence angle increases and a change in amplitude is observed. Experience has shown that when this effect is observed, there is a strong indication that gas is present. The pattern observed on the gather is referred to as the *hockey-stick effect* owing to its shape, and the effect is known as *amplitude versus offset* (AVO) (fig. 6–6). As the signal source is shifted, the same point in the subsurface will receive the incoming waveform at a different angle (fig. 6–2) for each CMP gather. The RC intensity at the various angles can be plotted directly against the incidence angle (θ), and a curve results.

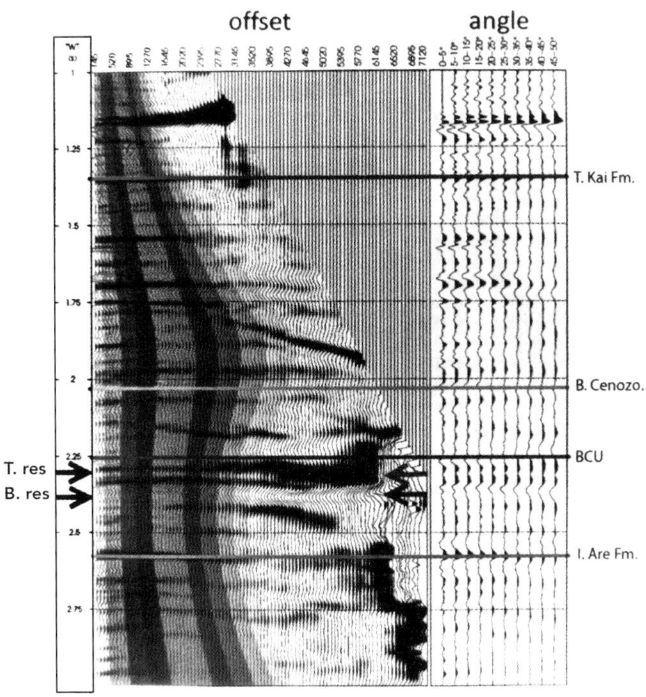

Fig. 6–6. AVO effect with incidence angle in color and offset distance increasing to the right. Band colors are scaled at 5° increments for the angle of incidence. (Reproduced from *The Leading Edge* with kind permission from the SEG; Avseth et al., 2014)

The observed amplitude $R(\theta)$ and the angle of incidence (θ), as $\sin^2 \theta$, are plotted (fig. 6–7). The result is an almost straight line relationship, whose slope is based on the response at angles less than 50°. The equation for the line is known as the *Shuey equation*, which is used to calculate the gradient (slope) and the *y*-intercept $R(\theta)$. Plotting the gradient versus the intercept allows general discrimination into specific classes of anomalies.

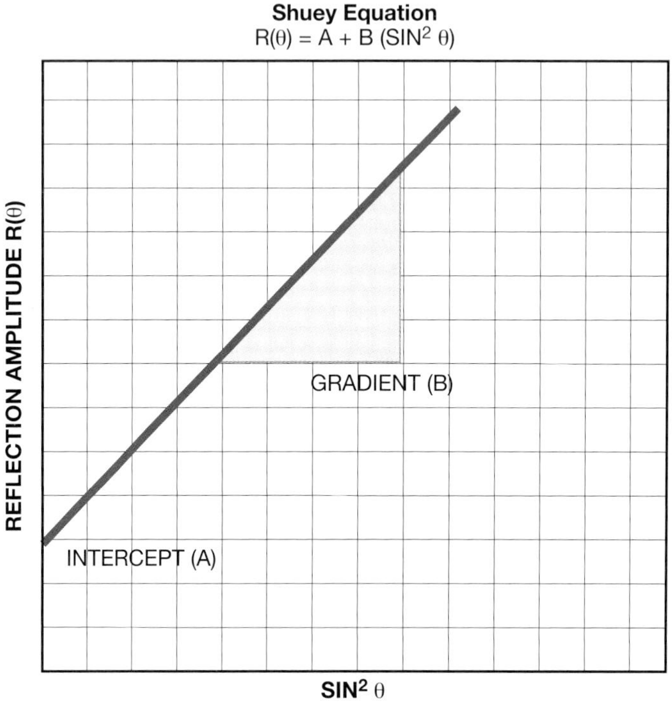

Fig. 6–7. Plot of reflection amplitude versus angle of incidence (Shuey equation)

The AVO anomaly has a proven track record as an indicator of gas in the subsurface, particularly in siliciclastic sediments. The most common seismic signatures for AVOs are classified as follows: the seismic bright spot (class III AVO), the lateral phase reversal (class II AVO), and the lateral dim-out of the reflector (class I AVO), where there is negative RC and decreasing amplitude with offset (class IV) (Rutherford and Williams, 1989; Castagna and Swan, 1997; Castagna et al., 1998). All of these AVO anomalies are indicators of the presence of potentially commercial gas, but they are dominant at different depths. Hilterman (2001) has suggested a depth range for these three classes, as shown in table 6–1. Schlumberger has also provided a summary of the description and applications of AVO anomalies (Chiburis et al., 1993).

Table 6-1. Depth range for classes of AVO anomalies

Class	Depth Range
III	<9,000 ft (<3,000 m)
II	9,000–14,000 ft (3,000–4,000 m)
I	16,000–30,000 ft (5,000–10,000 m)

Because the gradient and intercept data can be derived from the gathers, they can subsequently be displayed as conventional seismic lines. Instead of amplitude, either the gradient or the intercept values can be displayed to make the variations more visible. Russell (2010) has suggested that the difference between the gradient and intercept can provide a pseudo–shear reflectivity and that the sum of the two will provide a pseudo–Poisson's ratio. Hilterman (2001) has also suggested that the difference between the far trace and the near trace also approximates Poisson's ratio. The importance of Poisson's ratio will become more evident later in this chapter.

Another approach to quickly analyze the AVO is to display the seismic data based on the near and far traces. By subtracting the response of the far traces from the near traces, an approximation of the gradient can also be obtained (Ross and Kinman, 1995). Once again, what is being sought is a simple, cost-effective approach that can provide management with information to properly rank the opportunities and engineers with clues as to what to expect along the drilling path. Complex additional seismic processing is expensive and time consuming. Simple computations allow for a first look. It is important for the geophysicists to test various approaches with the well results and determine the most appropriate tool to resolve the problems at hand. As noted in the next section of this chapter, AVO has its limitations.

ROCK PHYSICS

If a synthetic seismic can be generated from petrophysical data, then the reverse can also take place; this is the essence of *recursive seismic inversion*. In other words, pseudo-sonic traces will be generated in place of the wiggle trace. Seismic inversion takes data from both compressional and shear velocity generated by the seismic and separates the velocity and density components. Given the AI (I_p = AI = $V_p\rho$) and SI (I_s = SI = $V_s\rho$), the following relation is apparent:

$$\frac{I_p}{I_s} = \frac{V_p\rho}{V_s\rho} = \frac{V_p}{V_s} \tag{6.6}$$

Laboratory experiments have shown that Poisson's ratio (ν) is directly related to the velocity of the material and, particularly, V_p/V_s:

$$\nu = \frac{(V_p/V_s)^2-2}{2[(V_p/V_s)^2-1} \tag{6.7}$$

Furthermore, Young's modulus (E) can be calculated as follows:

$$V_p^2 = \frac{E(1-\nu)}{(1+\nu)(1-2\nu)\rho} \tag{6.8}$$

where E is Young's modulus, V_p is the compressional velocity, ρ is the bulk density, and ν is Poisson's ratio (Sheriff, 2002.)

A major area of current research comprises the relationships of V_p/V_s to all rock physics parameters. Two of these relationships will be addressed now, and others will be discussed subsequently, once the main rock physics parameters are introduced.

The first set of plots that requires analysis is V_p/V_s related to AI for a specific reservoir and based on well data points (fig. 6–8). Here the response of fluids in the well is analyzed as a function of the variables. The second parameter that requires attention is the relationship of V_p/V_s versus porosity at the wellbore. The relationship between variations in V_p, derived from the sonic log, and porosity was discussed in the chapter 5; however, the effect of pressure must also be considered and is discussed below. In general, as the ratio V_p/V_s increases, so does the clay content. Importantly, the core data, petrophysics, and seismic must always have a feedback response relationship and provide similar results; otherwise, the results cannot be corroborated.

The Belgian engineer Maurice Biot authored several key studies into the relationship of porosity and elastic parameters in the late 1930s and 1940s. His pioneering early work on porosity and rock physics yielded the Biot compressibility constant (β_b). For a specific lithology and a specific facies, the relationship of the Biot compressibility constant to porosity is expressed as follows:

$$1 - \beta_b = (1 - \phi)^{3/(1-\varnothing)} \tag{6.9}$$

where ϕ is the porosity.

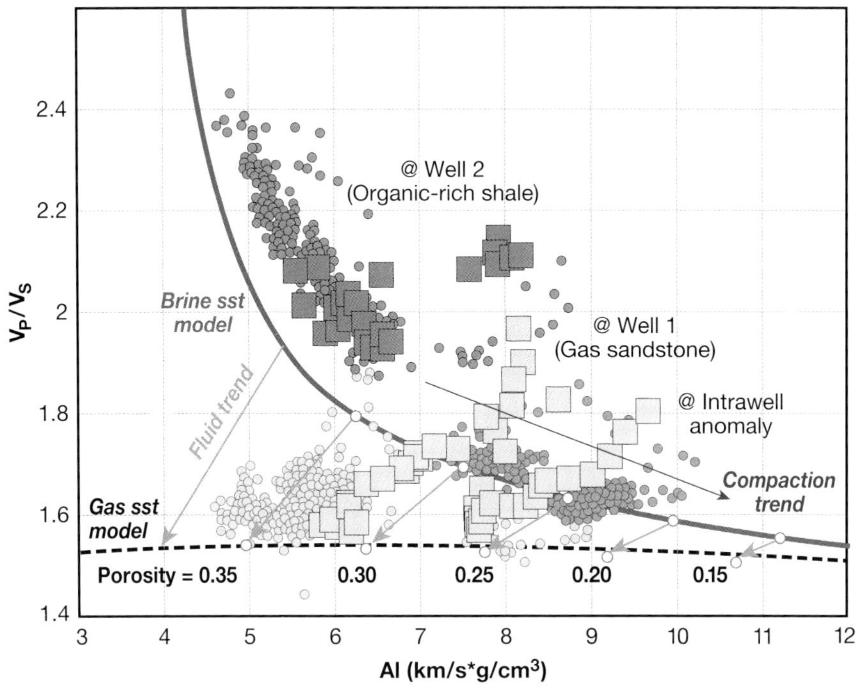

Fig. 6–8. Plot of V_p/V_s versus AI, showing characteristics of the reservoir. Data points shown in dark gray correspond to organic-rich shales. (Reproduced from *The Leading Edge* with kind permission from the SEG; Avseth et al., 2014)

There are issues with respect to equation (6.9) regarding the impact of pressure on the Biot compressibility constant and porosity in general. Sahay (2013) and Lee (2013) have brought attention to the problem of pressure differences affecting dry and wet samples and their effects on the Biot compressibility constant. Because we are dealing with compressible gases in shale resource assessment, the effects of pressure must be investigated further. Still, the basic relationship is important. Equivalent pressure conditions should be used when applying the Biot compressibility constant in project areas to validly predict trends.

There are two ways in which a rock is deformed in the subsurface without fracturing. The rock can be compressed or sheared (or both). If a rock is simply compressed, then the compressibility of the bulk rock is described in terms of the (pure) *compressibility modulus*, M_b, which is closely dependent on the rock's bulk modulus, K_b. The latter is the actual volume change due to compression and is affected by shear. If a rock is sheared, then the effect is described by the bulk rock *shear modulus*, μ_b, which is also the second Lamé parameter. The shear modulus is the easiest to evaluate. It is a function of the shear velocity, V_s, affecting the material and its bulk density, ρ_b, as follows:

$$\mu_b = V_s^2 \rho_b \tag{6.10}$$

For ease of computation, both sides of the equation are multiplied by ρ_b:

$$\mu_b \rho_b = V_s^2 \rho_b^2 = I_s^2 \tag{6.11}$$

Equation (6.11) implies that a shear impedance volume when squared will generate a $\mu_b \rho_b$ inversion.

A similar relationship is established for the pure compressional modulus, M_b:

$$M_b = V_p^2 \rho_b \tag{6.12a}$$

and, as a result,

$$M_b \rho_b = V_p^2 \rho_b^2 = I_p^2 \tag{6.12b}$$

The bulk modulus, K_b , is partially dependent on the shear velocity and the compressional velocity. An additional elastic property known as the first Lamé parameter, λ, is important to define and is widely applied in rock physics:

$$\lambda = \rho(V_p^2 - 2V_s^2) \tag{6.13a}$$

Multiplying both sides by ρ,

$$\lambda\rho = I_p^2 - 2I_s^2 \tag{6.13b}$$

Thus, seismic inversions can now be generated for $\lambda\rho$ and $\mu\rho$. Similar equations can be applied to generate additional inversions.

Bulk modulus is different. The relationship between bulk modulus and the Lamé parameters is known:

$$\lambda = K_b - \frac{2}{3}\mu \tag{6.14}$$

The bulk modulus can be determined in a laboratory. The definition of bulk modulus serves to describe the analysis. When a load is applied on a sample, a volume change is detected. As will be described in detail in chapter 8, the normal vertical load is σ_1 and is often the largest stress applied on a sediment. The applied stress directly affects the volume change, ΔV, and the bulk modulus, K:

$$K = \frac{\sigma_1}{\Delta V / V} \tag{6.15}$$

The Swiss mathematician Fritz Gassmann made a major step forward in the study of rock physics in the early 1950s. Gassmann observed the importance of differentiating the components of the bulk modulus and focused his attention on the *K* values for various rock types, fluids, and their relationship with porosity (see eq. [6.16]). Berryman (1999) and others later translated the original work by Gassmann, which serves as the basis for all fluid substitution modules currently used in the industry. It was shown that modifying the bulk modulus of the fluid affected the saturated bulk modulus, which was reintroduced into the equation. This is referred to as the *Biot-Gassmann relationship*. Berryman (1999) and Kumar (2006) have provided tutorials on the origin and application of the Gassmann equation:

$$\frac{K_{sat}}{K_o - K_{sat}} = \frac{K_{dry}}{K_o - K_{dry}} + \frac{K_{fluid}}{ø(K_o - K_{fluid})} \tag{6.16}$$

where K_{sat} is the saturated bulk modulus, K_{dry} is the dry sample bulk modulus, K_{fluid} is the fluid bulk modulus, K_o is the bulk modulus of the mineral, and ϕ is the porosity. Therefore, all the parameters in the Gassmann equation (eq. [6.16]) can be calculated except porosity.

The French petrophysicist Krief and colleagues (1990) further expanded on the Gassmann equation by incorporating the compressional and shear velocities. The Krief equation is as follows:

$$K_b = \rho_b(V_p^2 - \frac{4}{3}V_s^2) \tag{6.17a}$$

and

$$K_b\rho_b = V_p^2\rho_b^2 - \frac{4}{3}V_s^2\rho_b^2 \tag{6.17b}$$

Keeping in mind that one of the primary goals of the study of rock physics is to establish porosity values,

$$\frac{1}{M} = \left(\frac{\beta - ø}{K_m}\right) + \left(\frac{ø}{K_{fl}}\right) \tag{6.18}$$

where K_m is the bulk modulus of a mineral and K_{fl} is the fluid bulk modulus. The importance of the Biot compressibility constant (β) is evident in equation (6.18), and Lee (2008) has carried out extensive studies on the impact of the Biot constant on sediments.

Sharma and Chopra (2013a) have reformulated equation (6.17b) as follows:

$$K_b\rho_b = I_p^2 - \frac{4}{3}I_s^2 \tag{6.19}$$

and, in a similar calculation for Young's modulus,

$$E\rho_b = \mu\rho_b\left(\frac{3I_p^2 - 4I_s^2}{I_p^2 - I_s^2}\right) \tag{6.20}$$

These additional inversions can be generated to assist in determining shale properties such as brittleness.

Core data can be tested to obtain realistic values of V_p and V_s by simulating reservoir conditions. Once a porosity value is introduced, the value of K_b can be calculated. The more accurate the bulk modulus values are, the better the final results will be. The relationship made with the shear modulus μ can readily be established from shear velocity; thus, the Lamé constant λ can also be calculated. What will become evident in chapter 8 is that during fracking, the introduction of fluids under pressure modifies the saturated bulk modulus (K) as indicated in equation (6.15). As a result, the values of λ and μ in equation (6.14) will both decrease after fracking.

As can be observed, the seismic data for impedance—both AI (I_p) and SI (I_s)—can be combined and inverted into many of the of the principal rock physics parameters. Sharma and Chopra (2013*b*) have indicated that, by crossplotting the parameters $\lambda\rho$ versus $\mu\rho$, (V_p/V_s) versus I_p, and I_p versus I_s and calibrating with the Passey equation (see chap. 7), an estimate of TOC content can be made for the interval. Once again, calibration is crucial.

As a stand-alone tool, seismic is of moderate use. However, as the theme of this book, data integration is key to any successful exploration venture. For example, the LMR (lambda-mu-rho) plot in figure 6–9 demonstrates the application. The petrophysicist must first generate the same rock physics parameters described above. At the same time, data from core analysis—including all geological, geochemical, and rock mechanics aspects—must confirm the specific characteristics sought to determine the sweet spots for exploration. The log-derived rock physics data are graphed on LMR crossplots, and the characteristics of several sectors on the LMR plot are identified to rank prospective and nonprospective zones. The designated classified volume is transferred to the seismically derived crossplot for the zone of interest, and the most favorable attribute is highlighted on the new crossplot. The programs are designed to illuminate the LMR-designated classified volume and thus highlight areas of interest on the seismic view (fig. 6–9). Integration of data and detailed calibration are essential. Multiple parameters should be crossplotted and the results matched to confirm coincidence.

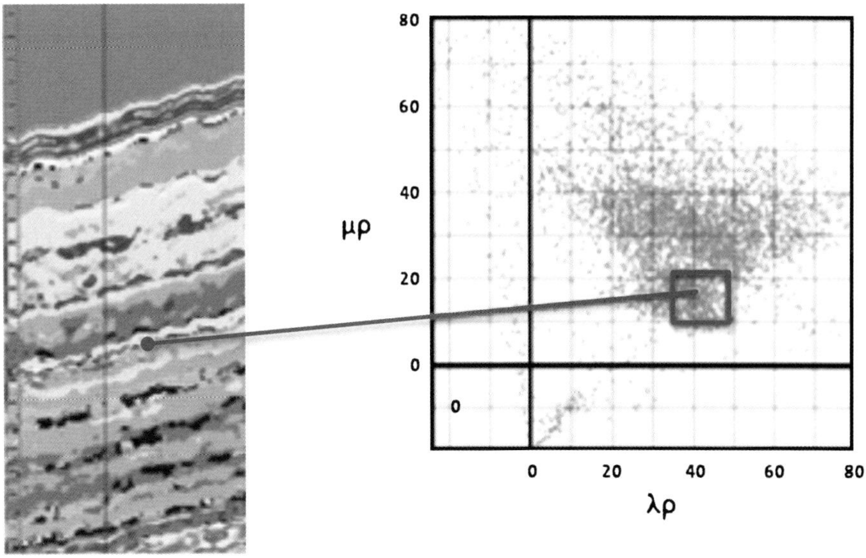

Fig. 6–9. LMR crossplot data with optimum area of interest. The log-derived sweet spot is designated a classified volume.

AVO analyses, in combination in combination with LMR studies, have proved valuable in improving our knowledge of rock physics and predicting reservoir rock parameters (Pengyuan et al., 2008). Figure 6–10 gives a general example of this framework. As data are introduced, the boundaries outlined should be modified. The potential for AVO studies in shales remains a subject of investigation, but the point of this discussion is that in the exploration of shales, many tools should be tested; moreover, those tools that do not reduce the exploration risk should be discarded, although they may apply to another project.

The LMR plot in figure 6–10 exemplifies the need to continuously reclassify the target parameters to fine-tune the exploration. In combination with data from figure 6–8, it is evident that organic shales have $V_p/V_s > 2.0$. Therefore, in the general shale region, they will tend toward the upper portion of the shaded shale dominant envelope (fig. 6–10). As the gas content increases, the value of λρ will tend to decrease. Indeed, as discussed above, the LMR values will decrease after fracking. If AVOs were not detected before drilling, is there a possibility they may be detected after drilling?

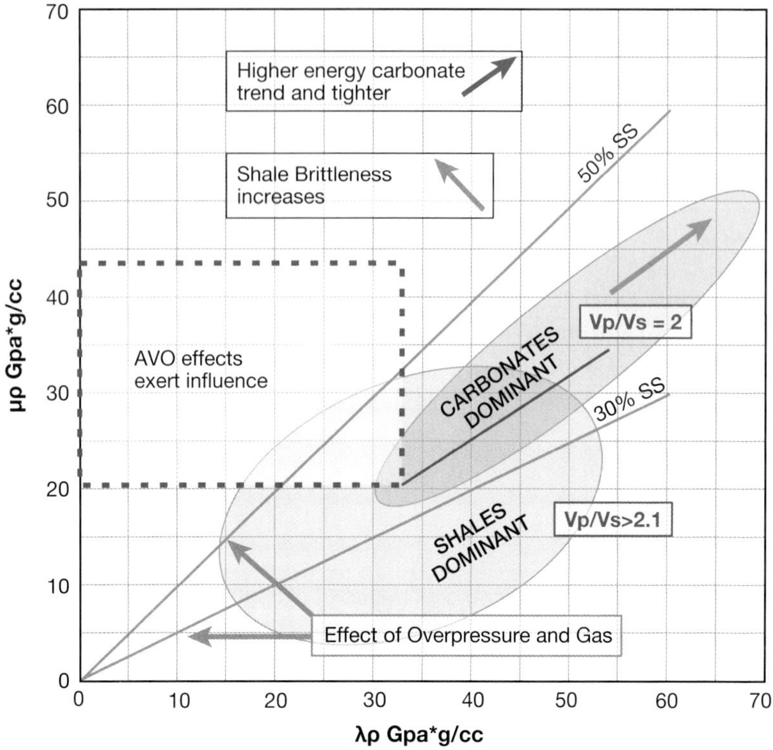

Fig. 6–10. LMR plot as a reference for AVO analysis. The boundaries for the lithologies serve as guides that should be refined for each individual project. (Modified from Perez and Tonn, 2003 and Perez, 2010)

In shale studies, two very important elastic parameters require further review: Young's modulus (E) and Poisson's ratio (ν). These are indicators of brittleness and ductility and must be determined in order to adequately fracture a sediment. So why not compute Young's modulus and Poisson's ratio directly? Why are the other parameters so important? As discussed above, once the Lamé parameters have been calculated, the remaining parameters can be generated. Originally, Gassmann and, later, Sheriff (2002) compiled a table now used by all rock physicists in which they related the elastic parameters to one another depending on the available data. The key is to determine which data one has available:

$$E = \mu \frac{3\lambda + 2\mu}{\lambda + \mu} \tag{6.21}$$

$$\nu = 0.5\lambda(\lambda + \mu) \tag{6.22}$$

$$E = 2\mu(1 + \nu) \tag{6.23}$$

BRITTLENESS

Experimental studies for the Barnett shale, using triaxial compresses among other techniques, have shown that Poisson's brittleness index (BI_v) and the Young's modulus brittleness index (BI_E) are as follows (Rickman et al., 2008):

$$BI_\nu = \left(\frac{\nu - 0.4}{0.15 - 0.4}\right) \times 100 \tag{6.24}$$

$$BI_E = \left(\frac{E - 1}{8 - 1}\right) \times 100 \tag{6.25}$$

or

$$BI_\nu = \frac{\nu - \nu_{min}}{\nu_{max} - \nu_{min}} \tag{6.26}$$

$$BI_E = \frac{E - E_{min}}{E_{max} - E_{min}} \tag{6.27}$$

Unfortunately, no single unique formula can be applied. The key is to apply several approaches and calibrate with observations in the specific area being worked on. Shales with the same composition but from different areas may yield different brittleness values. Even Rickman et al. (2008) have conceded that this formula does not apply to all areas and requires more detailed calibration. Their basic brittleness indices can be averaged between the two parameters. The overall brittleness index is

$$\frac{BI_\nu + BI_E}{2} \tag{6.28}$$

There are several ways of approaching brittleness using Young's modulus and Poisson's ratio. As Perez and Marfurt (2013) have indicated, a combined Young's modulus less than 5 GPa and Poisson's ratio of 0.25 suggests the presence of brittle rock, and the same Young's modulus along with higher Poisson's ratio suggests the presence of ductile material. The ultimate goal is to generate a plot that identifies the mechanical characteristics. The term used for this description is *mechanical stratigraphy*. Several contractors offer log curves of brittleness index. As in petrophysics (see chap. 5), defining the calculation and the cutoffs is critical.

ANISOTROPY

Gassmann was well aware that many sediments including shales are anisotropic. However, the above equations assume isotropic media. In isotropic media, there is no preferential flow. The opposite is the case for an anisotropic media.

The mathematics involved in the theoretical analysis of anisotropic media is *tensor mathematics*. Common mathematical problems are described in *scalar* quantities; for example, 2 + 2 = 4. *Vector* quantities involve a direction; for example, go 2 km to the northwest. Tensor quantities are of the form go 2 km northwest and when you get there go to third floor. Tensor mathematics requires computing matrices. Thomsen (1986) described the three directions of anisotropy, how to compute their values, and coined the terms ε, γ, and δ to describe them. These are used as standard references to describe anisotropic media. The theoretical aspects of anisotropic media are beyond the scope; suffice it to say that anisotropy results from originally random clay platelets becoming better oriented as overburden pressure is applied. In practice, anisotropy is considered in developing resource plays because of the significant presence of shales.

The approach used to establish the anisotropy in a sedimentary basin is based on seismic. Multicomponent seismic allows for the recording of the shear waves, which propagate in two directions: along the direction of the P wave, and perpendicular to the P wave. One is faster than the other. Therefore, by observation of the difference between the two, the direction of anisotropy and strain can be established. In a three-dimensional map display near the target of interest, vectors mark the overall trend and subsequently can be plotted on a rose diagram. In geophysical analysis of anisotropy, the *PP wave* refers to the analysis of the compressional to compressional reflected wave; the *PS wave*, sometimes referred to as the *c-wave*, is the reflection of the shear wave from the compressional wave and is used in the analysis of vertical transverse isotropic (VTI) analysis (Cai and Tsvankin, 2012). It is used to resolve the problems associated with the velocity anomalies related to anisotropy.

A quick review of the Thomsen (1986) anisotropy parameters is in order. Prior to multicomponent seismic acquisition, anisotropy was known but difficult to evaluate. The orderly stacking observed in shales causes physical properties to change in specific lateral and vertical directions, resulting in anisotropy. Velocity, among other physical properties, is affected by anisotropy. In particular, shear velocity is of significant importance because the horizontal shear has both a minimum and a maximum velocity. Thomsen (1986) focused on this difference by using the V_p, $V_{s\,max}$, $V_{s\,min}$, and density values to characterize his three parameters:

- δ: short offset effect, uses the NMO velocity correction to establish changes
- ε: long offset effect, measures the difference between the vertical and horizontal velocities
- γ: shear wave component

If a rock has horizontal anisotropy, it is said to be *vertical transverse isotropic* (VTI). VTI is commonly used in bedded shales. If the beds are dipping, the rock is *tilted transverse isotropic* (TTI), and if the feature is vertical, as with fractures and faults, it is *horizontal transverse isotropic* (HTI). VTI and HTI are used to establish the anisotropic effects of bedding, cracks, and fractures. Because the effects are three-dimensional, matrix algebra is used to allow the measurements to be compared simultaneously. The basic principle behind anisotropy, as noted on the Mohr diagram and discussed in chapter 8, for any stress applied (σ) in any given direction, there is a corresponding strain (often referred to as τ) that is a function of the elastic modulus (Young's modulus, E) of the medium. The relationship between AVO and anisotropy, particularly in shales, is an area of ongoing research (Li et al., 2003). Anisotropy analyses have importance in velocity studies, geohazard studies, improving location of landing zones, estimating brittleness, and establishing a proxy for horizontal stress by mapping azimuthal anisotropy.

Because there are three variables involved, it is interesting to analyze the relationship of the system of equations that describe the relationship. Matrix algebra provides a mechanism to analyze the linear equations and provide a description of the relationship between the parameters. The matrix determinant provides that information. (For a more detailed discussion on matrix algebra, see Campbell, 1977.) The determinant, $|D_{ia}|$, is the difference between the cross-products of a square matrix and allows it to define the relationships in equivalent matrices:

$$\begin{array}{lcc} & \text{Porosity} & \text{Permeability} \\ \text{A1} & \left[8 \right. & \left. 0.6 \right] \\ \text{A2} & \left[4 \right. & \left. 0.2 \right] \end{array}$$

$$|D_{ia}| = (8 \times 0.2) - (0.6 \times 4) = -0.8 \tag{6.29}$$

Given similar conditions, if the porosity and permeability of one well interval is given and the porosity is known in the adjacent well, then the determinant rule allows for the calculation of the permeability in the new well.

For a 3 × 3 square matrix, the determinant ($|D_{ic}|$) is slightly more complex:

$$\begin{bmatrix} a & b & c \\ d & e & f \\ g & h & i \end{bmatrix}$$

$$C = |D_{ic}| = a(ei - fh) - b(di - fg) + c(dh - eg) \tag{6.30}$$

Basically, the calculation is the sum of a series of 2 × 2 matrices.

The various stresses with reference to the Thomsen parameters (3 × 3 matrix) can be computed on a matrix and related to V_p/V_s. Perez et al. (2012) have shown that the determinant ($|D_T|$) with the two of the Thomsen parameters is as follows:

$$|D_T| = \frac{1}{4}\left(\frac{V_p}{V_s}\right)^2 \Delta\delta^{(v)} + 2\Delta\gamma \tag{6.31}$$

where V_p is the compressional velocity, V_s is the shear velocity, δ is the delta Thomsen parameter, ν is Poisson's ratio, and γ is the gamma Thomsen parameter. Perez et al. (2012) establishes the third Thomsen parameter ε to γ by relating them to the shear modulus.

Perez et al. (2012) have further defined the anisotropy gradient B_{ani}. The anisotropy gradient basically approximates the direction of greatest change. Other operators can be used for edge detection:

$$B_{ani} = \frac{1}{2}\left[\Delta\delta^{v} + 8\left(\frac{V_p}{V_s}\right)^2 \gamma\right] \tag{6.32}$$

Together, the two values can assist in comparing the results obtained from the various control points in the oil field being drilled. In addition, as will be discussed in chapter 8, a matrix determinant can be used to estimate the fracture closure index.

By careful calibration of the seismic-derived density against cores and logs, Sayers (2013) has demonstrated that the same tool can be used to study the organic richness in shales. By combined assessment of the PP waves, RC versus angle and the PS waves, and RC versus θ, Sayers established a relationship with TOC content. Sayers (2013) suggests that high TOC shales cause a reduction in the acoustic impedance and a reduction in amplitude that diminishes with offset, a typical AVO Class IV signature.

Rock physics is a relatively new science. Over the past half century, the basic tenets were studied and analyzed. With the advent of AVO in the 1980s, there was renewed interest in rock physics, and the study of rock physics increased exponentially over the past 20 years. This has led to considerable refinement of the knowledge base, with literally hundreds of publications on rock physics. The brief summary here serves as an introduction into the application of principal components of rock physics. A good starting point for interested readers is the *Rock Physics Handbook* (Mavko et al., 2009).

Importantly, there are three sources from which the rock physics parameters can be generated. The first and most accurate source is core data and must always be obtained for calibration. The second source is petrophysics, and the third source is the seismic data. Note that all three need to provide very similar mathematical results; that is, no single tool or parameter can provide perfect results alone. They must be calibrated, blind-tested, and redundant. The logs need to be adjusted to fit the cores, and the seismic needs to be adjusted to the petrophysical logs via the synthetics to verify that the model is within the bounds of the regional data. In chapter 8, a more detailed analysis showing how the geophysics, LMR, and pressure data are integrated.

SUMMARY

How seismic data of the project is acquired defines the limits of resolution of the data and determines the quality of the results necessary to implement a proper interpretation. Because the reflected ray travels both as a compressional wave and as a shear wave, these can be recorded with a multicomponent geophone to allow for advanced processing of the data. The initial seismic is displayed as either acoustic impedance (AI = I_p) or shear impedance (SI = I_s).

Synthetic seismic profiles, based on the petrophysical logs, are used to collect ground-truth data from the seismic and can be modeled to calibrate the extrapolated seismic parameters. By mathematical manipulation of the I_p and I_s cubes, the seismic can be inverted to provide parameters used to identify rock properties, such as dividing AI by SI to obtain V_p/V_s, other operations will yield the LMR coefficients, the Poisson's ratio and Young's modulus of the target horizons.

Using the acquired seismic gathers and plotting the reflection amplitude against $\sin^2 \theta$ generates a slope (gradient) and *y*-intercept used to define the AVO effect and thereby predict fluid types in the reservoir. Crossplotting the various parameters in combination with proper petrophysics, geochemistry, and geological data calibration can provide classified volumes to illuminate prospective targets with optimum porosity, density, pressure regimes, brittleness, and TOC content. Fluid substitution can be modelled using the Gassmann equation to improve lateral prediction.

Shales are anisotropic. The Thomsen anisotropic parameters help define the anisotropic effect. Matrix determinants for Thomsen parameters and anisotropy gradients provide valuable tools to describe the fracking effect and closure index.

REFERENCES

Anna, Lawrence O., John J. Miller, and Ofori N. Pearson. 2010. "Williston Basin Province—interpretation of 2-D seismic data, Mountrail County, North Dakota." Chapter 6 in *Assessment of Undiscovered Oil and Gas Resources of the Williston Basin Province of North Dakota, Montana, and South Dakota*. U.S. Geological Survey Digital Data Series DDS-69-W. pubs.usgs.gov/dds/dds-069/dds-069-w/contents/REPORTS/69_W_CH_6.pdf.

Avseth, Per, Tor Veggeland, and Frederik Horn. 2014. "Seismic screening for hydrocarbon prospects using rock-physics attributes." *Leading Edge*. 33: 266–274.

Berryman, James G. 1999. "Origin of Gassmann's equations." *Geophysics*. 64: 1627–1629.

Brown, Alistair R. 2011. *Interpretation of Three-Dimensional Seismic Data*. Tulsa: American Association of Petroleum Geologists and Society of Exploration Geophysicists.

Cai, Pengfei, and Ilya Tsvankin. 2012. "Joint migration velocity analysis of PP-and PS-waves for VTI media." In *SEG Technical Program Expanded Abstracts 2012*. Society of Exploration Geophysicists, 1–6.

Campbell, Hugh G. 1977. *An Introduction to Matrices, Vectors, and Linear Programming*. Englewood Cliffs, NJ: Prentice Hall.

Castagna, John P., and Herbert W. Swan. 1997. "Principles of AVO cross plotting." *Leading Edge*: 337–342.

Castagna, John P., Herbert W. Swan, and Douglas J. Foster. 1998. "Framework for AVO gradient and intercept interpretation." *Geophysics*. 63: 948–956.

Chiburis, Edward, Charles Franck, Scott Leaney, Steve McHugo, and Chuck Skidmore. 1993. "Hydrocarbon detection with AVO." *Oilfield Review*. 5: 42–50. www.slb.com/~/media/Files/resources/oilfield_review/ors93/0193/p42_50.pdf.

Dix, C. H. 1981. *Seismic Prospecting for Oil*. Boston: International Human Resources Development Corp.

Hilterman, Fred H. 2001. "Seismic amplitude interpretation." SEG and EAGE, Short Course no. 4.

Krief, M., J. Garat, M., J. Stellingwerf, and J. Ventre. 1990. "A petrophysical interpretation using the velocities of P and S waves (full-waveform sonic)." *Log Analyst*. 31: 355–369

Kumar, Dhananjay. 2006. "A tutorial on Gassmann fluid substitution: Formulation, algorithm and Matlab code." *Geohorizons*: 4–13.

Lee, Myung W. 2003. *Velocity Ratio and Its Application to Predicting Velocities: U.S. Geological Survey Bulletin 2197.* Reston, VA: U.S. Geological Survey. pubs.usgs.gov/bul/b2197/B2197-508.pdf.

———. 2008. "Comparison of the modified Biot-Gassmann theory and the Kuster-Toksöz theory in predicting elastic velocities of sediments." In *U.S. Geological Survey Scientific Investigations Report 2008–5196*.

Li, Y., J. Downton, and Y. Xu. 2003. "AVO modeling in seismic processing and interpretation, part 1." *CSEG Recorder*. 28: 43–52.

Mavko, Gary, Tapan Mukerji, and Jack Dvorkin. 2009. *Rock Physics Handbook: Tools for Seismic Analysis of Porous Media*. Cambridge, UK: Cambridge University Press.

Pengyuan, Sun, Lu Xiuli, Li Yanpeng, Yue Yuanyuan, and Chen Haifeng. 2008. "Elastic parameter AVO approximations and their applications." In *SEG Technical Program Expanded Abstracts 2008*. Society of Exploration Geophysicists, 523–527.

Perez, Marco. 2010. "Beyond isotropy—part 2: Physical models in LMR space." *CSEG Recorder*. 35: 36–43.

Perez, Marco, Bill Goodway, and David Close. 2012. "Geomechanics, anisotropy and LMR." Paper presented at GeoConvention 2012, Calgary, AB, May 14–18.

Perez, Marco A., and Rainer Tonn. 2003. "Reservoir modelling and interpretation with Lamé's parameters: A Grand Banks case study." Extended abstract. cseg.ca/assets/files/resources/abstracts/2003/404S0131.pdf.

Perez, Roderick, and Kurt Marfurt. 2013. "Calibration of brittleness to elastic rock properties via mineralogy logs in unconventional reservoirs." Paper presented at AAPG International Conference and Exhibition, Cartagena, Colombia, September 8–11. Search and Discovery article 41237. www.searchanddiscovery.com/documents/2013/41237perez/ndx_perez.pdf.

Rickman, Rick, Michael J. Mullen, James Erik Petre, William Vincent Grieser, and Donald Kundert. 2008. "A practical use of shale petrophysics for stimulation design optimization: All shale plays are not clones of the Barnett Shale." Paper SPE-115258-MS, presented at SPE Annual Technical Conference and Exhibition, Denver, September 21–24.

Ross, Christopher P., and Daniel L. Kinman. 1995. "Nonbright-spot AVO: Two examples." *Geophysics*. 60: 1398–1408.

Russell, Brian. 2010. "Making sense of all that AVO and inversion stuff." The Milton Dobrin Lecture, Houston, April 2010. www.agl.uh.edu/pdf/russell-2010.pdf.

Rutherford, Steven R., and Robert H. Williams. 1989. "Amplitude-versus-offset variations in gas sands." *Geophysics*. 54: 680–688.

Sahay, Pratap N. 2013. "Biot constitutive relation and porosity perturbation equation." *Geophysics*. 78: L57–L67.

Sayers, Colin M. 2013. "The effect of kerogen on the AVO response of organic-rich shales." *Leading Edge*. 32: 1514–1519.

Sharma, Ritesh Kumar, and Satinder Chopra. 2013*a*. "New attribute for determination of lithology and brittleness." Winner of the 2013 AAPG Jules Braunstein Award for best poster presentation at the AAPG 2013 Annual Meeting in Pittsburgh, May 19–22.

———. 2013*b*. "Conventional approach for characterizing unconventional reservoirs." *CSEG Recorder*: 35–38.

Schwenk, J. Tyler, Richard Miller, Julian Ivanov, and Jason McKenna. 2012. "Dispersion interpretation from synthetic seismograms and multi-channel analysis of surface waves (MASW)." In *SEG Technical Program Expanded Abstracts 2012*. Society of Exploration Geophysicists, 1–8.

Sheriff, Robert E. 2002. *Encyclopedic Dictionary of Applied Geophysics*. Tulsa: Society of Exploration Geophysicists.

Suarez, Gabriela M., and Robert R. Stewart. 2008. "Seismic source comparison for compressional and converted-wave generation at Spring Coulee, Alberta. Part I: Heavy vibe versus dynamite comparison." In *CREWES Research Report*. 20: 1–15.

Thomsen, Leon. 1986. "Weak elastic anisotropy." *Geophysics*. 51: 1954–1966.

7

GEOCHEMISTRY

TAKEAWAY

- Review the basics of organic chemistry of oil and gas in shales
- Analyze their starting and end products
- Establish the relationship between total organic carbon (TOC) and the hydrocarbon products
- Determine the effects of burial on source rocks and maturation
- Review petrophysical log TOC indicators
- Analyze how seismic can be incorporated to improve the prediction of subsurface thermal conditions

GENERAL GEOCHEMISTRY

Of all the aspects of the shale resource play, the geochemistry is among the best documented. Since the 1960s, a major push has been under way to improve our level of understanding on how source rocks function. The goal of the shale resource assessment is to produce oil and gas from precisely these formations. Before discussion of the geochemistry of shales, a review of basic organic chemistry is needed to understand the complex nature of crude oil compositions and the role played by total organic carbon (TOC).

Kerogen, immature amorphous organic material, is the source of hydrocarbons. Once heated to its initial cracking temperature, kerogen breaks down into simpler compounds. Pressure plays a secondary role. The produced oil is further cracked in the refining process to obtain the distilled products used commercially. The process of thermal alteration of the kerogen is known as *catagenesis*, and determining the thermal boundaries for each of its stages of cracking in the subsurface is of paramount importance. Aside from temperature, the stages are influenced by the available kerogen type in the shale.

On the basis of their composition, hydrocarbons are grouped into three categories—paraffins, naphthenes, and aromatics—which can be plotted on a ternary diagram (fig. 7–1). *Normal paraffins* include the simplest base compounds in organic chemistry. The core chemical structure is the methane molecule, CH_4, with a carbon atom at its center and four hydrogen atoms equally distributed on the carbon atom. As a molecule, CH_4 is unique. The ionic radius of carbon is 30 picometers (pm) (1 angstrom [Å] = 100 pm), and it has a relatively small *enthalpy*, which in thermodynamics is the energy necessary to dissociate a chemical bond. The difficulty lies in estimating how the bonds react after one hydrogen is removed and two carbons can bond with one another. Where thermodynamic conditions are appropriate, the higher bond energy of the carbon-carbon bond will prevail, and larger molecules can form (Ryan and Norris, 2014).

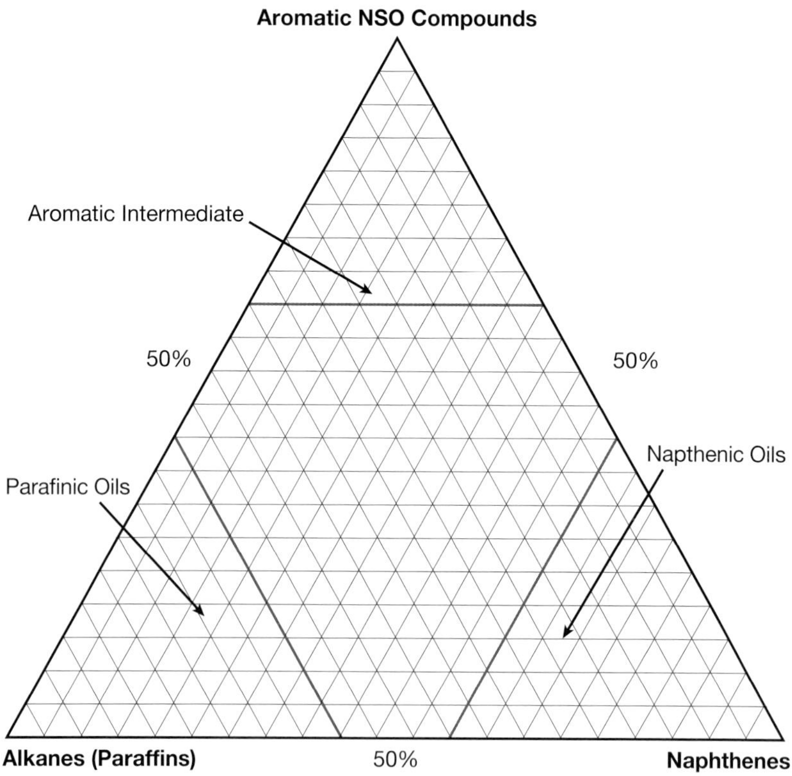

Fig. 7–1. Ternary plot classification of oils based on composition (Adapted from Tissot and Welte, 1984)

In organic chemistry, the terms *saturated* and *unsaturated* are used to describe molecules. Saturated molecules are those with single carbon-carbon bonds, and these are considered saturated with respect to their hydrogen atoms. The *normal alkanes* are saturated compounds. Unsaturated compounds have double or triple carbon-carbon bonds and therefore have lower hydrogen content. The higher the hydrogen index (HI) is, the better will be the quality of the hydrocarbon, and this point will be elaborated later in this chapter (see the section "Kerogen Typing"). The normal alkanes, also known as *paraffins*, are among the most common compounds in oil reservoirs. They are important constituents of gasoline, with heptane and octane being major classification tools. The octane used in the calculation is an isomer of the straight-chain alkane.

Naphthenes and *aromatics* are ring compounds. Naphthenes are single-bond rings, whereas aromatics have several double-bonded carbon atoms. The general formula for naphthenes is C_nH_{2n}. Type molecules include cyclopentane (C_5H_{10}), a pentagonal molecule, and cyclohexane (C_6H_{12}), a hexagonal molecule. Aromatics are double-bonded carbon ring compounds with the general formula C_nH_n. They are exemplified by benzene (C_6H_6). As their name implies, aromatics have a strong odor. Other common molecules in this class include naphthalene (mothballs). Naphthenic crudes are highly sought after because they are used to prepare lubricants.

Additional major constituents of hydrocarbons are the nitrogen, sulfur, and oxygen (NSO) compounds. These complex compounds are also called *hetero* compounds and, for classification purposes, are combined with the aromatics on the ternary diagram (fig. 7–1). The NSO compounds include all the organic acids and the *porphyrin* compounds. The latter are unique molecules because they can absorb cations. Note the distinction between absorbed cations, bonded to the molecule, versus adsorbed cations, which are attached but not bonded and are held in place by electrostatic van der Waals forces. Two well-known examples of molecules with a porphyrin core are hemoglobin and chlorophyll (fig. 7–2). They have similar structures, except that the heme porphyrin contains iron and not magnesium.

During catagenesis, the fatty acid tail is detached from the porphyrin head of chlorophyll and forms the alkanes. High concentrations of porphyrins indicate marine algal and bacterial source rock. Porphyrins in shales are also important because among the cations they bind are vanadium and nickel. Vanadium-to-nickel ratios in shales greater than 1 indicate a terrestrial source; by contrast, marine shales usually have a ratio less than 0.5. Because nickel is magnetic, crudes with a high nickel content provide a magnetic anomaly that may be detected with high-sensitivity magnetometer surveys.

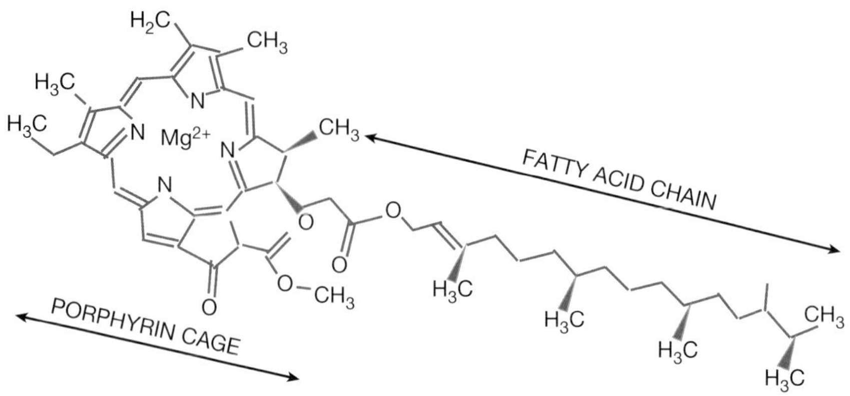

Fig. 7–2. Chlorophyll-type molecule with porphyrin core and fatty acid tail

Crude oil and gas densities represent an important physical property of hydrocarbons directly related to their chemical composition. Because most hydrocarbon densities are less than 1 g/cm^3, a convention was established whereby the density could yield a more manageable number. The formula for API number is as follows:

$$\text{API gravity} = \frac{141.5}{\text{Gravity of the oil at } 60°\text{F } (15.5°\text{C})} - 131.5 \qquad (7.1)$$

The API ranges used in classifying crudes are as follows:

- 0–14, heavy crudes
- 15–25, medium crudes
- 26–42, light crudes
- 43–52, condensate

Higher-API crudes have greater value, are less viscous, and are lighter. For exact determination of composition of the crude, gas chromatography and mass spectrometry ideally should be applied to the samples.

Methane is the most common hydrocarbon in space and on Earth, where it is ubiquitous on continents and oceans. In the deep oceans, methane combines with water and freezes in deep ocean sediments as methane clathrates, $CH_4 \cdot 6\ H_2O$. There is great interest in discovering economic means of producing methane clathrates; however, with current production techniques, this remains uneconomical.

Basic organic chemistry begins with the natural gases. As a group, they form part of the normal alkanes beginning with methane, CH_4, as the initial starting block. As they link along a straight chain, the general formula reflects the relationship by means of the standard formula (*n*-alkanes) is C_nH_{2+2n}. Other examples are ethane, C_2H_6; butane, C_3H_8; and so forth. These are referred to in short as C1, C2, C3, C4, and so on, based on the number of carbons in the molecule. Carbon numbers can continue from C1 through C34+.

A common analysis of any oil sample is a *gas chromatogram* of the normal alkanes (figs. 7–3 and 7–4). To identify the various carbon numbers for the normal alkanes, chemists place an "n" before the "C" followed by the carbon number. Isomers have the same formula as the *n*-alkanes, except that they have branched molecular structures. Isomers are distinguished from their *n*-alkane counterparts by their contractions having an "i" (in place of "n") before "C." The *x*-scale on the gas chromatogram is retention time, and the y-axis is the intensity (fig. 7–3); usually, nC12 and higher carbon numbers are detectable on a gas chromatogram. The results must be referenced to proper calibration samples.

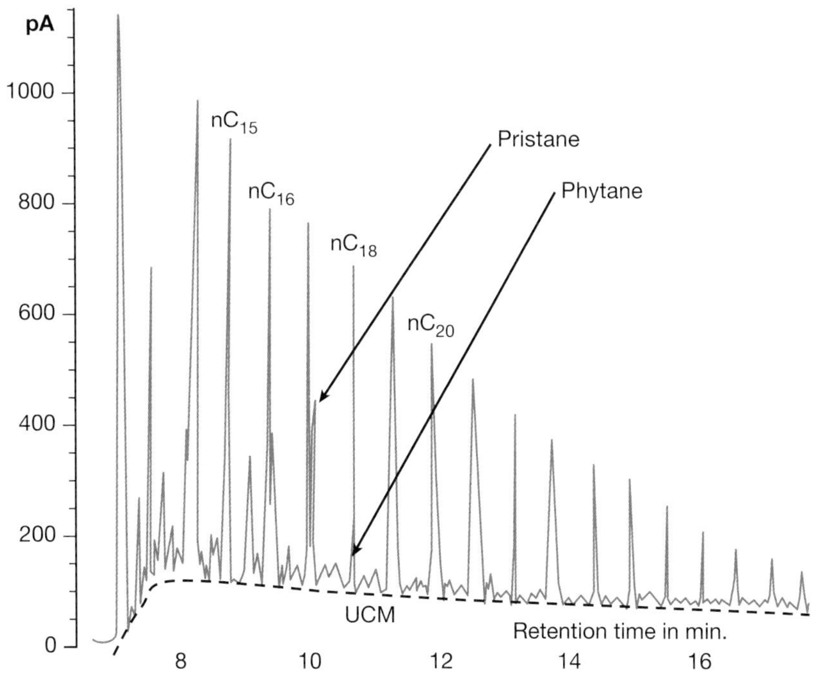

Fig. 7–3. Gas chromatogram for Cretaceous source rocks in South Texas. The sample is heated, and the retention time is plotted on the x-axis. The intensity is measured along the y-axis. "UCM" stands for unresolved complex mixture. Note the pristane/phytane (Pr/Ph) peaks; a very high Pr/Ph ratio suggests terrestrial input. In general, carbon numbers 1–4 are gases, 4–14 are gasoline, 15–23 are gas oil, and 24–35 are wax distillates. (Reproduced with kind permission of K. K. Bissada; Bissada et al., 1990)

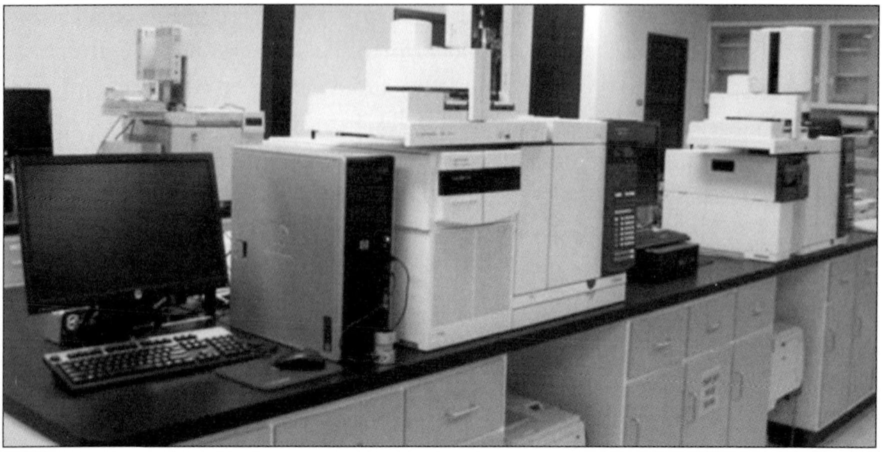

Fig. 7–4. Combined gas chromatograph and mass spectrometer (University of Houston geochemistry laboratory, Agilent 7000B GC-QQQ-MS)

Several techniques exist to measure the intensity on the gas chromatograph, the most common of which is based on flame ionization. The liquid hydrocarbon is placed in a column and slowly heated. The light compounds vaporize first, and the remainder are retained in the column. The relative abundance of each of the compounds can be estimated by calculating the area under each curve and establishing what percentage they represent (similar to the method outlined for area calculation by x-ray diffraction [XRD] in chap. 2). Early interpretation of the gas chromatogram establishes whether odd- or even-numbered carbons predominate; this is also referred to as the *carbon preference index* (CPI). Often, marine shales have an odd or even predominance. The CPI can be used to distinguish various oils from one another.

There are two special isomer peaks on the gas chromatogram: one near the nC17, *pristane* (Pr); and another near nC18, *phytane* (Ph) (fig. 7–3). The origin of these two molecules has been studied in great detail, and the Pr/Ph ratio has been documented as an indicator of terrigenous (predominantly plant derived) versus strictly marine (bacterial and marine organism derived) source. Tissot and Welte (1984) indicated that when Pr/Ph is significantly greater than 3.0, the source rock has a strong terrigenous component. It is clear the Cretaceous source studied by Bissada et al. (1990) documented a significant terrigenous input. Interestingly, the Bakken oils, at Williston Basin, also have relatively high Pr/Ph ratios, ranging from 0.76 to predominantly 1.45–2.0 (Price and LeFever, 1994), indicating a significant terrestrial contribution mixed with the marine source.

Furthermore, it has been suggested that the Pr/Ph ratio may provide valuable information on the redox conditions during deposition. Didyk et al. (1978)

indicated that when Pr/Ph is less than 1, the environment of deposition is anoxic; hence, the chances of a well-preserved organic carbon are high. By contrast, high Pr/Ph values suggest oxidation; therefore, the chances of preserving the organic carbon may be lower. Low Pr/Ph values are sought by the shale resource explorationist. There remains debate as to the validity of using Pr/Ph as indicator of oxidation state; moreover, as hydrocarbons mature, their Pr/Ph ratios change. A modified version of this relationship is currently being applied to source rock studies in which the Pr/nC17 ratio is correlated against the Ph/nC18 ratio. The upper quadrant of the plot corresponds to anoxic conditions, and the lower quadrant corresponds to oxidation conditions (Jarvie, 2001). Before a conclusive association is made, there should be several strands of evidence that confirm deductions based on Pr/Ph; alone, they cannot provide a reliable answer.

Usually, gas chromatography is run concurrently with mass spectrometry (fig. 7–4). The latter is used to record the biomarkers in the crude. Important biomarkers include hopanes, steranes, and terpanes. The biomarkers also appear as small peaks above UCM (fig. 7–3) envelope of the gas chromatogram. By magnifying the plot, their relative abundance can be estimated. If needed, they should be studied with very sensitive equipment. For additional information on the use of biomarkers, refer to the work of Peters and Cassa (1994), who have noted a second problem with biomarkers: namely, they evolve in a similar manner as biota do, because that is their origin, and as a result, various age units will contain different biomarkers. For example, the hopane oleanane is mainly present in source rock shales from the Late Cretaceous and Tertiary.

If C1 through C4 are in a gas phase under normal atmospheric conditions and if they are the only hydrocarbons in the reservoir as a group, they are classified as a dry gas reservoir. Because C1 through C4 are gases even under reservoir conditions, their behavior is subject to the *universal gas law*:

$$PV = nRT \tag{7.2}$$

where P is the pressure, V is the volume, T is the temperature, R is the gas constant, and n is the number of moles. Most often in dry gases, C1 exceeds 90%. C5 and higher are liquids at one atmosphere pressure. In all reservoirs, a sample is taken to a laboratory and is subjected to a test called a *PVT* (pressure, volume, and temperature) analysis.

The most common means of sample collection are the *drill stem test* (DST) and the *repeat formations test* (RFT). Both approaches require isolating the interval to be tested, opening a chamber, collecting the sample, and sealing the chamber. The RFT allows for up to three samples to be taken in one log run. A more recently developed tool, the *modular formation dynamics tester* (MDT), allows for more sampling and collection of other reservoir data.

Figures 7–5 and 7–6 give an example of a partial PVT analysis. PVT is a basic assay requirement for any crude transported to a refinery. One problem with simply using the universal gas law is that, under subsurface conditions, gas does not behave as an ideal gas; thus, a compressibility factor, Z, needs to be introduced, which is used in the estimation of gas reserves. The PVT analysis also provides Z values for the reservoir fluids (fig. 7–7). In chapter 9, we will discuss this topic in more detail.

Table 10–1 (cont.)

Page 2 of 15

File RFL 76000

Company Good Oil Company — Formation Cretaceous

Well Oil Well No. 4 — County Samson

Field Productive — State Texas

HYDROCARBON ANALYSIS OF Reservoir Fluid SAMPLE

Component	Mol Percent	Weight Percent	Density @ 60°F Grams per Cubic Centimeter	°API @ 60°F	Molecular Weight
Hydrogen Sulfide	Nil	Nil			
Carbon Dioxide	0.91	0.43			
Nitrogen	0.16	0.05			
Methane	36.47	6.24			
Ethane	9.67	3.10			
Propane	6.95	3.27			
iso-Butane	1.44	0.89			
n-Butane	3.93	2.44			
iso-Pentane	1.44	1.11			
n-Pentane	1.41	1.09			
Hexanes	4.33	3.97			
Heptanes plus	33.29	77.41	0.8515	34.5	218
	100.00	100.00			

Fig. 7–5. Partial PVT analysis on a Cretaceous oil sample formation in Texas. Normal gas chromatograms taken at the well site record C1–C8 only and are reported as ppm. (Based on McCain, 1990.)

Table 10–1 (cont.)

Page 5 of 15

File RFL 76000

Well Oil Well No. 4

Differential Vaporization at 220 °F.

Pressure PSIG	R_{sD} Solution Gas/ Oil Ratio (1)	B_{oD} Relative Oil Volume (2)	B_{tD} Relative Total Volume (3)	Oil Density gm/cc	Deviation Factor Z	Gas Formation Volume Factor (4)	Incremental Gas Gravity
2620	854 = R_{sDb}	1.600 = B_{oDb}	1.600	0.6562			
2350	763	1.554	1.665	0.6655	0.846	0.00685	0.825
2100	684	1.515	1.748	0.6731	0.851	0.00771	0.818
1850	612	1.479	1.859	0.6808	0.859	0.00882	0.797
1600	544	1.445	2.016	0.6889	0.872	0.01034	0.791
1350	479	1.412	2.244	0.6969	0.887	0.01245	0.794
1100	416	1.382	2.593	0.7044	0.903	0.01552	0.809
850	354	1.351	3.169	0.7121	0.922	0.02042	0.831
600	292	1.320	4.254	0.7198	0.941	0.02931	0.888
350	223	1.283	6.975	0.7291	0.965	0.05065	0.988
159	157	1.244	14.693	0.7382	0.984	0.10834	1.213
0	0	1.075		0.7892			2.039
		@ 60°F = 1.000					

Gravity of residual oil = 35.1°API @ 60°F.

(1) Cubic feet of gas at 14.65 psia and 60°F per barrel of residual oil @ 60°F.

(2) Barrels of oil at indicated pressure and temperature per barrel of residual oil at 60°F.

(3) Barrels of oil plus liberated gas at indicated pressure and temperature per barrel of residual oil at 60°F.

(4) Cubic feet of gas at indicated pressure and temperature per cubic foot at 14.65 psia and 60°F.

Fig. 7–6. Partial PVT analysis tabulating gas chromatography data to provide information on oil and gas density and Z factor. (Based on same data as fig. 7–5; McCain, 1990.)

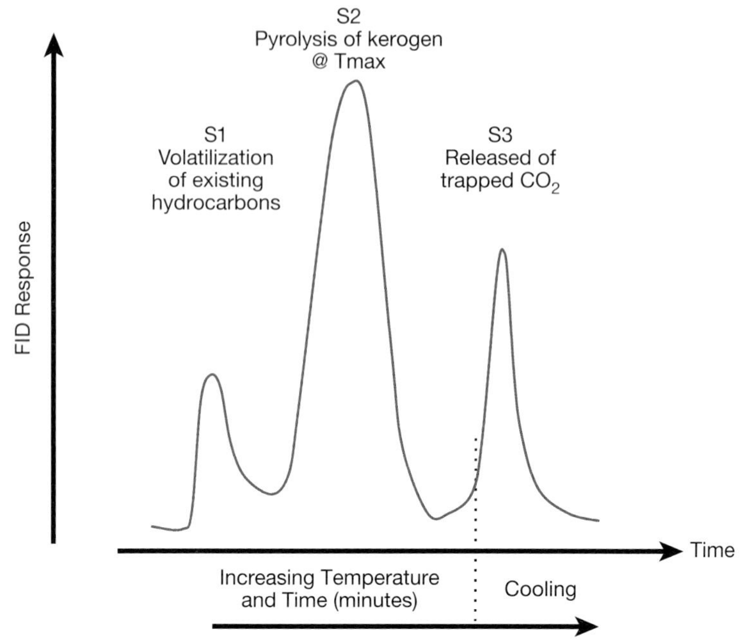

Fig. 7–7. Rock-Eval plot defining the characteristics of each peak. "FID" denotes flame ionization detector. (Based on Tissot and Welte, 1984)

While drilling all exploration and production wells, conventional gas chromatography should be carried out at the well site. Baroid and Core Laboratories have developed different formulas for the analysis of well site–derived gas chromatography data. The Baroid method shows why calculating an estimate of the response is important. In all cases, the gas must be corrected for background gas and trip gas. The Baroid method (Pixler, 1969) uses the corrected ratios of C1/C2, C1/C3, C1/C4, and C1/C5+ to provide information on the wetness of the gas, as well as its producibility. In general, if the ratios are below 2 units or above 100 units each, the interval is considered nonproductive; if the ratios are between 20 and 80, the zone is considered to be a gas producer; and if the ratios are between 2 and 20, the interval is considered to be an oil producer.

Figures 7–5 and 7–6 show a PVT analysis for an oil well. However, as an exercise, if we were to consider the separator gas as the only fluids in the well, then on the basis of the data provided by the PVT analysis (fig. 7–5) and using the mole percent yield from that chart, the Pixler (1969) ratios (C1/C2, C1/C3, etc.) can be applied as follows:

- C1 = 36.47
- C2 = 9.67

- C3 = 6.95
- C4 = 3.93
- C5 = 1.41
- C6 = 4.33
- C1/C2 = 3.77
- C1/C3 = 5.25
- C1/C4 = 9.28
- C1/C5+ = 6.35

Note that any proportional gas chromatography volume data can be used. As a well is being drilled, the gas chromatogram provides information to assist drilling to stay on target (Unrau and Nagel, 2012).

A second method to determine the wetness of the interval was proposed by Unrau and Nagel (2012), based on the *wetness ratio* (WR):

$$WR = \frac{C2 + C3 + C4 + C5}{C1 + C2 + C3 + C4 + C5} \times 100 \tag{7.3}$$

Therefore, using the same values from figure 7–6, the PVT values yield, WR = (21.96/58.43) × 100 = 38. WR of 0.5–17.5 indicates productive wet gas, and WR of 17.5–40 indicates productive oil. Further, the *balance ratio* (BR) is defined as follows:

$$BR = \frac{C1 + C2}{C3 + C4 + C5} \times 100 \tag{7.4}$$

Again by use of the values from the PVT analysis, BR = (46.14/12.29) = 3.75. BR of more than 100 indicates that the dry gas is probably a nonproducer. The lower the WR is, where the BR is less than 100, the wetter will be the produced gas. Very low wetness and BR above 20 indicate that liquid oil production is likely. If the BR is less than the WR, then wet gas production is likely. Both methods indicate oil production.

Other techniques for evaluating if a well will produce gas, oil, or mixed gas and oil include the *gas-to-oil generation index* (GOGI), originally proprietary is now commercially available from several vendors (e.g., Intertek and Zetaware). GOGI uses the data from the gas chromatogram and generates a ratio (C1–5/C6+). If the GOGI ratio is less than 0.35, the yield is oil; if it is 0.35–0.5, the yield is mixed oil and gas; and if it is above 0.5, the yield will be predominantly gas. If shale oil is produced, a second ratio needs to be analyzed: the *gas/oil ratio* (GOR). GOR is usually measured in standard cubic feet of gas per barrel (scf/bbl). Again, higher GOR values make the products less economically viable and indicate overpressure. It is important to monitor the GOR in the field because it changes over the life of the field. The rate of change allows for an improved production forecast.

KEROGEN TYPING

Thus far, we have reviewed the basic chemical composition of the kerogen and the potential by-products kerogens can produce. The next parameter that requires analysis is typing of the kerogen, to determine the richness and establish the level of maturity of the source rock shale. Crude samples and source rock extracts are collected and submitted to a laboratory for testing in a now-traditional tool known as *Rock-Eval pyrolysis*. The reason a specific type of study is selected is that the analysis must be reproducible and readily comparable with other crudes. It is often recommended that an immature source rock interval be analyzed for comparative studies. On such samples, determining the critical temperatures associated with the stages of catagenesis of the kerogen as well the hydrogen index (HI) and the original TOC content are critical for the exploration process. Determining the activation energy of the kerogens involved in the source rock shale is critical to establish the critical moment of generation. Once again, analysis of immature kerogens is critical.

The basic principle of Rock-Eval pyrolysis is to heat a sample in an inert atmosphere (fig. 7–7). A sample is placed in a container and progressively heated to 550°C, and then the products are measured. Over time, four distinct peaks are observed, which Tissot and Welte (1984) have labeled S1, S2, S3, and S4 (fig. 7–7); S1 and S2 measure hydrocarbons, S3 is a measure of the CO_2 produced after the breakdown of the kerogen, and S4 is the residual carbon in the system. S1 is produced from the free hydrocarbons in the sample; in other words, the amount of free oil in the sample pore space can be computed. The S2 peak is the hydrocarbon potential in the source rock and is used in estimating the generation potential remaining in the source interval. It is not the original generation potential. This is important because as the source rock matures, the S2 peak decreases, and consequently the S1 peak increases. Also, remember that not all hydrocarbon produced is trapped in the reservoir. Therefore, the sum of S1 + S2 is *not* the original generation potential of the source rock. The best value for an original S2 peak or generation potential should be estimated where the source rock is in an immature state. It may be interesting in the future to combine Rock-Eval with differential thermal analysis/thermal gravimetry (see chap. 2) to improve the determination of the products of catagenesis.

Total organic carbon

The observed TOC content in the system is determined using Rock-Eval:

$$TOC = \frac{0.82(S1 + S2) + S4}{10} \tag{7.5}$$

where the result is weight percent based on milligrams of hydrocarbon per gram of rock. This computed value yields the observed TOC. The original TOC was likely higher because portions of the hydrocarbons have likely migrated away from the sample. Once again, it is important to study the immature source rock and establish an original TOC value. A second problem that needs to be considered is that TOC values measured on cuttings from the same interval may be as much as 50% lower than those from core samples (Jarvie and Hill, 2011).

Implication of T_{max}

As the rock is heated beyond 350°C in Rock-Eval, the S2 peak is generated at a specific temperature, T_{max} (fig. 7–7). T_{max} has been compared with vitrinite reflectance (R_o) in common tests and found to be closely proportional to R_o; hence, it is used as a maturity index (fig. 7–8). However, caution should be taken to calibrate the T_{max} value in a specific area of study against other thermal indicators to ensure accuracy; it should not be used in isolation to determine maturity values.

MATURITY	Ro	Tmax (C°)
Immature	0.2–0.60	< 435
Early	0.6–0.65	435–445
Peak	0.65–0.90	445–450
Late	0.90–1.35	450–470
Post mature	>1.35	>470

Fig. 7–8. Maturity indicator comparisons from Peters and Cassa (1994). Other maturity indices are shown in figure 7–9 and will be summarized in the next section ("Thermal Effects Applied to Burial History").

Current research has questioned the validity of a simple translation between T_{max} and vitrinite reflectance (Wust et al., 2013). The two should be calibrated in each area of study. The vitrinite reflection (R_o) index is a coal rank reference for maturity. Vitrinite serves as a guide only for Upper Paleozoic, Mesozoic, and Tertiary rocks. It is basically the intensity of light that reflects from one specific polished maceral—namely, vitrinite. This maceral is found only in plants. Caution should be taken to separate the reflectance of other macerals (e.g., liptinite, inertinite) from that of vitrinite. Older Paleozoic rocks have no vitrinite because plants bearing vitrinite did not exist yet. Several other maturity indices, including pollen coloration index (themal alteration index, TAI) and conodont alteration index (CAI) (see chap. 3), can be used in place of the vitrinite reflection index. Work is also being carried out on an illite crystallinity index (the Kubler index), as well applying Fourier transform infrared (FTIR) spectrometry to determine maturity (see chap. 2).

Additional important computations from the Rock-Eval results are the HI, oxygen index (OI), and the production index (PI) or transformation index. HI = (S2/TOC) × 100; OI = (S3/TOC) × 100; and PI = S1/(S1 + S2) . A plot of HI versus OI, known as a *van Krevelen plot* (fig. 7–9), is used to classify the kerogen type from predominantly marine (high HI and low OI) to terrigenous (low HI and high OI) source rocks. The latter group of kerogens are more gas prone. A modified *van Krevelen plot* is used because it allows for better integration with the ultimate resource potential calculation (Nordeng, 2012). The original *van Krevelen plot* used hydrogen to carbon and oxygen to carbon ratios.

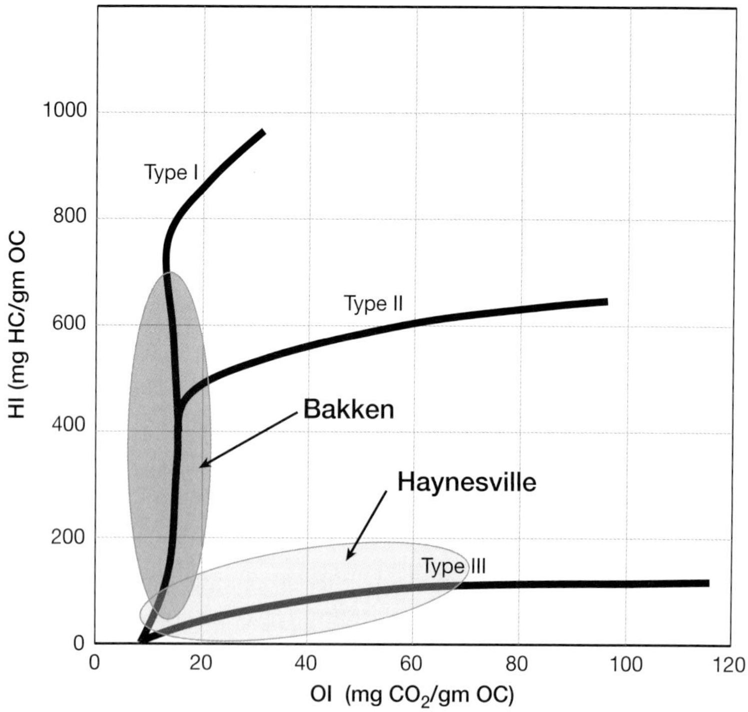

Fig. 7–9. Modified van Krevelen Plot for the Haynesville Formations, North Louisiana (light gray), and low-maturity Bakken samples (dark gray).

In general, when both HI and OI are low, the source rock is considered mature and burned up, as is the case with the Haynesville crudes in figure 7–9 (Goddard,

2016). The high OI suggests very-high-maturity source rocks and or terrigenous input—hence, very gas prone. By contrast, the Bakken is more oil prone. In the Bakken, at higher maturities, the HI is reduced. These calculations are critical for the estimation of reserves of the resource play and will be reviewed in chapter 9 (see the section "Reserve Estimate").

Importantly, the generation potential for a specific area in a basin must be understood. Calculation of the potential requires comparison of the original HI (obtained from an immature interval) against the observed HI. Note that in the immature sample, the S1 peak will be significantly lower. Hence, the original TOC can be estimated. As hydrocarbons are generated, the TOC is reduced. TOC measurements also include organic carbon in bitumen residual that does not play a role in movable hydrocarbons reserves.

THERMAL EFFECTS APPLIED TO BURIAL HISTORY

In conventional hydrocarbon exploration, the principal source rock *kitchen* is rarely drilled. The kitchen usually lies near the basin axis in a synclinal position. However, in the case of the unconventional resource play, the source rock is the target, and, whereas general estimations are acceptable for traditional exploration source kitchens, more precise calculations are needed for the resource play. Moreover, the target must be within a specific temperature range and is often drilled in the hydrocarbon kitchen as the kerogen transforms into the gas phase, thus providing wet gas. The burial history also includes the thermal history of the shale. As a result, the moment at which the source rock enters the oil window can be defined together with the paragenetic sequence (diagenetic history) of the source interval. The results are subsequently summarized on a critical moment chart.

Kitchen maps, with their maturity levels contoured, can be generated and serve an important function for any conventional or unconventional hydrocarbon play. In particular, they provide the basis for estimating the hydrocarbon generation volumetrics. In the case of unconventional resource assessment, they enable focusing on the sweet spots for exploration and field development, as well as improving the definition of the target risk segments. Bissada et al. (1990) have applied an interesting map using the critical moment of generation; specifically, they restored the structural map to the time of the critical moment for specific areas and demonstrated how the local conditions might have modified the sweet spot by indicating length of stay in the optimum temperature range.

The steps in analyzing the thermal history of a basin are as follows:

- Establish the present-day heat flow in the area of interest

- Determine the geothermal gradients above and at the target horizon
- Predict past and present thermal conductivities (k, in watts/m•°C)
- Model the thermal history of the area on the basis of the area's tectonic history

The equation for heat flow, Q (commonly in units of mW/m^2, although eq. [7.6] uses W/m^2), is relatively simple, using the geothermal gradient (GG), dt/dz (in °F/100 ft or °C/100 m, with a conversion factor of °C/100m = °F/100 ft × 1.82). Heat flow is constant at a given time:

$$Q = -k\frac{dt}{dz} \tag{7.6}$$

Among the most important elements of a resource play is its thermal history. The heat flow in a region is constant over short periods of time; the more ephemeral variables are the GG and the thermal conductivity, and they are inversely proportional. The heat flow does vary over long time periods, and it should be modeled with care. Because the principal objective of the shale resource play is to produce wet gas, the temperature conditions must be ideal at the present time and during the latter thermal history of the shale. Burial history provides the only tool to determine the degree of the maturity history of the target formation. Once established, a map of the maturation indices provides information on the timing of entry into the oil and gas window and the present-day thermal history prior to drilling. For a detailed discussion on the topic, refer to the work of Waples (1980).

Translations are available for time temperature index (TTI) to R_o, and level of organic maturity (LOM) used elsewhere in this book (fig. 2–9). Commercially available modeling tools are commonly used. The concept of graphically describing the burial history and estimating the degree of maturation was first developed in 1970 by the Russian geochemist Lopatin and expanded by Waples (1980, 1985). An explorationist should have a clear understanding of the input parameters and how specific problems are resolved. As such, a simple one-dimensional model is described. Once the fundamental parameters are understood, more sophisticated, three-dimensional models are recommended.

Before discussion of the burial history of the region of study, it is important to establish the GG in the area of interest. The primary source for temperature information in the subsurface is from wells drilled. All logging runs must record *bottom-hole temperature* (BHT) at the bottom of the logging point. The information is reported on the log header (fig. 7–10).

Log Measured From RKB 24.2 Ft.
Drilling Measured From SAME

Date	7-1-89	8-13-89
Run No	ONE	TWO
Depth–Driller	11900	15500
Depth–Logger (Schl.)	11872	15473
Btm. Log Interval	11868	15470
Top Log Interval	3066	11891
Casing–Driller	13 3/8 @ 3067	9 5/8 @ 11900
Casing–Logger	3066	11891
Bit Size	12 1/4	8 1/2
Type Fluid in Hole	LIGNOSULF.	LIGNOSULF.
Dens. / Visc.	11.5 / 38	14.4 / 42
pH / Fluid Loss	11.0 / 5.4 ml	10.5 / 7.4 ml
Source of Sample	MUD PIT	PIT
Rm @ Meas Temp	.742 @ 94 F	.602 @ 94 F
Rmf @ Meas. Temp	.754 @ 76 F	.460 @ 64 F
Rmc @ Meas Temp	1.110 @ 94 F	.903 @ 94 F
Source Rmf / Rmc	M / C	M / C
Rm @ BHT	.384 @ 188 F	.228 @ 260 F
Circulation Stopped	0400/7-1	1900/8-13
Logger on Bottom	1030/7-1	0300/8-14
Max. Rec Temp.	188 F	260 F
Equip. / Location	8411 / OPEL.	8411 / OPEL.
Recorded By	GROUMS	LOUVIERE
Witnessed By	IMMEL-LUKAVSKY	IMMEL-LUKAVSKY-WYCHE-

Fig. 7–10. Typical well log header for South Louisiana well. BHT for each log run is recorded twice: "Max. Rec. Temp." and "R_m @ BHT." In addition, the time between "Circulation Stopped" and "Logger on Bottom" is critical. The mud progressively warms as time passes. Other important points are measured depth (corrected to subsea) and density of the mud (in ppg).

The BHT is recorded with the Rm resistivity (near the bottom of fig. 7–10). Occasionally, the logger will repeat the maximum observed temperature. In the well shown in figure 7–10, there are two log runs: 7-1-89, with a logging interval depth of 11,872 ft; and 8-13-89, with a logging interval depth of 15,470 ft. Note that the measurement of the kelly bushing on the rig floor is 24 ft above sea level. Therefore, subsea depths are –11,848 ft and –15,446 ft, respectively. The maximum temperatures recorded are 188°F and 260°F, respectively. The calculations should be carried out in meters and degrees centigrade. Therefore, logging info should read as follows:

- –3,611 m, at 81°C, for the first run (7-1-89)
- –4,708 m, at 127°C, for the second run (8-13-89)

As discussed earlier in chapter 5, drilling mud at surface temperature is introduced into the wellbore to cool and lubricate the bit. As a result, the BHTs must be corrected. When a production well is used with multiple log runs, a method known as a *Horner plot* is used to determine the correction. The plot takes into account the lag in circulation time between when drilling stopped and when the logging run began. Mud circulation stopped at 4:00 a.m., and logging commenced at 10:30 a.m. on July 1, for the first run; and mud circulation stopped at 7:00 p.m. on August 13, and logging commenced at 3:00 a.m. on August 14, for the second. Thus, the time since circulation was 6.5 h for the first run and 8.0 h for the second run. In the Horner plot, the temperature differences for each subsequent log run at the same depth are compiled, and the formation temperature is extrapolated. In most cases, this information is not available. A best approximation is achieved by use of Harrison et al.'s (1983) estimation. Based on hundreds of samples, Harrison et al. arrived at a best-fit second-order polynomial to establish the correction factor:

$$\Delta T = -16.51 + 18z - (2.34 \times 10^{10}) - 6z^2 \quad (7.7)$$

where z is the depth.

By use of equation (7.7), the BHT should be corrected 18°C for the first run and 16°C for the second run. Therefore, the corrected values should be –3,611 m, at 98.5°C, for the first run, and –4,708 m, at 143°C, for the second run. The temperature gradient is as follows:

- Run 1: (99°C – 20°C)/36.1 = 2.2°C/100 m, where 20°C is the surface ambient temperature
- Run 2: (143°C – 99°C)/(47.08 – 36.11) = 4°C/100 m

The surface ambient temperature may change depending on latitude. It may be higher or lower than 20°C. The lower section has a GG that is based on the difference between the two BHTs in the well. There is an obvious increase in the lower section. If the computation were performed using only the deeper value, the GG would be (143°C – 20°C)/47 = 2.67°C/100 (i.e., significantly higher than 2.2°C/100 m).

Why are the two runs so different? Recall equation (7.6), which gives the heat flow. The well is located in South Louisiana, where the heat flow has been measured at 54 mW/m^2 = 0.054 W/m^2. If the GG is 2.2°C/100 m or 4°C/100 m, then the thermal conductivity is 0.054 W/m^2/(2.2°C/100 m) = 2.45 W/m/°C; or, 0.054 W/m^2/(4°C/100 m) = 1.35 W/m/°C.

The second question is, "Why is the thermal conductivity lower in the deeper section?" The answer is overpressure. Returning to the well log header (fig. 7–11), note that the mud weight for the upper section was 11.5 pounds per gallon (ppg), and that for the lower section was 14.4 ppg. Given that 1 psi/ft = 0.433 g/cm^3 = 1 ppg/19.27 psi/ft. Thus, the conversions are 11.5/19.27 = 0.60 psi/ft, and 14.4/19.27 = 0.75 psi/ft; note that 0.75 psi/ft is significant overpressure.

Importantly, the mud weight provides the pore pressure in the reservoir (1/19.27=.052):

$$\text{Pore pressure (psi)} = 0.052 \times \text{mud weight} \times \text{depth} \tag{7.8}$$

Thus, the pore pressure for the first run is 0.052 × 11.5 × 11,848 = 7,085 psi = 49 megapascals (MPa). For the second run, it is 0.052 × 14.4 × 15,446 = 11,566 psi = 80 MPa.

The mechanical aspects of overpressure and the consequences of high pressure will be discussed in chapter 8. Prediction of overpressure is a critical issue in all predrill analysis, including shale resource assessment. Note that the presence of overpressure requires that a good seal be present, allowing the overpressure cell to be preserved. This brings us back to the factors associated with overpressure:

- The temperature within the overpressure compartment will always be higher
- The porosity is better
- The interval velocity is slower

In the discussion on smectite-to-illite conversion (see chap. 2), the topic of overpressure was mentioned. As the illite structure collapses, it greatly improves the seal capacity of the shale. Within the affected overpressure cell, fluids are inhibited from moving, hence preserving porosity. Solidity was defined previously as 1 – porosity. The term is commonly used in the study of thermal conductivity.

UNDERSTANDING THERMAL CONDUCTIVITY

Thermal conductivity is the ability of the material to transmit or transfer thermal energy. Each material has a unique thermal conductivity value, k_f. From the standpoint of the current study, a few important k_f values (in W/m/K) (K in Kelvin) are the following:

- Water, 0.58
- Sandstone (dry), 1.4
- Sandstone (wet), 2–4
- Limestone, 1.26–1.33 or higher
- Gasoline, 0.15

- Ethane, 0.018
- Coal, 0.2
- Bitumen, 0.18
- Mica, 0.71
- Kimmeridge clay, 0.87–2.01 (Midttømme et al., 1998)
- Smectite, 1.8 (Midttømme et al., 1998)
- Illite, 1.8 (Midttømme et al., 1998)

What is evident from the preceding list is the effect of lithology and mixture present in a rock system. Also, keep in mind that older sediments will always have higher thermal conductivities than will their younger counterparts. Given that the net thermal conductivity is the sum of 0.7 k_m + 0.3 k_f, if a sandstone has a porosity of 30% with a matrix conductivity (k_m) of 1.4 and the pores are filled with either water, ethane, bitumen, or smectite, then

- For water (k_f = 0.58), (0.7 × 1.4) + (0.3 × 0.58) = 1.154
- For ethane (k_f = 0.018), (0.7 × 1.4) + (0.3 × 0.018) = 0.985
- For bitumen (k_f = 0.18), (0.7 × 1.4) + (0.3 × 0.18) = 1.034
- For smectite (k_f = 1.8), (0.7 × 1.4) + (0.3 × 1.8) = 1.52

In other words, if the overlying layer is 100 m of a sand/shale mixture with zero porosity and has a seal, the heat flow is 0.05 W/m^2, and the underlying layer is 100 m of an ethane-filled pore space, as indicated above, then the GG will be 3.3°C/100 m for the upper tight sand and 5°C/100 m for the porous unit. Because thermal conductivity and GGs are inversely proportional, unique conditions may occur, as is the case with the Bakken of North Dakota. The overlying Lodgepole Formation (Madison Group) has a thermal conductivity of 3.45 W/m/K. The underlying Bakken has a thermal conductivity ranging from 2.77 W/m/K (Nordeng and Nesheim, 2011) to significantly less, 1.1 W/m/K (Gosnold et al., 2010), depending on the study area (fig. 7–11). Three Forks has a thermal conductivity of 3.5 W/m/K. Given a heat flow of 65 mW/m^2 for the first area of study and 51 mW/m^2 for the second and by use of equation (7.6), the GGs for the two formations are 1.89°C/100 m for Lodgepole and a GG of 2.35–4.6°C/100 m for the Bakken. Three Forks has a GG of 1.45°C/100 m (assuming a heat flow of 51 mW/m^2).

Figures 7–12 and 7–13 illustrate a study performed in South Louisiana, where mud weights were compared to the thermal conductivity for 1,000 wells. The deeper the well was, the lower was the thermal conductivity value. Where mud weights are above 15 ppg—in other words, overpressure—the thermal conductivity decreases. It is an indication of greater porosity and more fluid retention as the mud weight continues to increase. It also suggests that the temperature is higher at greater mud weights. The reason for the spread in the values is that fluid content was not considered.

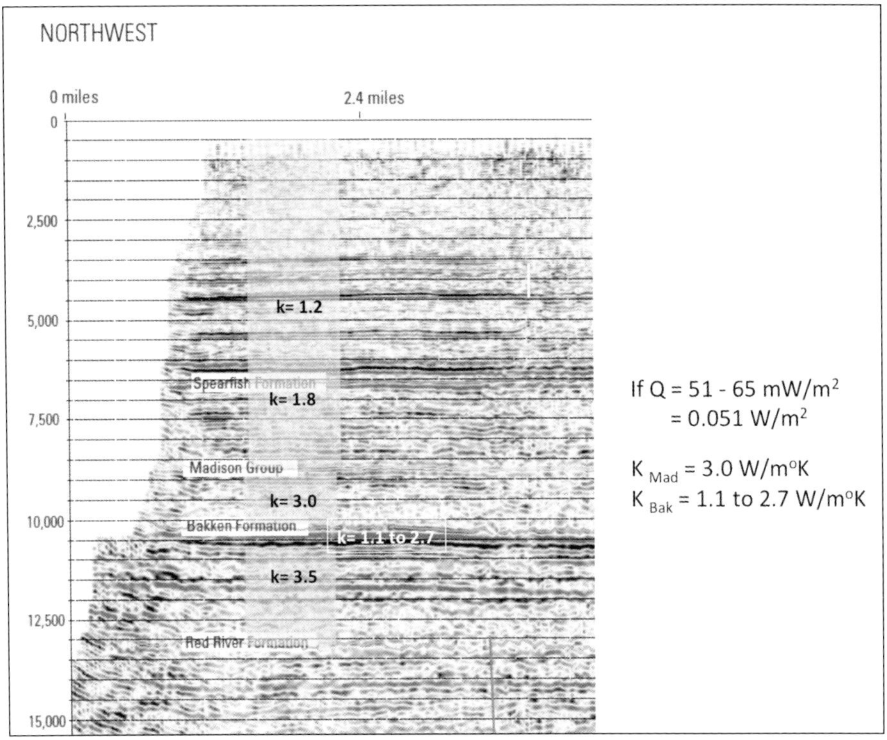

Fig. 7–11. Seismic line from Williston Basin (Mountrail County, North Dakota) showing variations in thermal conductivity values of the strata. (Reproduced with kind permission from the U.S. Geological Survey. Thermal conductivity data are from Gosnold et al., 2010.)

The next question that needs to be answered is the timing of the overpressure. In the discussion on smectite diagenesis (see chap. 2), it was mentioned that the activation energy for the smectite/illite reaction attains a maximum at 80°C. Therefore, in the burial history of the sample, the timing for the entry into overpressure can be established. It is then possible to introduce the new GG at that time and continue the burial history.

Burial history of a horizon is first a function of the rate of sedimentation of each sedimentary unit overlying the exploration target horizon, which in turn is a function of the amount of accommodation space created and sediment supply. As discussed previously, if a stratigraphic zone that is affected by neither faulting nor erosion is chosen and the interval thickness is corrected for dip, then the interval, bracketed by faunal control, will yield the rate of deposition. The rate of sedimentation should be compared in the basin to ensure that reasonable values are input. Second, by application of the thermal history of the applicable segment of the basin under investigation, the level of maturation of hydrocarbons in the system can assist in predrill definition of hydrocarbon type and the definition of potential sweet spots. Third, determine if unconformities are present and significant. Fourth, if the amount of compaction can be estimated, then it can be applied to determine its impact.

Before the burial history of a region can be described, its basic geologic history must be understood. In a basin undergoing initial exploration, it is important to determine the dominant megasequences and uplift history and establish their impact on the thermal history. The uplift history can produce a cooling effect owing to significant erosion. Cooking organic material is analogous to cooking meat. Once a degree of cooking is achieved, when heating is discontinued, the meat retains the same degree of cooking. The uplift history freezes the maturation until heating is resumed beyond the previous heating cycle.

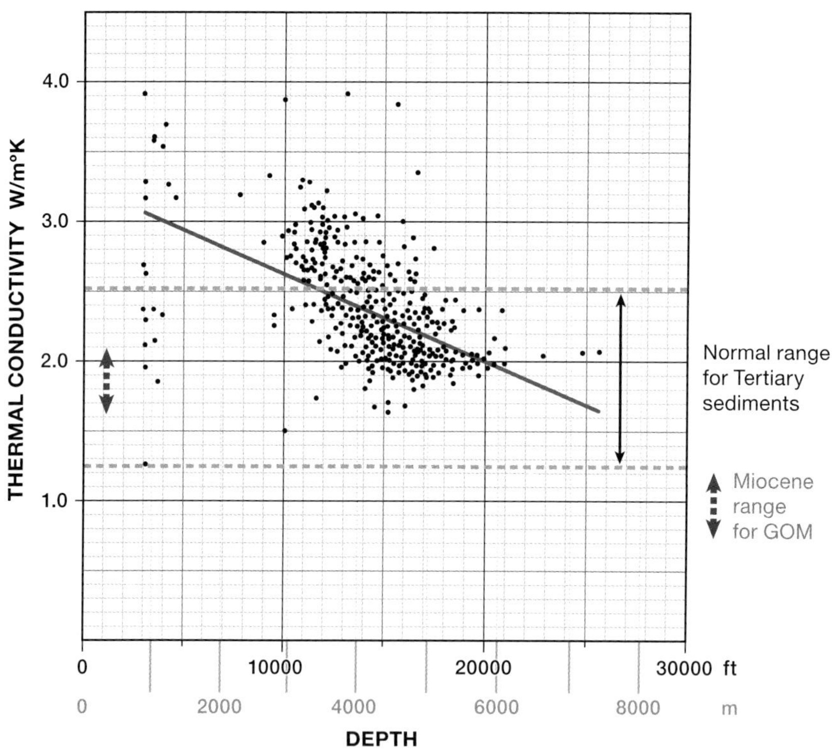

Fig. 7–12. Plot of thermal conductivity versus depth in South Louisiana. A decrease is seen with depth, and GGs rise with depth (in ft) "GOM" denotes Gulf of Mexico. (Data by the author)

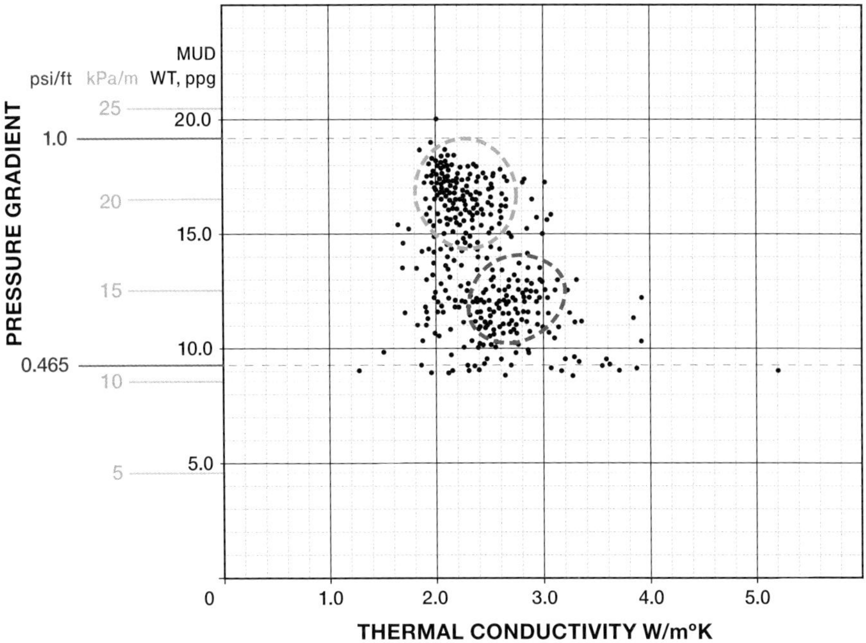

Fig. 7–13. Plot of mud weight versus thermal conductivity. As mud weight increases, indicating overpressure, the thermal conductivity is reduced and the GG rises. (Data by the author)

Most common sedimentary basins formed in one of the following five tectonic settings:

- Rifting to passive margin
- Foreland basin associated with plate collision and located in a back-arc setting
- Intracratonic basins
- Forearc basins
- Pull-apart basins associated with strike-slip

Individual basins may have different overprinted tectonic characteristics. Each tectonic phase has a different geothermal history. Normally for uplift to take place, heat provides the energy allowing for the displacement. Global heat-flow maps are available for most countries. They not only provide the present-day heat flow but also provide examples of heat flows associated with tectonic events that can be applied to the past.

One example is the Red Sea Rift, in Northeast Africa. Detailed heat-flow data from the region provide a model for heat flow associated with rifting. The present-day Afar triangle (fig. 7–14) along the southern Red Sea and the Sudan/

Eritrea border is the most active current rifting. Heat flows in excess 180 mW/m^2 have consistently been reported. Hence the current exploration interest in offshore Sudan. Rifting is episodic, and the central portion of the Red Sea was likely dominated by Oligocene (± 29 Ma) rifting. The central Red Sea has heat flows in the range of 70–80 mW/m^2. The cooling effect is consistent with McKenzie (1978) curves (fig. 7–15; Allen and Allen, 2013), with a crustal stretching factor of 6+.

Most studies of thermal histories of graben regions assume a very simplistic, uniform thermal model. The evidence from the Red Sea demonstrates the opposite. Thus, as additional thermal information is modeled in a particular basin, the original suppositions may need to be updated as well.

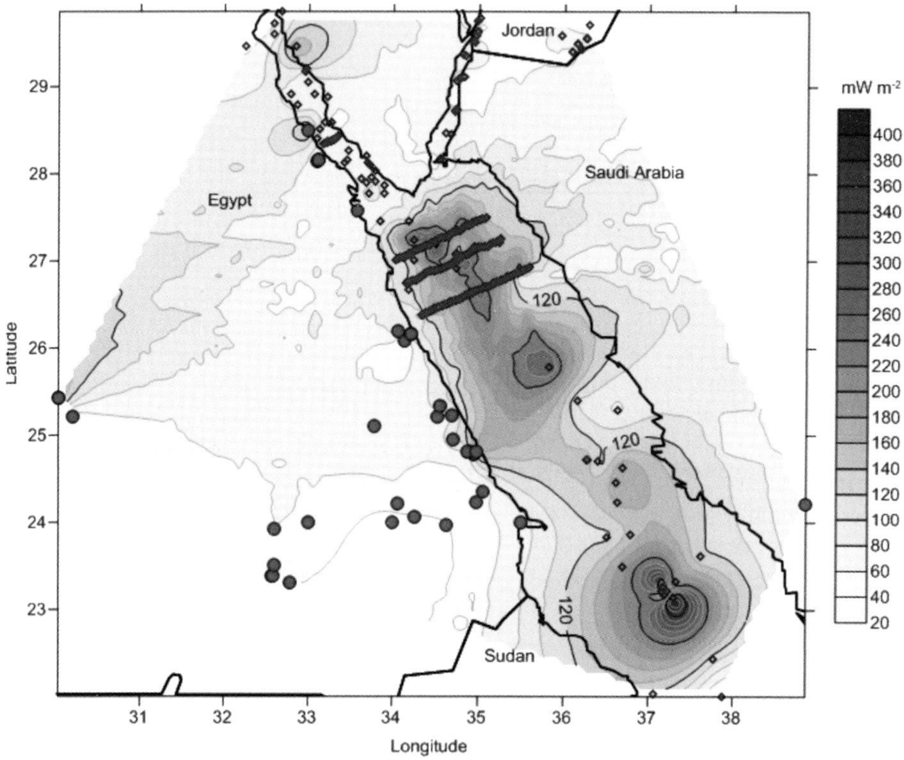

Fig. 7–14. Present-day heat flow map in the Red Sea (Reproduced with kind permission from Professor Gosnold, University of North Dakota, 2017)

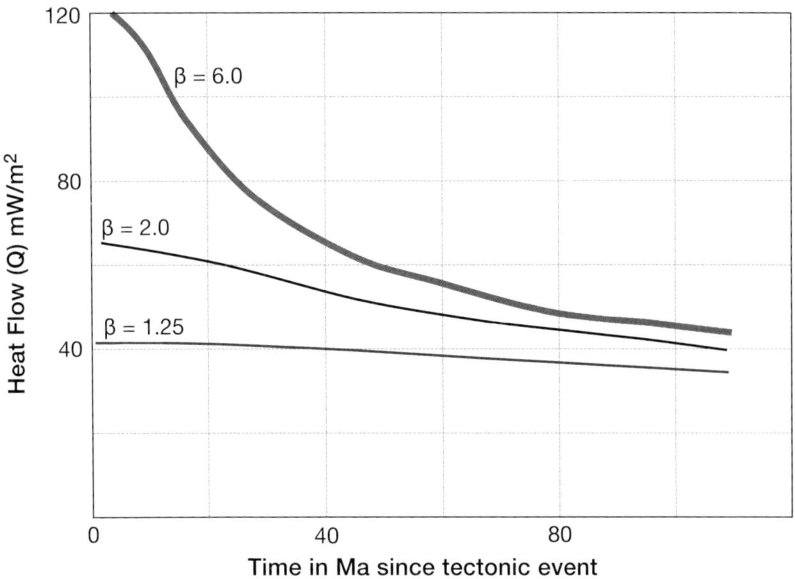

Fig. 7–15. Red Sea crustal stretching factor (β). The Red Sea likely has a β factor of 6+. Within 20 million years, the heat flow can drop from 180 to 80 mW/m^2. The top line represents possible incremental heat flow due to Jurassic rifting in the South Texas Gulf of Mexico region. (Adapted from Allen and Allen, 2013)

The Haynesville cross-section (fig. 4–7) can be used as a model of the effects of the two rift phases in the Gulf of Mexico: the Norian, 210 Ma (Late Triassic) (Bartok, 1993), and the Bathonian, 166 Ma (Marton and Buffler, 1994; Pindell, 1985). The heat-flow history of these two rift phases affected the Kimmeridgian (153 Ma) Haynesville Formation of East Texas and North Louisiana as well as the Late Cretaceous (95 Ma) Eagle Ford Formation of South Texas, but in different ways. Both formations were deposited along a passive margin setting overlying a preexisting rift sequence.

If a simple shear model is assumed for both rifts, the heat-flow decline curve in figure 7–15 can be applied. In general, the Triassic rifting is considered more intense, resulting in higher heat flows ($\beta \leq 6$). The Jurassic rifting was less intense in the region in terms of crustal stretching ($\beta \leq 2$ (Buffler and Sawyer, 1985). By use of the plot in figure 7–15, the Norian to the Bathonian transpired over 40 million years.

Therefore, the heat flow by the Bathonian was approximately 73 mW/m^2 along the β-6 line from an original heat flow of 94 mW/m^2 (thick line in fig. 7–15). This line represents a possible model for the heat flow affecting the region after the Jurassic rifting event. The Kimmeridgian was deposited approximately 10 million years after the rifting event. Therefore, the heat flow may have been as high as 80–90 mW/m^2 at the time of deposition. Today, the heat flow in the region is 70 mW/m^2. Given that Haynesville was less consolidated, its thermal conductivity was also lower than at the present day. Therefore, the GG was significantly higher than in the present day, and this value needs to be input in the geothermal model for the formation maturation study. For Eagle Ford, no correction is required because the heat flow at the time of deposition was nearly the same as it is today. In any study, the paleo heat flow should be estimated and its impact determined using the best models available. Once drilling begins, the models should be adjusted to fit the data. An additional example for the Eagle Ford (Cretaceous) will be discussed later in this section.

The effects of erosion may further impact the thermal history of a basin. For the current discussion, a simple passive-margin model containing a mild uplift is described. Two examples will be used here: first, a model Tertiary shale with a simple burial history and a high GG is plotted; next, the same model is repeated with a significant erosion event. These examples demonstrate the impact of changes in burial history to provide an understanding the mechanism at work. It is evident that younger sediments required more heating than older sediments do because they had less time to cook. The average heat-flow value is 50 mW/m^2 (0.05 W/m^2), and GG is 2.5°C/100 m (0.025°C/m). Therefore, the average thermal conductivity is 2 W/m/°C. (Note that either Kelvin or centigrade can be used because thermal conductivity is measured per degree.)

Robertson (1988) reviewed thermal conductivities for various rock types, including sediments at varying porosity (solidity) values. Shales with 20% sandstones have thermal conductivities of 2.2–2.5 W/m/°C. In general, older sediments tend to be more compacted and have higher k values.

Heat-flow maps of present-day South Texas show that 60 mW/m^2 is a reasonable number today (fig. 7–16). However the prognosticated heat flow for the area shows that during the Cenomanian (100 Ma), the heat could have been as high as 70 mW/m^2. It is also evident that significant variations in heat flow occurred within the main target area, marked by gray in figure 7–16. As discussed earlier (see chap. 2), during illitization, water is expelled. Studies carried out by Beziat et al. (1988) demonstrated that thermal conductivity in clay minerals increases with increasing density and water content. Because during illitization the density of illite is higher and the water content is reduced, the two effects may cancel out the change in thermal conductivity.

Present-day heat-flow maps of Texas are also available from Blackwell et al. (2011) and the Bureau of Economic Geology of Texas. Note the proximity of the Cretaceous reef trend (black lines in fig. 7–16) to the heat-flow high. The reef trend is on the Angelina Caldwell High, most likely a horst block associated with the Triassic rifting. The nature of the high is still a topic of discussion. However, behind the high lie a series of sag basins known as the Karnes and Atascosa Troughs, which contain some of the best producing trends in the Eagle Ford shale. The setting is very similar to the Haynesville cross-section (fig. 4–6).

As stated earlier, sediments can also be sources for radiogenic heating. A recent study in the Gulf of Mexico Basin demonstrated that the Frio Formation shales (Oligocene) can generate 1.8–2.2 $\mu W/m^3$ (McKenna and Sharp, 1998). Hot shales such as the Eagle Ford would be expected to generate 2.5–3.0 $\mu W/m^3$. Additional work is required in this area to determine its impact. A 1 m^2 block with a thickness of 50 m hot shale can generate 0.15 mW/m^2 and therefore does not appear to have a significant influence on the GG of the region.

Regional cross sections should be constructed, and the corrected BHTs should be plotted on the cross-section, to resolve the impact of tectonics on heat flow and, therefore, on temperature gradients. This is exemplified in the region affected by the Rome trough, Marcellus play, in West Virginia and Pennsylvania (Frone et al., 2015). There, the Precambrian Rome graben still exerts a slight influence by increasing the heat flow over the graben. The strongest effect is associated with the west verging thrust of the Appalachian, causing the Marcellus to be in the dry gas window. More than 1,600 BHTs were corrected and integrated to obtain a good map of the heat-flow data in West Virginia.

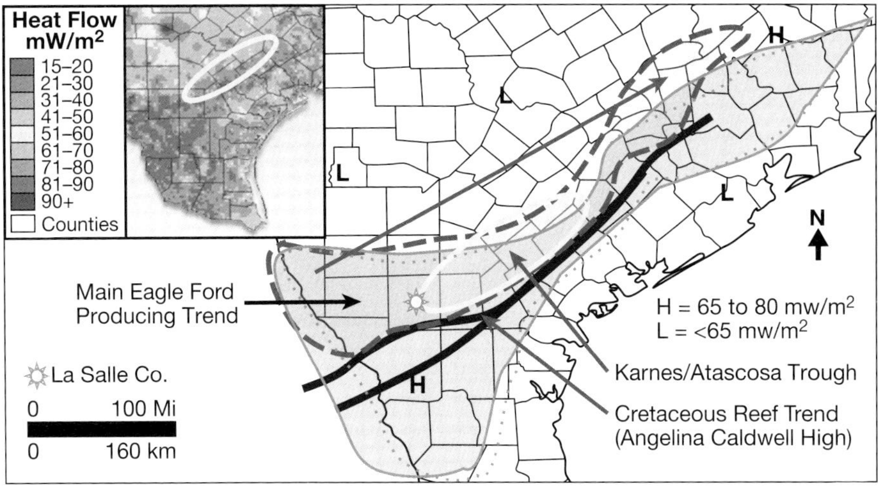

Fig. 7–16. Simplified present-day heat-flow map of Texas overlain with the Eagle Ford producing trend. Note higher heat flow over parts of the trend and variability over the trend. (Inset heat-flow map is reproduced with kind permission from the Bureau of Economic Geology of Texas; any subsequent reproduction requires their permission.)

DETERMINING AMOUNT OF EROSION

Several techniques estimate the amount of erosion in a given region. For erosion to be considered important, it must have a significant impact on the burial history, it should be relatively late in the burial history, and it should probably be in excess of 500 m. The first approach, which is easiest to visualize and fastest to compute, is to estimate the amount of erosion by analyzing seismic lines. By observation of the presence of angular unconformities and extrapolation of the thickness from the adjacent sedimentary wedge, a reasonable minimum value can be estimated for the erosion. Paleontology (as described in chap. 3) can identify the time gap near the objective.

Another common technique relates the vitrinite reflectance (R_o) gradient (fig. 7–17). In general, vitrinite reflectance of the coal-bearing fragments in the cuttings samples increases with depth in a gradual manner. If there is an abrupt shift, an unconformity is suspected. The normal curve is projected, and the difference between the observed and projected trend lines provides an estimate for the amount of erosion.

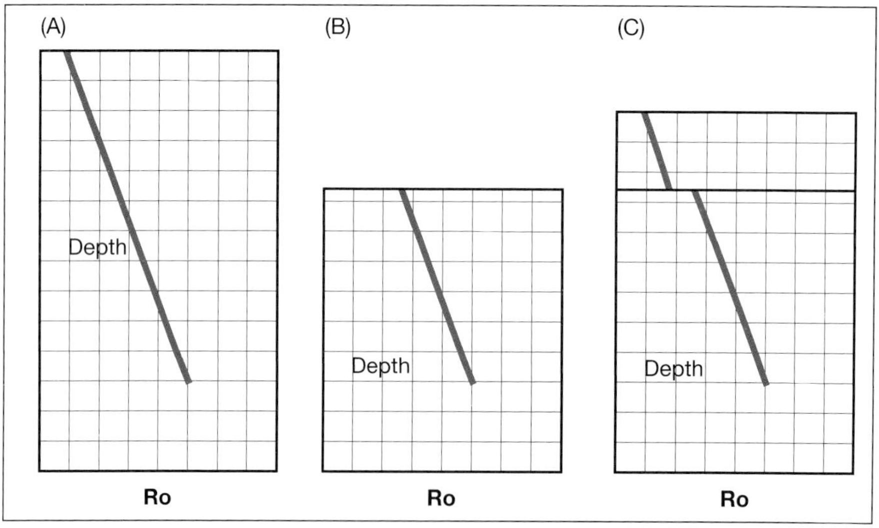

Fig. 7–17. Vitrinite reflectance gradient. Column A shows a normal increase in vitrinite reflectance with depth; column B shows the effect of erosion; and column C shows subsequent deposition.

In all three cases shown in figure 7–17, the exact same GG is used. The separation of the lower line to the upper line in column C is a function of the amount of erosion. This is known as the *depth-equivalence erosion* estimate. In general, the same rule holds for any porosity tool, except that higher porosities correspond to shallower depths. Sonic logs are often used, given that once a higher interval velocity is attained, even when subjected to erosion, the higher velocity is retained in a manner similar to the curve in figure 7–16.

A problem arises if, in addition to erosion, there is also overpressure. The two processes, erosion and overpressure, interfere with the overall signal. In all cases, it is imperative to generate a detailed burial history for several control points by using different techniques. The ages should be controlled with biostratigraphy.

Several mathematical techniques are also available, one of which is described below in detail. When analyzing sonic logs, care must be taken to study similar lithologies only. Usually shales are chosen. A moving average of the sonic log with a bin size of 20 ft and sampling every 2 ft is selected to avoid analyzing noise. The result is plotted with a depth chart along the x-axis and a sonic log response on the y-axis. Because of the hyperbolic decline of porosity in the first 1,000 ft, these data are ignored, and only the general straight-line segment is used.

Magara (1976) presented a similar approach using sonic log data (fig. 7–18). He compared different wells. In the present study, the same well is used, and the Magara computation is applied. It is suggested that the following steps be taken in sequence:

- First, edit the sonic log to eliminate cycle skips and washouts.
- Second, from gamma-ray or SP log (or both), establish a cutoff for the intervals dominated by shales (e.g., $V_{sh} > 50\%$). Note that the section should not be faulted and eroded, because these two effects will interfere with one another.
- Third, generate a moving average for the shale fraction of the well. In Excel, “offset” is 2 ft and “count” is every 20 values (bin size). This generates a curve that is relatively smooth and has a single value as the moving average.
- Fourth, plot the result on a graph with depth on the abscissa and log-derived Δt from the sonic log on the ordinate. As mentioned previously, the section below 1,000 ft has a porosity decrease that is nearly a straight line; as a result, the Δt from the sonic log will also increase in a linear pattern. If there is no unconformity, then the slope of the line will be continuous. If there is no unconformity but a section with overpressure is encountered, then there will be a break at the overpressure and the slowness will increase, meaning that the interval velocity will decrease; the section beneath the overpressure will eventually normalize to the trend line of the upper section. If there is an unconformity, the value beneath the overpressure zone will decrease in slowness. In figure 7–18, there are three families of lines in the plot: 1,200–2,500 ft, normal burial; 4,000–5,000 ft, overpressure; and 5,000–7,000 ft, subunconformity. At the break point, the dashed line in figure 7–18 is the zone of overpressure. The deeper zone

(lower oval in fig. 7–18) has its slope parallel but offset from the upper section. The slope of both lines is M_i.

- Finally, extrapolate both trend lines to the *y*-intercept. Indicate the trend for the underlying curve and extend the trend to the *y*-intercept. Because the interval was buried deeper, the slowness is lower than the overlying section. The slope and y-intercepts will define the equations; the difference between y-intercepts divided by the slope, M_i, of the original sequence provides an indication of possible erosion. Care must be taken because overpressure will also cause a reduction in velocity. The example in figure 7–18 shows the effect of overpressure that should be ignored in erosion analysis.

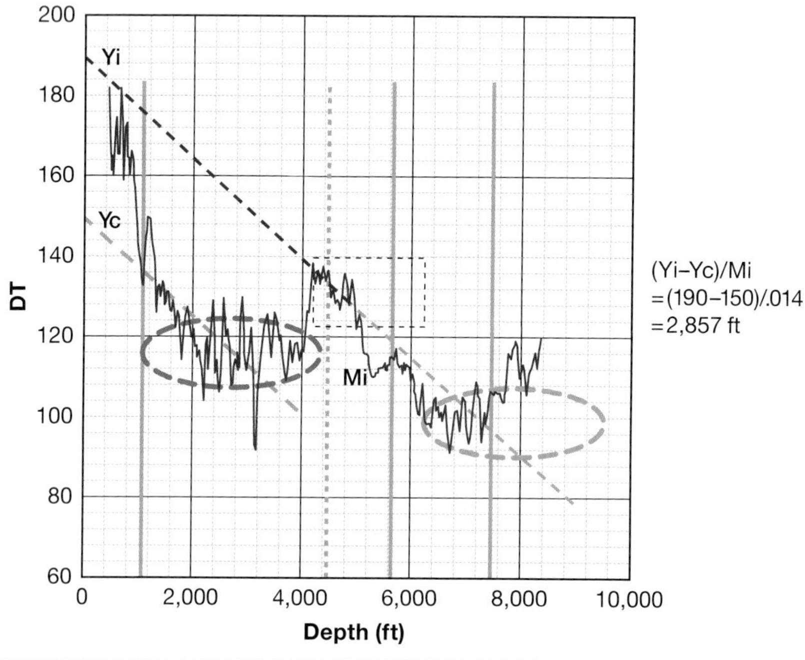

Fig. 7–18. Method used to estimate erosion, based on a single well, using moving average for sonic shale slowness. The dashed line is the zone of the unconformity.

At least two wells should be analyzed using the same technique. Note that long shale sections provide good slope data. Where there is intercalated silt and shale, the velocity data are noisy, both ovals, and should be ignored. In figure 7–18, the calculated amount of erosion is 2,857 ft (870 m). Seismic suggested nearly 2,500 ft of erosion in the same region. If the unconformity (dashed line in fig. 7–18) is significant, then the seismic velocity contrast should be noticeable. Several wells will provide an improved regional slope and thus improve the erosion estimate.

BURIAL HISTORY

A typical tabulation includes the age (in Ma) of the top and base of each rock unit. The general lithology should be noted along with an erosion estimate if available. The premise of the burial history plot, demonstrated graphically, is that sediments were deposited at specific rates. These deposition rates can be determined from the slope of the line on the burial history and should be independently corroborated with biostratigraphy (see chap. 3).

The plot is initiated at the base of the youngest unit. Its age is identified and connected to the depth corresponding to the base of the unit. It is apparent that the underlying units must have been buried at a similar rate; therefore, subsequent lines are parallel over the same time intervals. Each subsequent unit follows the same construction, except that each starting point at depth corresponds to the onset age of the sedimentation and the base corresponds to its age.

Lopatin (1971) and, subsequently, Waples (1980) described an approximation of the rate of increased maturation and average rate of activation of the kerogen as a function of residence time and corresponding increasing increments of temperature. The simplified increase is exponential by a factor of 2^n, where *n* begins with –7 and increases by a factor of 1 at every 10° increment. Depending on the ambient surface temperature, the plot can begin at either 20°C or 30°C in continental areas. In the model described, 30°C is used as a starting point (fig. 7–19A and B). All values on the plot are in SI units. The n factor at 30°C–40°C is –7 (2^{-7}). At each increment of 10° the index adds +1; therefore, 40°C–50°C is 2^{-6}, etc. The factor is multiplied by the retention time or residence time within that temperature bracket and the cumulative value is summed (Waples, 1980). For an initial, manual model, a simple GG could be used: 2°C/100 m yields a 10° change at every 500 m, 2.5°C/100 m every 400 m, 4°C/100 m every 250 m. A global average GG is 2.5°C/100 m. In the described model, an Early Paleocene source rock is analyzed. In the first case, there is continued sedimentation. In the second case, a 500 m erosion event took place during the Late Eocene to Early Oligocene. It is important to run a model where there is no erosion to ensure that the proper rates of deposition are used (slope of the lines). The downspout on the burial history (fig. 7–19B) reflects that the erosion event is bracketed by the age of the last appearing fauna under the unconformity and the age of the first appearance over the unconformity. The effects on the well are significant. For detailed burial histories all of the factors mentioned above should be considered: heat flow changes through time, thermal conductivity changes through time, the impact of overpressure, and using the appropriate kerogen activation energy model.

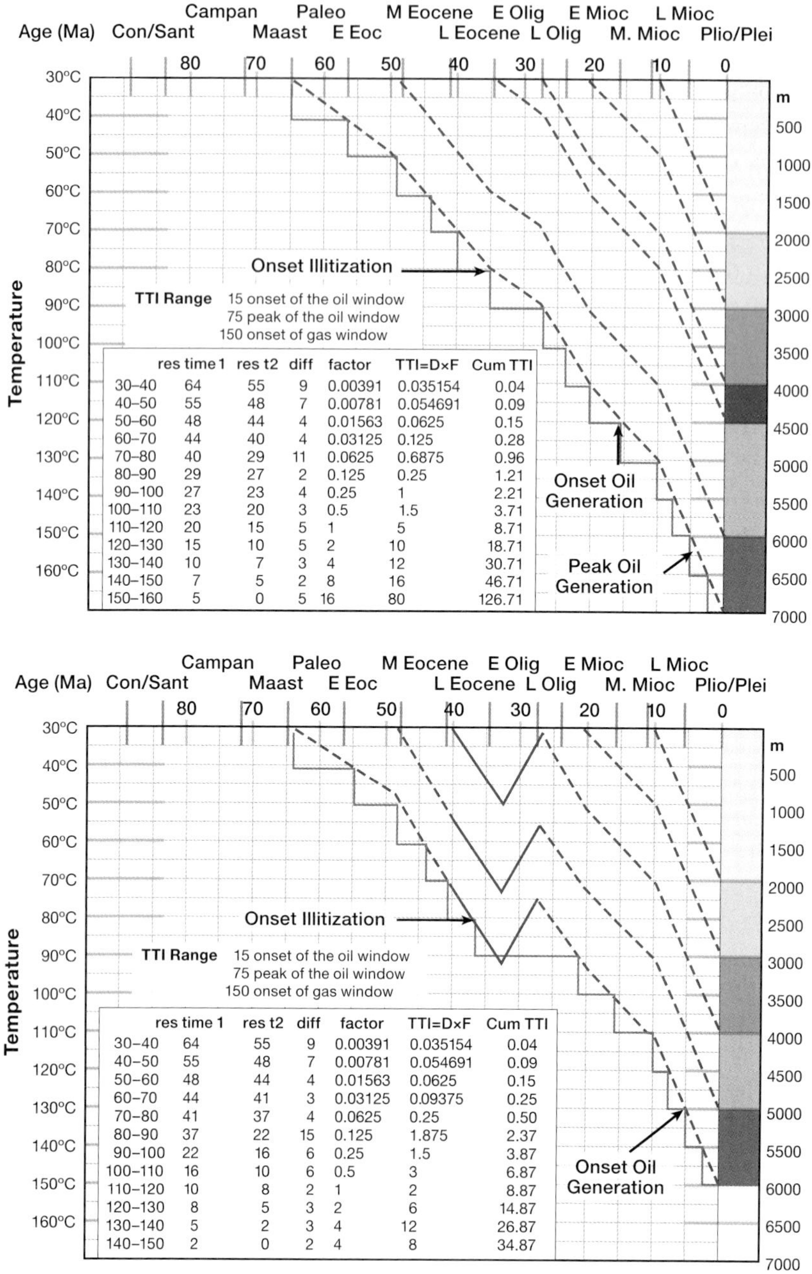

Fig. 7–19. Burial history diagrams. The same data set is used in panels A and B; however, B has a 500 m erosional event during the Late Eocene. Cumulative TTI values are listed in each chart. For both A and B, illitization occurs 35 Ma. For A, onset oil is 26 Ma, and peak oil is 4 Ma; for B, onset oil is 5 Ma and peak oil is not reached. Reference data: Illitization at 80°C, onset of oil at TTI 15 (equivalent R_o, 0.65), peak generation at TTI 75 (equivalent R_o, 1.0), and onset of gas generation at TTI 150–160 (equivalent R_o, 1.35).

For the initial model, a constant GG of 2°C/100 m was used. It is evident that illitization has occurred; hence, overpressure is very likely in the region at depths greater than 2,500 m. With knowledge of the lithology and sediment characteristics above and below the overpressure zone, the thermal conductivity of each can be determined. As previously discussed, an improved estimate can be made of the GG prior to the illitization and the calculation rerun after illitization.

As described previously, heating kerogen is an irreversible reaction. The same is true of illitization. Once a sediment has been buried, it retains its maximum attained maturity. If the sediment is exhumed partially or completely, the cooking of the kerogen will reinitiate once the previously attained maturation level is surpassed. If BHT data are available in overpressure intervals, then the new GG can be introduced after illitization, and the calculation can be continued.

The results of the burial diagram should always be summarized on a *critical moment chart* (fig. 7–20). The critical moment chart enables any reader to understand all the timing events affecting the reservoir.

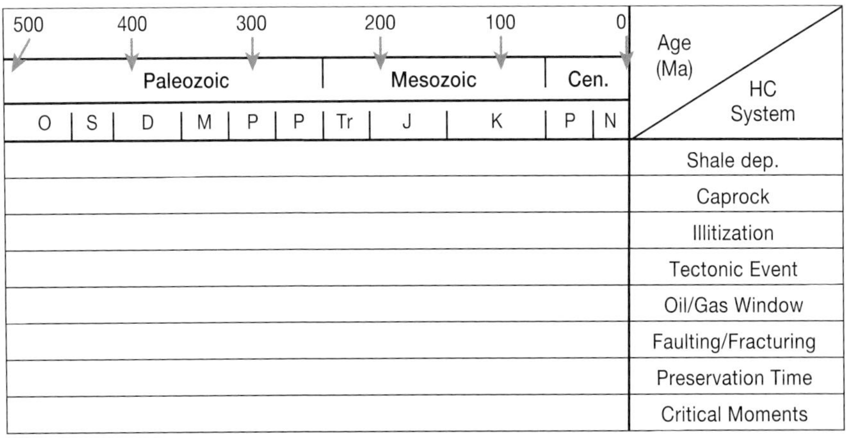

Fig. 7–20. Critical moment chart summarizing all the events affecting the reservoir. Additional categories can added, such as time of illitization and timing of overpressure. (Based on Magoon and Beaumont, 1999)

CATAGENESIS

In chapter 2, we discussed the diagenesis of smectite as it is converted to illite and briefly mentioned that carbonates and other labile sediments such as ash beds can alter diagenetically. As a general rule, diagenesis is significant and evident down to the oil generation window. Below that threshold, we refer to the alteration of organic matter entering the catagenesis phase, which is the focus of the following discussion.

As mentioned previously, as a kerogen cracks, the molecular composition of the products change. Also, as the mineralogy is being altered, the organic products react with the chemicals and changes in water chemistry. When hydrocarbons begin to migrate into organic pores, there is fractionation differentiating the chemistry of fluids as the hydrocarbons migrate (Jarvie, 2017). It is important to consider the alteration products when comparing hydrocarbons from different wells along a trend. Loucks et al. (2009) have described spongy porosity (fig. 7–21) and suggested that it might be a result of catagenesis. Many of the products described thus far in the present chapter are the results of catagenesis.

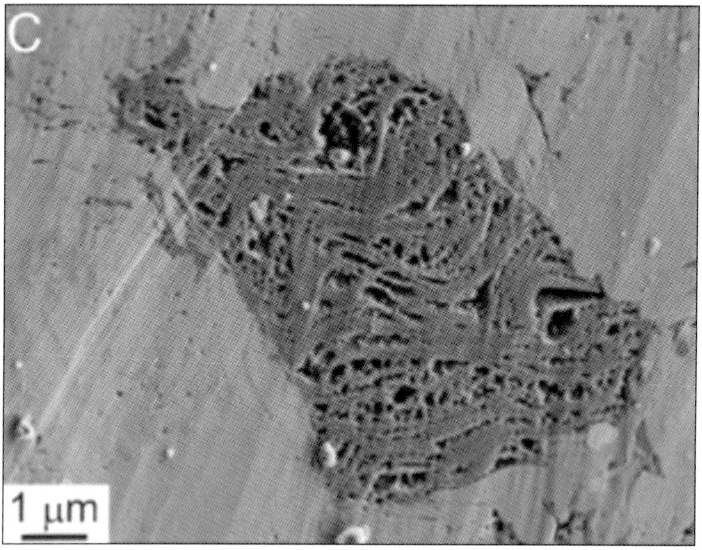

Fig. 7–21. Micropores in kerogen creating spongy porosity during maturation (Reproduced with kind permission of the Bureau of Economic Geology of Texas; Loucks et al., 2009)

GEOCHEMICAL PETROPHYSICS

Once the organic chemistry of a reservoir has been confirmed from oil samples and cores, the next logical step is to evaluate the petrophysical response of the oil shale followed by an extrapolation to the seismic section. Previous discussions on petrophysics (see chap. 5) have shown that richness in organic material should be detectable on logs. The basic logs that can provide information on organic content of shales are gamma-ray logs to establish the shale content, density determination, sonic logs, and resistivity estimates. Other logs can also be used.

The thermal maturity of the shales also plays a key role in understanding the log response. Passey et al. (1990) have considered the effects of resistivity, sonic, and maturity. Hester and Schmoker (1985) performed a comparative evaluation of density and gamma ray between proximal organic rich and organic poor sediments. In all cases, the results must be calibrated with a selection of wells and outcrops in the area of study. Calibration is critical because, for example, density and sonic logs respond to the overall mineral composition of the sample, which may include three or more constituents. Differentiating the significance of the TOC component may be difficult. Thermal maturity plays a key role in quantification of TOC. As discussed previously, as the kerogen matures and migrates, the TOC content is reduced. The volume of free hydrocarbons directly influences the resistivity response. Hot shales may be the result of high concentrations of uranium and not necessarily only TOC or maturity. Borehole effects—and their correction—pose an additional problem, particularly in organic-rich shales. This is especially problematic with logs that require borehole wall contact.

Hester and Schmoker (1985) derived the following empirical relationship:

$$TOC = \frac{100\rho_o(\rho_b - 0.9922\rho_{mi} - 0.039)}{R\rho_b(\rho_o - 1.135\rho_{mi} + 0.675)} \tag{7.9}$$

where ρ_b is the measured bulk density, ρ_o is the density of the organic matter, ρ_{mi} is the matrix and pore fluid density, and $R\rho_b$ is the density ratio of organic matter to organic carbon.

Equation (7.9) has been used widely. However, the indices are based on Hester and Schmoker's (1985) calibration for the Woodford shale of Oklahoma. These indices should thus be adjusted to fit the data for each individual region. Several independent calculations should be compared. Furthermore, equation (7.9) does not consider the maturation effect.

The most commonly used log-derived TOC estimation is the Exxon approach, using the Passey equation (fig. 7–22) (Passey et al., 1990). The premise of the Passey equation is that an organic-rich shale, given a critical LOM (level of organic maturity) and an appropriate density, will generate a resistivity response that can be related to TOC (fig. 7–23). As noted in chapter 5, the sonic log response can be related directly to density (Gardner et al., 1974). The approach used by Passey is to overlay the resistivity and sonic logs on the same plot and relate the overlap to the estimated TOC in the well. In all areas, the log response must be calibrated with known TOC values and the indices adapted to ensure an accurate estimation. The Passey equation is also known as the $\Delta \log R$ plot (fig. 7–24).

$$\Delta \log R = \log_{10}\left[\frac{R}{R_{ns}}\right] + K \cdot (P - P_{ns})$$

$\Delta \log R$	is curve separation
R	is measured formation resistivity
R_{ns}	is resistivity of non-source shales
P	is porosity log reading
P_{ns}	is the porosity log reading in non-source shales, and
K	is a scale factor dependent on the porosity log measurement units

TOC is calculated using: $TOC = \Delta \log R \cdot 10^{(2.297-.1688 \cdot LOM)}$

Fig. 7–22. TOC estimation using Passey equation. The result is commonly known as Δ log *R*. LOM has equivalency to R_o and TTI (Adapted from Passey et al., 1990; reproduced with kind permission from the AAPG)

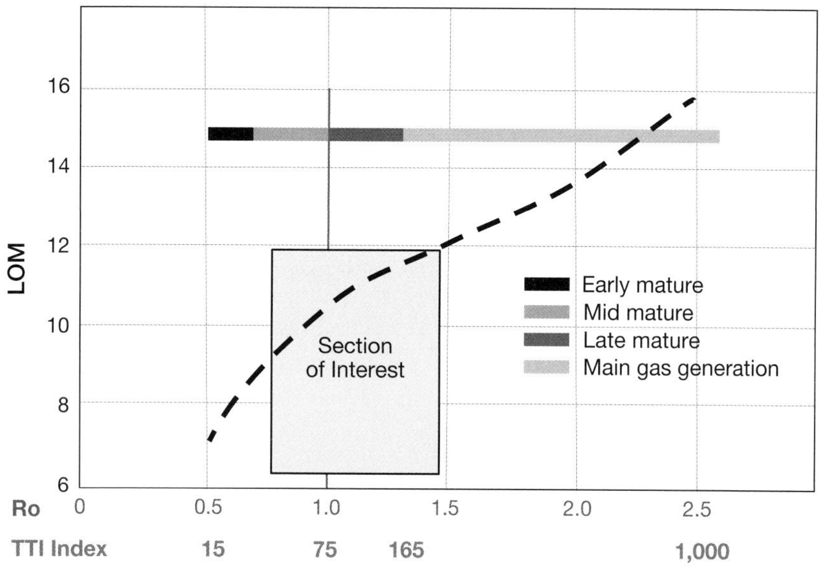

Fig. 7–23. Correlation chart relating LOM and TTI and R_o (Adapted from Passey et al., 1990)

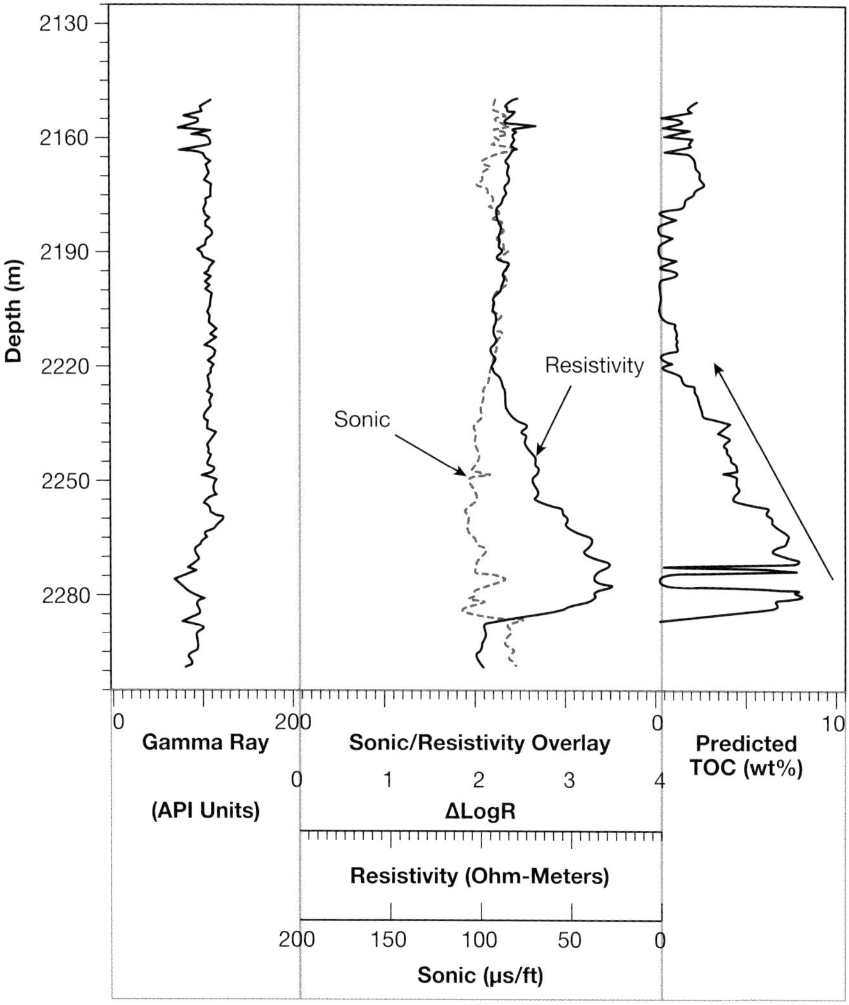

Fig. 7–24. Sample logs showing typical Δ log *R* plots and correlation gamma-ray log (Adapted from Passey et al., 1990; reproduced with kind permission from the AAPG)

A conversion chart provides the appropriate LOM value for the well in question (fig. 7–23). Once the TOC has been estimated in a region, maps can be generated for each interval of interest. For example, the corrected TOC distribution for the Lower Bakken is provided by the U.S. Geological Survey and Jin and Sonnenberg (2013) (fig.7–25). Note the wide variation in values for TOC; this will be critical in estimating reserves.

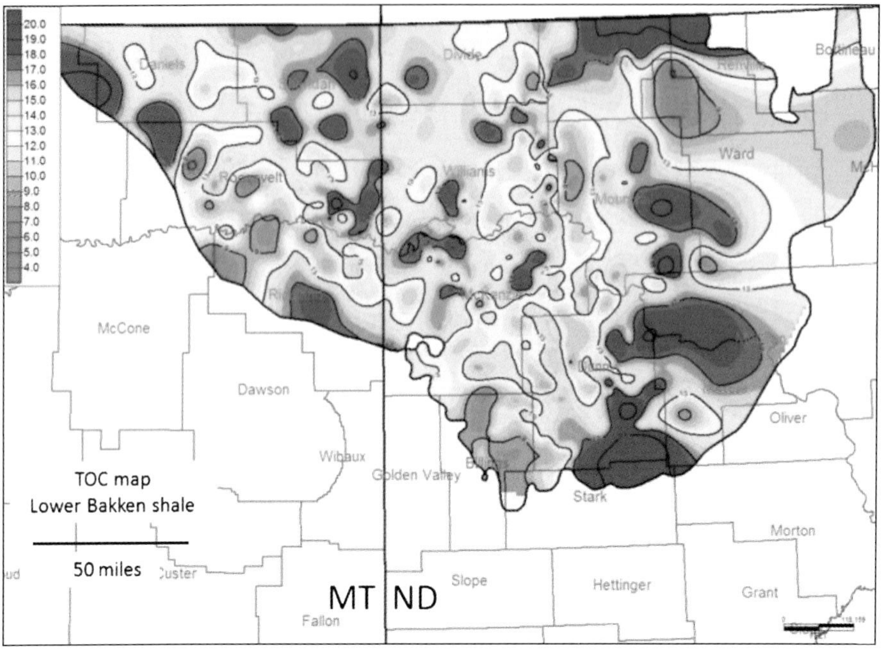

Fig. 7–25. TOC distribution map for the Lower Bakken Formation, Williston Basin, Eastern Montana/Western North Dakota (Jin and Sonnenberg, 2013). Note the uneven distribution of the higher TOC values. (Adapted from Jin and Sonnenberg, 2013; reproduced with kind permission of Stephen A. Sonnenberg, Colorado School of Mines)

Once the regional TOC is available, the porosity and thickness of the organic-rich strata can also be mapped, and the sweet spot can be established. However, it is very difficult to estimate average TOC for a region. In fact, within a 50-mile (80 km) radius in any county, the range may vary from 4% to 20% TOC.

CHEMOSTRATIGRAPHY

Studies of elemental composition differences along a wellbore can further assist in correlations, as well as provide information on the presence of source rocks (fig. 7–26). Specific elements can be diagnostic for anoxic conditions, as is the case with the *enrichment factor of vanadium* (EFV) track. High values of EFV are indicative of anoxic conditions. EFV can also be used to ensure that the horizontal well is drilling along the correct pathway (Ratcliffe et al., 2012). High uranium concentration is an indication of more shale and potential for organic material to trap uranium in porphyrin molecules. In addition, because it is common to observe the presence of ash layers, titanium concentrations may serve as correlation markers even though the ash layer is thin (Calvin et al., 2017).

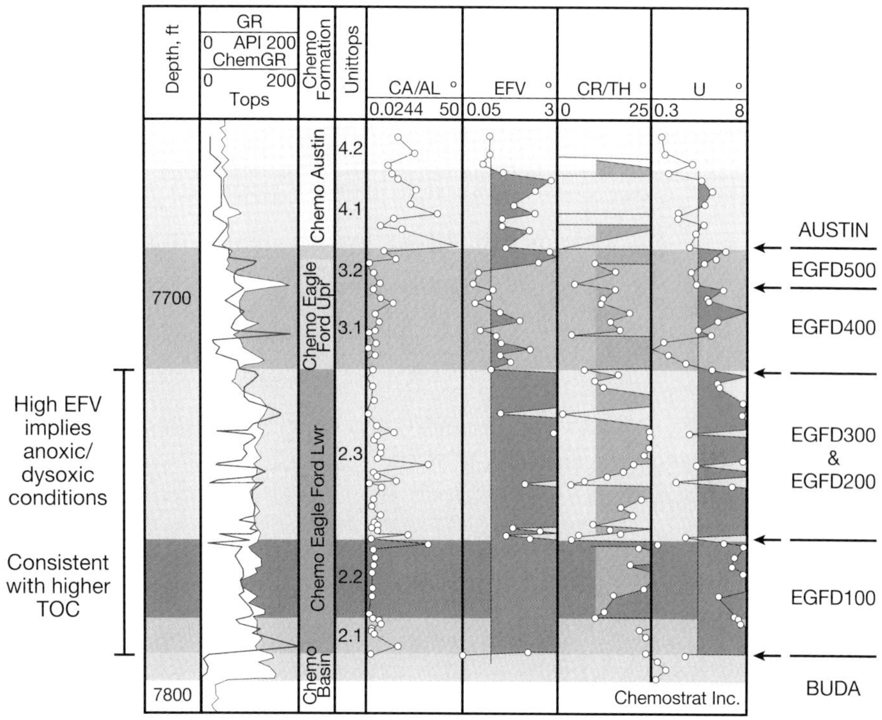

Fig. 7-26. Application of chemostratigraphy to correlations and regional geologic information (Reproduced with kind permission of John A. Breyer and the AAPG; Breyer et al., 2011)

SEISMIC APPLICATION TO GEOCHEMISTRY

The application of seismic to geochemistry is a topic that requires further investigation. The concept of thermal resistivity studies is well known. It states essentially that, if the interval thickness is known and the thermal conductivity of the unit is known and its previous history can be estimated, then, given a specific heat flow, temperature isotherms can be computed. However, this can be carried out only on depth-converted seismic. For example, Gosnold et al. (2012) have provided detailed thermal conductivity values for various stratigraphic units in the Williston Basin. With knowledge of the heat flow in the area, isotherms can be generated and overlain on the seismic. If the tuning thickness is adequate, then the density inversion seismic can be calibrated to the TOC from wells, enabling extrapolation of similar conditions on the three-dimensional seismic cube.

SUMMARY

A general introduction to organic chemistry—in particular, as applied to hydrocarbons—is essential to understand the terminology used in hydrocarbon analysis and in reports provided by contractors. Catagenesis describes the diagenesis of kerogen (source material for hydrocarbons) in the subsurface.

Oils are classified as alkanes, naphthenes, and aromatics. Alkanes have the highest hydrogen index (saturated). Gases are low-carbon-number alkanes: methane is C1, ethane is C2, butane is C3, and so forth. Isomers are branched alkanes. Pristane and phytane are important isomers that help define aspects of the kerogen.

Gas chromatography, mass spectrometry, and Rock-Eval pyrolysis are used to study oils, gases, and kerogen. The last of these allows for TOC calculations and definition of kerogen types. PVT analyses are a basic assay tool for all hydrocarbon evaluations.

Heat flow and burial histories can be used to determine timing and level of maturity attained by the source rock. Techniques have been described for determining the amount of erosion affecting the area of study.

The important role of changes in thermal conductivity is reviewed. The potential role of catagenesis in porosity creation is described. The Passey equation to determine TOC from logs is explained graphically. TOC is not distributed in a uniform pattern over a basin.

Chemostratigraphy improves the definition of potential source intervals and aides in correlation. Seismic data can be used to assist lateral prediction of geochemical conditions in the area of study.

REFERENCES

Allen, Philip A., and John R. Allen. 2013. *Basin Analysis: Principles and Application to Petroleum Play Assessment.* 3rd ed. Chichester, UK: Wiley-Blackwell.

Bartok, Peter. 1993. "Prebreakup geology of the Gulf of Mexico-Caribbean: Its relation to Triassic and Jurassic rift systems of the region." *Tectonics.* 12: 441–459.

Beziat, Alain, Michel Dardaine, and Victor Gabis. 1988. "Effect of compaction pressure and water content on thermal conductivity of some natural clays." *Clays and Clay Minerals.* 36: 462–466.

Bissada, K. K., B. J. Katz, S. C. Barnicle, and D. J. Schunk. 1990. "On the origin of hydrocarbons in the Gulf Of Mexico Basin—a reappraisal." In *Gulf Coast Oils and Gases—Their Characteristics, Origin, Distribution, and Exploration and Production Significance: Proceedings of the Ninth Annual Research Conference, Gulf Coast Section Society of Economic*

Paleontologists and Mineralogists Foundation, Dietmar Schumacher and Bob F. Perkins (eds.). Austin, TX: Bureau of Economic Geology, 163–171.

Blackwell, D. D., M. C. Richards, Z. S. Frone, J. F. Batir, M. A. Williams, A. A. Ruzo, and R. K. Dingwall. 2011. "SMU geothermal laboratory heat flow map of the conterminous United States, 2011." https://www.smu.edu/Dedman/Academics/Programs/GeothermalLab.

Breyer, John A., Richard Denne, Jonathan Funk, Tobi Kosanke, and Joan Spaw. 2013. "Stratigraphy and sedimentary facies of the Eagle Ford shale (Cretaceous) between the Maverick Basin and the San Marcos Arch, Texas, USA." Paper presented at the AAPG Annual Convention and Exhibition, Pittsburgh, May 19–22. Search and Discovery article 50899. www.searchanddiscovery.com/pdfz/documents/2013/50899breyer/ndx_breyer.pdf.html.

Buffler, Richard T., and Dale S. Sawyer. 1985. "Distribution of crust and early history, Gulf of Mexico basin." *Gulf Coast Association of Geological Societies Transactions*. 35: 333–344.

Calvin, Christina, H. Gamero-Diaz, L. Mosse, R. Malpani, C. K. Miller, J. Xu, and K. Fisher. 2017. "Diagenesis of ash beds in mudrocks and their impact on production." Paper presented at Houston Geological Society Applied Geoscience Conference, The Woodlands, TX, March 7–8.

Didyk, B. M., B. R. T. Simoneit, S. C. Brassell, and G. Eglinton. 1978. "Organic geochemical indicators of palaeoenvironmental conditions of sedimentation." *Nature*. 272: 216–222.

Frone, Z. S., D. D. Blackwell, M. C. Richards, and M. J. Hornbach. 2015. "Heat flow and thermal modeling of the Appalachian Basin, West Virginia." *Geosphere*. 11: 1279–1290.

Gardner, G. H. F., L. W. Gardner, and A. R. Gregory. 1974. "Formation velocity and density—the diagnostic basics for stratigraphic traps." *Geophysics*. 39: 770–780.

Goddard, Donald. 2016. "Bossier-Haynesville shale, North Louisiana." *LSU Center for Energy Studies*. Personal communication, correspondence, June 12, 2016.

Gosnold, William D. 2017. The global heat flow data base of the International Heat Flow Commission. Personal communication with author by correspondence (4/2/17).

Gosnold, William D., Mark R. McDonald, Robert Klenner, and Daniel Merriam. 2012. "Thermostratigraphy of the Williston Basin." *Transactions Geothermal Resource Council*. 36: 663–670.

Harrison, W. E, K. V. Luza, M. L. Prater, and P. K. Cheung. 1983. "Geothermal resource assessment in Oklahoma." *Oklahoma Geological Survey*, Special Publication 83-1.

Hester, T. C., and J. W. Schmoker. 1985. "Oil and gas investigations chart." Report #OC-0126 in *Selected Physical Properties of the Bakken Formation, North Dakota and Montana Part of the Williston Basin*, T. C. Hester and J. W. Schmoker (eds.). Reston, VA: U.S. Geological Survey.

Jarvie, Daniel M. 2001. "Williston Basin petroleum systems: Inferences from oil geochemistry and geology." *Mountain Geologist*. 38: 19–41.

———. 2017. "Interaction of organic and inorganic matter: Impact on composition and fractionation of petroleum." Paper presented at Houston Geological Society Applied Geoscience Conference, The Woodlands, TX, March 7–8.

Jarvie, Daniel M., and R. J. Hill. 2011. "Understanding unconventional resource potential by conventional petroleum system assessment." Paper presented at the West Texas Geological Society Fall Symposium, Midland, TX, September 28–30. Search and Discovery article 40840. www.searchanddiscovery.com/documents/2011/40840jarvie/ndx_jarvie.pdf.

Jin, Hui, and Stephen A. Sonnenberg. 2013. "Characterization for source rock potential of the Bakken shales in the Williston Basin, North Dakota and Montana." Paper presented at the Unconventional Resources Technology Conference (URTeC), Denver, August 12–14.

Lopatin, N. V. 1971. "Temperature and geologic time as factors in coalification" [in Russian]. *Izvestiya Akademii Nauk: Seriya Khimicheskaya*. 3: 95–106.

Loucks, Robert G., Robert M. Reed, Stephen C. Ruppel, and Daniel M. Jarvie. 2009. "Morphology, genesis, and distribution of nanometer-scale pores in siliceous mudstones of the Mississippian Barnett Shale." *Journal of Sedimentary Research*. 79: 848–861.

Magara, Kinji. 1976. "Thickness of removed sedimentary rocks, paleopore pressure, and paleotemperature, southwestern part of Western Canada Basin." *AAPG Bulletin*. 60: 554–565.

Magoon, Leslie B., and Edward A. Beaumont. 1999. "Petroleum systems." Chapter 3 in *Treatise of Petroleum Geology/Handbook of Petroleum Geology: Exploring for Oil and Gas Traps*. Tulsa: American Association of Petroleum Geologists.

Marton, György, and Richard T. Buffler. 1994. "Jurassic reconstruction of the Gulf of Mexico Basin." *International Geology Review*. 36: 545–586.

McCain, William D. 1990. *The Properties of Petroleum Fluids*. Tulsa: PennWell.

McKenna, Thomas E., and John M. Sharp Jr. 1998. "Radiogenic heat production in sedimentary rocks of the Gulf of Mexico Basin, South Texas." *AAPG Bulletin*. 82: 484–496.

McKenzie, Dan. 1978. "Some remarks on the development of sedimentary basins." *Earth and Planetary Science Letters*. 40: 25–32.

Midttømme, K., E. Roaldset, and P. Aagaard. 1998. "Thermal conductivity claystones and mudstones of selected from England." *Clay Minerals*. 33: 131–145.

Nordeng, Stephan H., Timothy O. Nesheim. 2011. "Determination of subsurface temperatures and the fraction of kerogen converted to petroleum within the Rauch Shapiro Fee #21-9, Billings County, North Dakota." North Dakota Geologic Investigation No. 146.

———. 2012. "Basic geochemical evaluation of unconventional resource plays." *Geo News*. North Dakota Department of Mineral Resources. January, 2012.

Passey, Q. R., S. Creaney, J. B. Kulla, F. J. Moretti, and J. D. Stroud. 1990. "A practical model for organic richness from porosity and resistivity logs." *AAPG Bulletin*. 74: 1777–1794.

Peters, Kenneth E., and Mary Rose Cassa. 1994. "Applied source rock geochemistry." Chapter 5 in *The Petroleum System—from Source to Trap*. AAPG Memoir 60. Tulsa: The American Association of Petroleum Geologists, 93–120.

Pindell, James L. 1985. "Alleghenian reconstruction and subsequent evolution of the Gulf of Mexico, Bahamas, and Proto-Caribbean." *Tectonics*. 4: 1–39.

Pixler, B. O. 1969. "Formation evaluation by analysis of hydrocarbon ratios." *Journal of Petroleum Technology*. 21: 665–670.

Price, Leigh C., and Julie LeFever. 1994. "Dysfunctionalism in the Williston Basin: The Bakken/mid-Madison petroleum system." *Bulletin of Canadian Petroleum Geology*. 42: 187–218.

Ratcliffe, K. T., J. Woods, and C. Rice. 2012. "Determining well-bore pathways during multilateral drilling campaigns in shale resource plays: An example of chemostratigraphy from the Horn River Formation, British Columbia, Canada." Paper presented at the Eastern Australian Basins Symposium IV, Brisbane, September 10–14.

Robertson, Eugene C. 1988. *Thermal Properties of Rocks: Open-File Report 88-441*. Reston, VA: U.S. Geological Survey.

Ryan, Lawrie, and Roger Norris. 2014. *Cambridge International AS and A Level Chemistry Coursebook with CD-ROM*. Cambridge, UK: Cambridge University Press.

Tissot, Bernard P., and Dietrich H. Welte. 1984. *Petroleum Formation and Occurrence*. Berlin: Springer.

Unrau, Sean, and Bill Nagel. 2012. "Using compositional gas data for horizontal drilling and practical detection of natural gas liquids (NGL)." Paper presented at GeoConvention 2012, Calgary, AB, May 14–18. www.cspg.org/cspg/documents/Conventions/Archives/Annual/2012/104_GC2012_Using_Compositional_Gas_Data_for_Horizontal_Drilling.pdf.

Waples, Douglas W. 1980. "Time and temperature in petroleum formation: Application of Lopatin's method to petroleum exploration." *AAPG Bulletin*. 64: 916–926.

———. 1985. *Geochemistry in Petroleum Exploration*. Boston: International Human Resources Development Corp.

Wust, Raphael, A. J., Brent Nassichuk, Ron Brezovski, and Paul Hackley. 2013. "Vitrinite reflectance versus pyrolysis Tmax data: Assessing thermal maturity in shale plays with special reference to the Duvernay shale play of the Western Canadian Sedimentary Basin, Alberta, Canada." SPE Asia-Pacific Unconventional Resource Conference and Exhibition-. Society of Petroleum Engineers.

8

ENGINEERING

TAKEAWAY

- Predrill preparation—anticipating problems
- Static model is used to design geosteering, properly model faults and preexisting fractures, and assist in the selection of target fracs
- Pilot well—confirmation of stratigraphy, logging for rock properties, minifrac, acidizing, and use as recorder for microseismic
- Overburden stress determination
- Vertical and horizontal stress, Poisson's ratio, and Young's modulus,
- Application of uniaxial and triaxial compress
- Pore pressure
- Fracture gradient, effective stress, and frac pressure
- Rock physics
- Rock mechanics
- Global stress map and local effects on well orientation
- Mohr circle
- Leak-off test
- Microseismic
- Hydraulic fracking process
- Capillary pressure and wettability
- Postdrill review
- Produced water disposal and aquifer management

WELL PLANNING

It is important not only to explore and produce the shale resources but also, based on an understanding of their geology, to drill them efficiently and hydraulically fracture (frac) them so as to maximize production. It is in the engineering phase of drilling and production that the fully integrated geoscience effort brings forth its results. Physical and chemical conditions can be predicted. Seismic data have allowed, with a high degree of accuracy, extrapolation of reservoir characteristics from known conditions to the target reservoir. In any drilling activity, there are three basic stages:

- Predrill preparation
- Drilling activity
- Postdrill and production analysis

Geoscientists will have greatest impact in the pre- and postdrill stages. During the drilling and completion of the well, the drilling engineer is the primary driver, and the geoscientists are focused primarily on sample description, geosteering of the well, and interpreting the microseismic events.

The focus of this chapter is on the predrill stage, although there will also be some discussion of postdrill production. The reservoir engineer, in conjunction with the geoscientists, manages the postdrilling analysis. Postdrilling also includes excess-water management. For a more detailed discussion on drilling procedures for oil wells, refer to the work of Bommer (2008). In all cases, before a horizontal well is drilled in an area, it is important to drill vertical pilot wells. Aside from the obvious depth confirmation to direct the horizontal well, several other data sets provided by the vertical well are essential to achieve economic success.

In shale resource plays, the form of stimulation most commonly applied is hydraulic fracking in a horizontal well. The very nature of fracking requires an in-depth understanding of pressure and stress regimes affecting the reservoir target. Before any exploration venture, two basic steps must be taken: first, do a historical review of the exploration in the area; and second, analyze and synthesize all the available data. It is important to integrate literature on the region with a review of the exploration experience in the region. Drilling problems, completion issues, expected production potential and other concerns should be anticipated. Objectives must be clearly defined; otherwise, even if detected, issues may still be overlooked. The data analysis should be sequential, focusing first on the most critical elements.

The previous chapters dealt with the geological information, the geochemistry of the expected target, and its geophysics and petrophysics. The current chapter will focus on the engineering data required prior to drilling. What needs to be done before drilling is to integrate and analyze all the information described in the previous chapters to generate a comprehensive drilling strategy.

PREDRILLING THE HORIZONTAL WELL—INCLUDING THE PILOT WELL

Prior to drilling the horizontal well, at a minimum, studies should include the following steps, each of which will be described later in this chapter:

- Develop an adequate static model for the reservoir to ensure an optimum target
- Evaluate pressures and stresses in surrounding wells with emphasis on determining overburden stress, pore pressure, and overpressure
- Estimate fracture gradient
- Determine the matrix stress coefficient, K_{mc}, that applies to your project
- Apply the Matthews-Kelly stress coefficient, K_{mk}
- Apply the Eaton equation to improve pore pressure prediction and frac gradient
- Evaluate drilling conditions in the area
- Analyze the drilling exponent, d_e
- Apply modifications of the Eaton equation to integrate the seismic data
- Understand the direction of regional and local stress regimes to ensure proper orientation of the wells
- Propose the required coring and logging suite to maximize the evaluation of the target horizon—in particular, fullbore formation microimaging (FMI) to identify the intensity and orientation of open fractures, and dipole sonic to calculate the geomechanical properties and seismic response for the interval under investigation
- Drill the pilot well
- Propose a minifrac (also known as a *leak-off test*) to establish fracture gradient and fracture orientation, as well as their geometries and permeability estimates, and integrate results with those of previous minifracs
- Apply the Mohr circle to better understand the effect of the stresses on the sediments under investigation
- Determine rock strength using uniaxial and triaxial compression on core samples

- Define a proper work flow to ensure that subsequent wells are drilled more efficiently
- Prepare an adequate well design
- Estimate the development costs (see also chap. 9)
- Decide on the economic viability of the project

STATIC MODEL

Prior to drilling of an initial horizontal well, a static reservoir model should be generated. By definition, this model assumes that physical and chemical properties of a reservoir can be extrapolated from known control points to the target objective based on sets of rules determined from experience in the area and constrained by geostatistical tools. The static model integrates all the parameters outlined in the previous chapters and includes a detailed analysis of the structure, lithology, facies, deposition, and diagenesis (integrating surface geology, core data, and log facies analysis). The data are described in terms of both a sequence stratigraphic and rock physics framework and are integrated with the seismic interpretation. An important additional display includes detailed fault geometries and potential fracture zones. Both should be avoided when targeting frac targets.

Once production is in place, the same parameters are updated into a dynamic flow model (not dealt with in this study). The dynamic model is evaluated by the reservoir engineer and involves flow characteristics, fluid types, fluid phase, and pressure decline. The seismic data provide the detailed mapping that defines the framework for the static model and are used to define the geosteering parameters. Any potential drilling issues, such as preexisting faults and fractures, can be highlighted.

The previously described model provides engineers with an upscale geocellular model to better plan the upcoming well, generate a production forecast, and estimate the economics of the project. An example of a program used for a projection analysis is Schlumberger's Eclipse program, which has been applied in the industry for over 25 years. With the advent of extended horizontal hydraulic fracturing, there is a need to understand how the preexisting fractures behave and determine their potential size, areal density, orientation, and intensity of fracturing. The geocellular model should include a prediction on how the potential fractures will enhance drainage. Once the fractures are modeled, it becomes possible to define the dimensions of the individual drainage blocks and avoid unnecessary fracs.

Previous experience in the region is vital in developing an appropriate model, and neural networks provide the mechanism to evaluate the model. At the very least, the geocellular model contains cell sizes equivalent to the proposed drainage block—or, better still, a few cells for each proposed drainage block. As the data are analyzed, the model can be modified, and the relationship with each adjacent cell can be indicated as either corner-to-corner or edge-on-edge (fig. 8–1). The cells may also be deformed if subsequent drilled wells confirm the necessity for such a change.

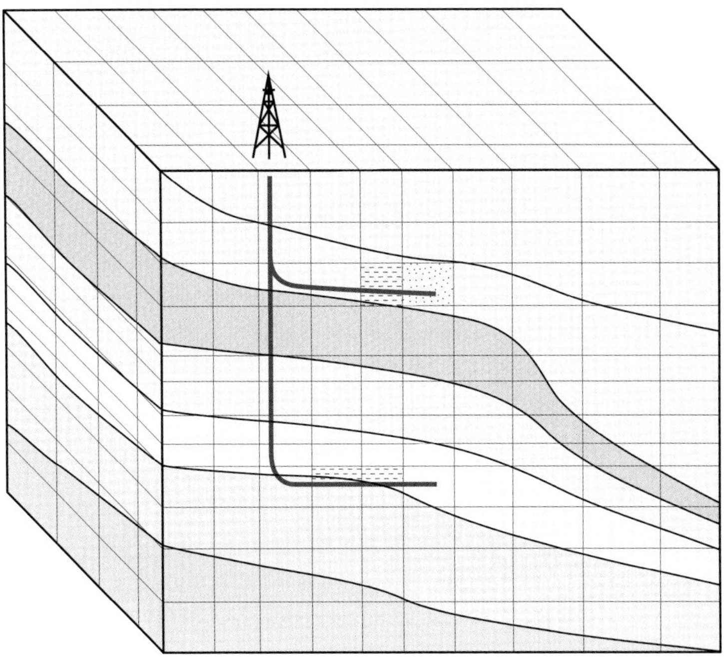

Fig. 8–1. Geocellular model. The model can also be applied to a lithofacies map or any other specific property and can be in cross-section or three-dimensional. In this example, dashed cells are more favorable than dotted. Drilling and completions are based on the interpretations.

From the geocellular model, a detailed cross-section is generated where, as discussed above, each cell has the dimensions of the flow unit. Thus, the precise target for the optimum hydraulic fracture location can be programmed (fig. 8–2). Specific attributes can be color-coded; cells to be avoided due to faulting or other issues can be indicated as well.

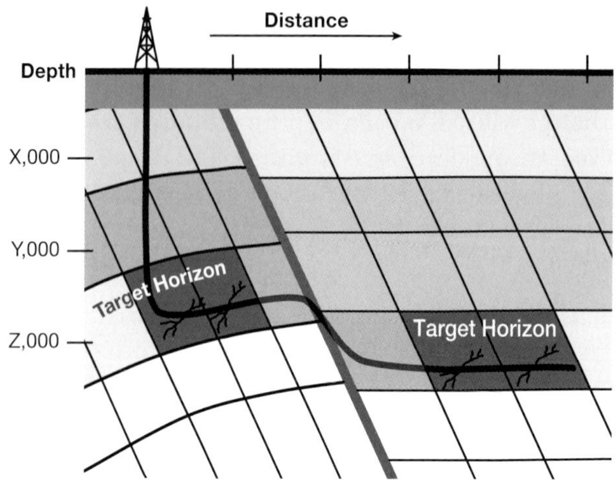

Fig. 8-2. Detailed cross-section generated from the static model used to guide the horizontal well and select cells to be fracked and cells to avoid

Once the seismic and geological model is established, it is important to assign risk to the individual cells. Neural networks provide such a tool, and the results can be displayed on radar plots. An example from medicine is useful to explain neural networks. Data from patients of a particular age group and sex can be collected at a cardiology department in a hospital. All patient data are input into two groups. The first group is analyzed for randomness, and the second is studied for patterns from which models can be developed to predict efficiency of procedures. Differentiating between the two is vital in developing improvements. In geoscience, the same process can be applied.

The initial step in a neural network study is to first study *randomness*. This is defined as a set of events that when repeated will always have a 50% chance of occurrence (i.e., the same probability as a coin toss). The next step is to determine the presence of linear logic: that is, if "a" and "b" are present, then "c" occurs. Neural networks generally assume a nonlinear path and focus on a condition where either "a" or "b" can occur and its consequence(s). What is being sought is recognition of a pattern. Once observed, can it be followed by subsequent events? Do the data indicate a normal or skewed distribution, and if so, why (or why not)? The next step is to weigh influences: if "a" is three times greater than "b," will "c" occur faster or slower? Then, determine if there is any self-organizing pattern. Is there a pattern to the input response on the model? How well can it predict events? For additional information, refer to the work of Nikravesh et al. (2003). Probabilities will be discussed further in chapter 9. An additional step is to ensure that signals are separated from noise.

Multiple parameters can be displayed on a radar plot (fig. 8–3) to better visualize the neural network results for a specific interval of interest. Each parameter may be totally independent, and the radii scales may vary to become more illustrative. Software is available to assist in creating these plots. On the basis of experience in the area, the probability of success can be overlain on the map (fig. 8–3).

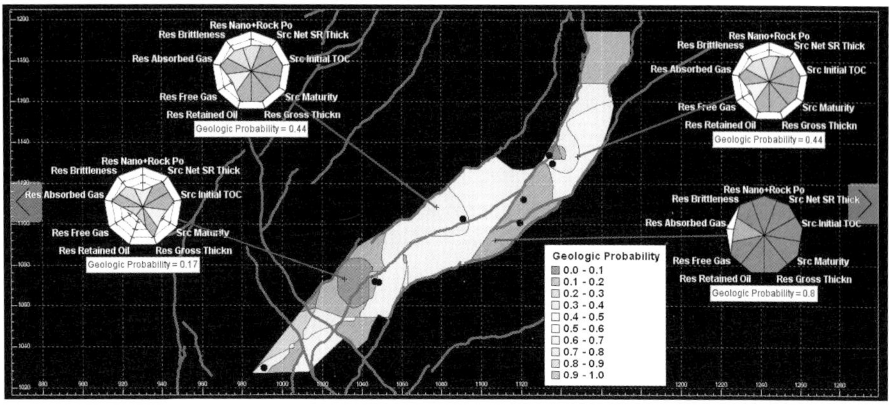

Fig. 8–3. Radar plot for specific parameters applied to a specific formation. The legeld shows the likelihood of occurrence. (Reproduced with kind permission of Platte River Associates)

Both the static model and the radar plot are designed to predict subsurface conditions prior to drilling. Once the geological model is described, the stress conditions in the subsurface must be analyzed.

PRESSURE REGIMES

Among the most important initial calculations in a given basin are determinations of the pressure and the stress regime near the reservoir. The overburden pressure and pore pressure information must be included to make these calculations. The basic purpose for analyzing overburden is to determine both the weight or gravitational force of rock overlying the target horizon and its effects on the target horizon. Because it is relatively straightforward to estimate the overburden stress, it can be carried out early in the exploration phase and subsequently recalibrated as more information is provided. It is evident that the sediments at depth are subjected to significant pressures and temperatures. Although the following discussion will focus on the significance of the pressure regime, the temperature information will be reintroduced, following on the heels of geochemistry temperature analysis.

Fracking requires accurate knowledge of all aspects of the pressure and stress regimes on the target subsurface section. Even simple maps, such as depth maps of casing points, may be of great help in identifying changes on pressure regimes. Note, however, that the first two sets of casing—namely, the *conductor* and *surface casing*—are not pressure dependent. All others are mainly pressure dependent.

General stress regimes observed in geology must be reviewed prior to discussion of overburden. There are three possible basin scale stress regimes: normal extensional; reverse or thrust faults, both grouped as compressional; and strike-slip regimes. For the purpose of the current study, only stresses applied to extensional systems will be considered because most shale resource plays fall into this category. For compressional regimes, a similar logic is used, except that the values have to compensate for the higher horizontal stresses associated with compression. An exception occurs in shallow burial conditions (<1,500 m), where horizontal stresses often exceed vertical stresses, even in extensional regimes.

Two tools are used to analyze stresses: Stress can be calculated on the basis of known physical properties of the rocks, or the reservoir conditions may be simulated in a laboratory by using core data from the reservoir and applying both uniaxial and triaxial compresses on the samples. The general stratigraphy, density, and thicknesses are needed to compute the vertical stress on a rock. The second tool comprises application of uniaxial and triaxial compress to determine how the rock fractures. For over 100 years, both sets of data have been incorporated and analyzed using the Mohr stress circle to define the mechanisms of fracturing, joints, and joint orientation.

Subsurface stress fields in a normal extensional system tend to be orthogonal in nature, with the principal stress, σ_1, at depths greater than 1,500 m as vertical. This stress (σ_1) is also equivalent to the overburden stress, σ_{ob}. By definition, the minimum stress is denoted as σ_3 and the intermediate stress as σ_2, which is then by definition the maximum horizontal stress. Importantly, the effective overburden is influenced by fluid content; for the initial estimation of overburden stress, the fluid effect, or pore pressure, is disregarded. Also, for simple calculations, it is assumed that $\sigma_2 = \sigma_3$; in triaxial compression, this is referred to as the *confining stress* and is equal on both sides. The vertical, overburden stress on a reservoir is the weight of rock volume overlying the target horizon and is a function of the *overburden gradient* (OBG), also known as the *lithostatic gradient*. It is common to assume that for young sediments the lithostatic gradient is 1 psi/ft (22.6 kPa/m). For computation of the true vertical stress, σ_1, the density of each rock unit from i to n (ρ_i) is multiplied by the thickness of each interval (z_i) and the acceleration of gravity ($g = 9.8$ m/s^2):

$$\sigma_1 = (\Sigma_i^n \rho_i z_i)g \tag{8.1}$$

For 2,000 m of section, density of 2,500 kg/m^3, and acceleration of gravity of 9.8 m/s^2 equation (8.1) yields 49 10^6 $kg/(m\ s^2)$ = 49,000 kPa. Thus, the pressure gradient is 49,000 kPa/2,000 m = 24.5 kPa/m, which is very close to 1 psi/ft.

The mass density in the overburden equation (8.1) is given in kg/m^3. A density of 2.65 g/cm^3 translates to 2,650 kg/m^3. In geoscience, the overburden is measured in pascals (SI units; 1 Pa = 1 N/m^2) or psi (English units). One difficulty in drilling engineering is this mixed use of units. Wherever possible, metric units will be used; however, for clarification and ease of computation, the English system will be invoked, because most drilling parameters and equations are based on the English system.

At the reservoir itself, the actual stress is the overburden stress minus the pore pressure, because the fluids in the pore act to support the stress. Hydraulic fracking requires exceeding the highest horizontal stress observed at the target horizon. In essence, the frac pressure is close to the sum of the overburden stress, σ_1. A modern adaptation to the classic triaxial compress allows for a determination of the effect of σ_2 and σ_3 by providing an additional independent confining stress. Further research in this area is required (Vachaparampil, 2017).

For quick calculation of pore pressure, the drilling mud density is determined by weighing one gallon of mud at the drill site. This is done on a regular basis to ensure proper mud balance. The result is in pounds per gallon (ppg), which is the common measurement used in all wells: 1 ppg = 19.7 psi/ft; 1 psi/ft = 22.6 kPa/m; and 1 MPa = 145.3 psi.

For a typical sequence in the East Texas Basin, σ_1 is summed as follows:

- Tertiary and Upper Cretaceous sandstone/shale, 1,000 m × 2,400 kg/m^3 × 10 m/s^2 = 24 MPa
- Cretaceous shale/anhydrite and carbonates, 500 m × 2,650 kg/m^3 × 10 m/s^2 = 13.3 MPa
- Eagle Ford/Woodbine ("Eaglebine"), 500 m × 2,500 kg/m^3 × 10 m/s^2 = 12.5 MPa
- Total overburden (σ_1) is the sum of the three = 49.8 or 50 MPa

When the effect of porosity is disregarded, a margin of error is introduced in the overburden stress equation. As noted in chapter 5, porosity decreases with depth as a consequence of differential compaction and is a function of sediment load. Shallow sediments have very high porosities and are supported by the fluids in the pore space. Once a sufficient load is attained, the fluids are progressively expelled, and the porosity is reduced. Logging the well during the drilling phase allows for determination of the porosity at specific intervals and, consequently, the preparation of a porosity decline function with depth. While a well is being drilled, curves generated from adjacent wells should be compared to the current well to observe local anomalies. Two logs are commonly used to determine porosity, the sonic log and density log. Of particular significance is the density log. The bulk density of the rock is a function of the matrix density and the porosity.

Based on the analysis of multiple data sets, Hubbert and Willis (1957) proposed an initial estimate for the relationship between vertical (σ_1) and horizontal (σ_3) stress as follows:

$$\sigma_1 = 3\sigma_3 \tag{8.2}$$

Therefore, in the above example, as a first approximation, $\sigma_3 = 17$ MPa .

The minimum confining stress required prior to fracturing is partially a function of the Poisson's ratio (ν) and the Young's modulus (E) of the material. By definition, as a rock is compressed vertically, it is forced to deform laterally in accordance to the minimum stress regime and strain. For this relationship to be true, $\sigma_2 = \sigma_3$. Laboratory studies have demonstrated the following:

$$\frac{\sigma_2}{\sigma_1} = \frac{\nu}{(1-\nu)} = K_{mc} \tag{8.3}$$

This relationship is a fundamental rock mechanics property and will be introduced in several subsequent discussions as the rock matrix stress coefficient, K_{mc} (fig. 8–4).

The following common ranges for Poisson's ratio (ν) in sediments have been given by Ikelle and Amundsen (2005):

- Consolidated shales, 0.25–0.35
- Consolidated sandstone, 0.15–0.25
- Poorly consolidated shale, 0.35–0.45
- Poorly consolidated sandstone, 0.30–0.35
- Gas charged sandstone, 0.10–0.20
- Limestone, 0.25–0.35
- Coal, 0.35–0.45
- Steel, 0.30

Therefore, most shales at depth are consolidated and have an approximate Poisson's ratio (ν) of 0.3, although the Poisson's ratio for highly organic-rich shale may be closer to 0.35. If $\nu = 0.3$ is assumed, then

$$\frac{\sigma_2}{\sigma_1} = \frac{0.3}{0.7} = 0.43 \tag{8.4}$$

In the calculation of equation (8.4), K_{mc} will be rounded to 0.4. In nature, σ_3 is not equal to σ_2 . As discussed above and as various studies have confirmed, σ_2/σ_1 is closer to 1/3. Hoek and Brown proposed the term *K*, subsequently renamed K_{mc}, to relate the average horizontal to vertical stress, which is the abscissa in figures 8–4 and 8–5 (Hoek, 1988). Hoek and Brown pointed out that as a rule, rocks behave in a simple and predictable manner when stress is applied. Note that where stress and strain are applied, by definition Young's modulus (*E*) applies.

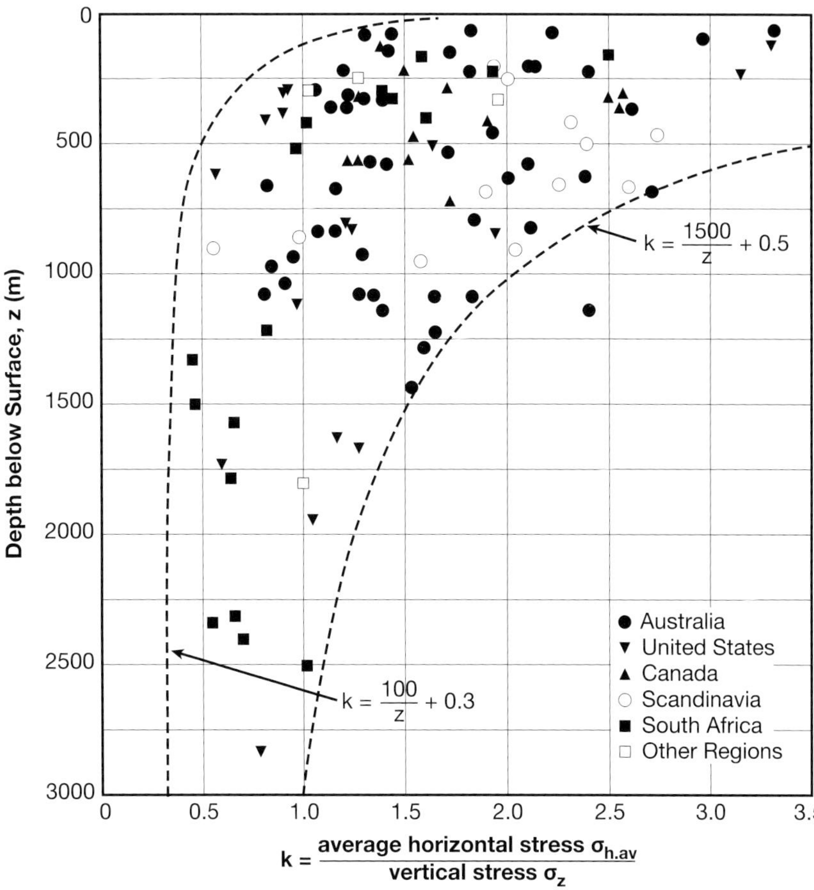

Fig. 8–4. Matrix stress, *K* or K_{mc}, as function of depth (Hoek, 1988). In the reference diagram, $\sigma_{Ha} = \sigma_2$ and $\sigma_v = \sigma_1$. As noted previously, at depths less than 1,500 m, the stress field associated with the horizontal stress is dominant. (Figure provided by and reproduced with the kind permission of Professor Evert Hoek)

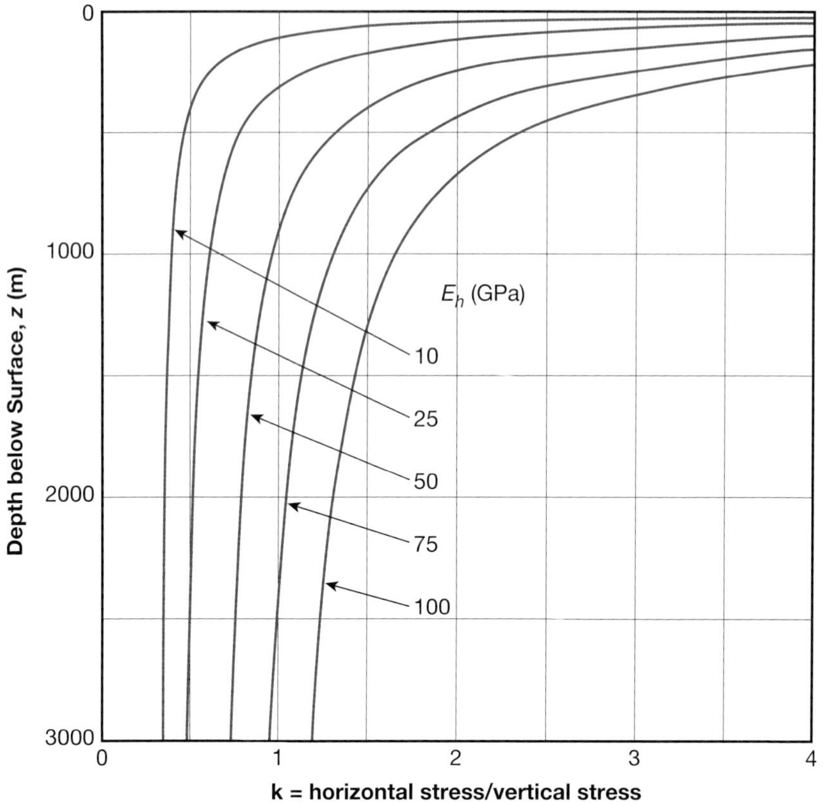

Fig. 8–5. Same plot as in figure 8–4 except that Young's modulus is also considered. (Figure provided by and published with kind permission of Professor Evert Hoek; Sheorey, 1994)

Sheorey (1994) modified the original Hoek-Brown relationship (figs. 8–4 and 8–5) to reflect variations in Young's modulus (E). Most siliciclastic sediments have Young's moduli of 13–25 GPa (giga pascal); carbonates range from 50 to 90 GPa. Because the common depth ranges for sediments in oil reservoirs range from 2,000 to 3,000 m (or more), K_{mc} values of 0.5–0.8 are common (fig. 8–5). Recalling the previous estimate for K_{mc} (eq. [8.4]), based on the Poisson's ratio, the value was 0.43; this is considered a low value. The lithostatic gradient is usually near 1 psi/ft, and the K_{mc} value is nearly constant for a specific rock type. Minifracs (leak-off tests) allow for an improved estimation of the minimum horizontal stress (σ_2) and is discussed in detail later in this chapter (see the section "Leak-Off Test (Minifrac) and Permeability"). Studies in the Eagle Ford and Bakken have shown that the actual pressure gradients at the reservoir level range from 0.75 to 0.9 psi/ft, with an average value of 0.8 psi/ft (Xiong et al., 2015).

From the previous discussion, if σ_1 = 50 MPa, then a corresponding maximum horizontal stress, σ_2 , with a K_{mc} of 0.6 yields an average horizontal stress of 30 MPa. This can be rearranged to give the following relationship:

$$K_{mc} = \frac{\text{average horizontal}}{\text{vertical stress}} \tag{8.5}$$

For K_{mc} = 0.8 (an average K_{mc} for a rock with a Young's modulus of 50 GPa) the resulting maximum horizontal stress, σ_2 , would be 40 MPa. If the pressure gradient is 0.8 psi/ft, as noted above, or 18 kPa/m at 2,000 m, then σ_1 = 36 MPa. For K_{mc} = 0.8, the average σ_2 = 29 MPa. As can be seen, the initial estimate using only the Hubbert-Willis ratio yields σ_2 = 17 MPa. This is lower and conforms to an average Young's modulus of 8–9 GPa in figures 8–4 and 8–5 and is very near the minimum line on the Hoek-Brown relationship (figure 8–5). As noted before, the simple relationship shown in equation (8.5) represents the absolute minimum computed ratio. We will elaborate on this topic further once the Mohr diagram is described later in this chapter.

ROCK QUALITY DESIGNATION

In 1967, Deere coined the term *rock quality designation* (RQD) to define rock quality by observing cores (Deere and Deere, 1989). In essence, RQD looks at cores taken in various wells and compares how well the cores are preserved. A subjective measure of the core as a percent, RQD is determined as follows: first, measure the length of the core; next, measure the length of each piece greater than 10 cm, and divide this by the total length; finally, multiply the result by 100% to give the RQD.

How the RQD is related to the matrix stress coefficient, K_{mc}, is not clear; nevertheless, intuitively, there should be a relationship. As K_{mc} increases, the sediment will become unstable and fracture. To the drilling engineer, the data have significance because if the rock has open fractures, it will have a more fragmented core. The more fragmented the rock is, the lower will be the tensile strength. A similar logic can be applied on outcrop descriptions. The concept is important. For exploration within North Dakota's Bakken Formation, the University of North Dakota has registered the RQD descriptions for many cores into their basinwide geomechanical database.

Hoek and Brown also used the number of joints per unit area as an indication of tensile strength (Hoek, 1988). More recently, Aadnøy and Looyeh (2011) have suggested that, because rocks in the subsurface have undergone micro- to macrofracturing, the tensile strength of the rock is 0. As will be observed later in this chapter (see the section "The Mohr Circle"), a tensile strength of 0 has important implications; it means that the Mohr-Coulomb envelope must pass through the origin and may imply zero cohesion.

PORE PRESSURE

An initial estimate for pore pressure is always based on the mud weight used in the well. Pore pressure prediction is vital in all drilling, including shale resources.

By definition, pore pressure corresponds to the pressure within the fluids in the rock pore space. Several redundant techniques are applied to ensure accurate estimate of pore pressure. Among the most commonly used are those based on the mud weights used in the wells. They are often followed by a more robust determination such as the Matthews and Kelly (1967) method and several variations of the preferred Eaton (1975) method. The information is subsequently integrated into seismic interpretation and extrapolated to future well locations to anticipate potential problems. Currently, it is a standard procedure to evaluate each well within the immediate radius of the proposed well in great detail by analyzing all mud weights used and reviewing any problems that may have been encountered. For example, the U.S. Geological Survey has compiled pressure and temperature information on 590 wells in South Texas alone (fig. 8–6). The report includes an open-file spreadsheet with the data for each well (Burke et al., 2011).

As depth increases, porosity decreases, and density and velocity increase (fig. 8–7). Because velocity data are readily available in most areas, they should be used in the initial evaluations of pressures. Average normal velocity curves can be generated, and a well that diverges from the normal can be analyzed to determine the causes. More detailed analyses of velocity data will be presented later in this chapter (see the section "Eaton Method"). Overpressure effects (fig. 8–7) and unusually high TOC content, among other characteristics, can result in deviations from the norm. Moreover, as shown in a Bakken interval, more than one rock type can coexist—namely, a ductile shale with high TOC content and low density and a brittle tight sand or carbonate Middle Bakken unit. The two have different interval velocities; therefore, the velocity is also a function of which member is analyzed and possible overpressure effects (fig. 8–8). An example of velocity versus depth for the Bakken interval in the eastern Williston Basin shows high variability irrespective of depth (fig. 8–8). This is common in shale resource plays owing to the effects of organic richness. Coal beds have a similar response. The velocity data can also be used to estimate the reservoir pressure and therefore must be considered in pressure analyses (again, see "Eaton Method").

As discussed previously, the pore pressure is estimated from the drilling mud weight (in ppg). In the English system, commonly used in drilling, the pore pressure (P_p) is calculated as follows:

$$P_p = 0.052 \times \text{mud weight (ppg)} \times \text{depth (ft)} \tag{8.6}$$

The maximum conventional mud weight that can be used is 18.7 ppg, yielding a pressure at 8,728 ft of 8,487 psi for that depth, or 0.97 psi/ft. Normal hydrostatic (water column) pressure gradients are 0.45–0.47 psi/ft, soft overpressure is 0.47–0.70 psi/ft, hard overpressure is 0.7–0.99 psi/ft, and hyperpressure is greater than 1 psi/ft. For a sample region in South Texas, an actual mud weight at 6,728 ft (2,057 m) is 9.3 ppg. By use of equation (8.6), it yields a pressure of 3,254 psi (22.4 MPa). Note that the estimated hydrostatic pressure at that same depth would be 6,728 ft × 0.465 psi/ft = 2,200 psi (15.2 MPa), indicating that the pressure in the well has soft overpressure. Regional studies providing pressure information are common (e.g., see Burke et al., 2011).

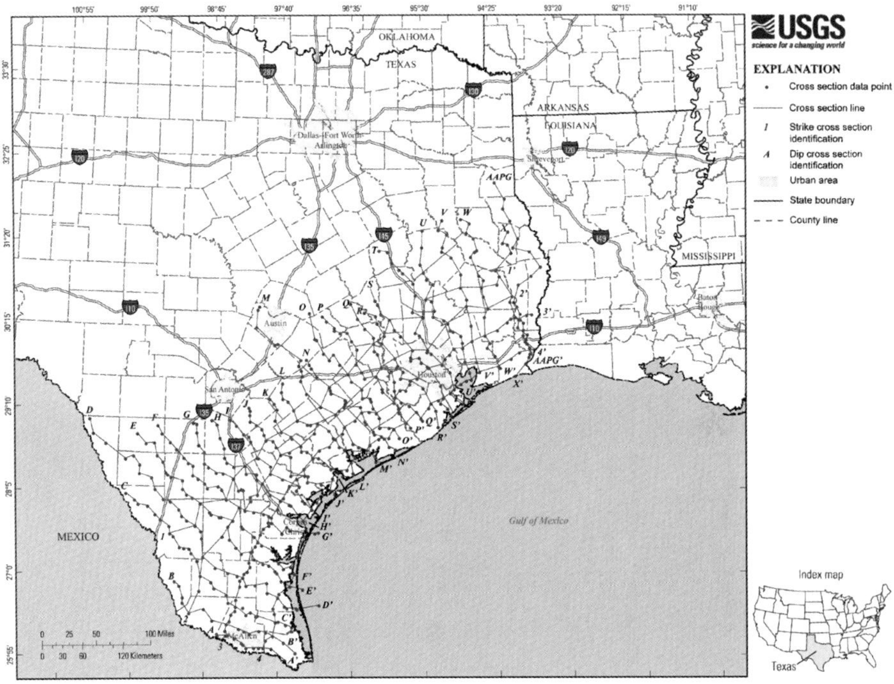

Fig. 8–6. Map of bottom hole temperature and pressure for South Texas. (Reproduced with kind permission from the U.S. Geological Survey. Both the report and spreadsheets with data are available online at the U.S. Geological Survey website; see also Burke et al., 2011.)

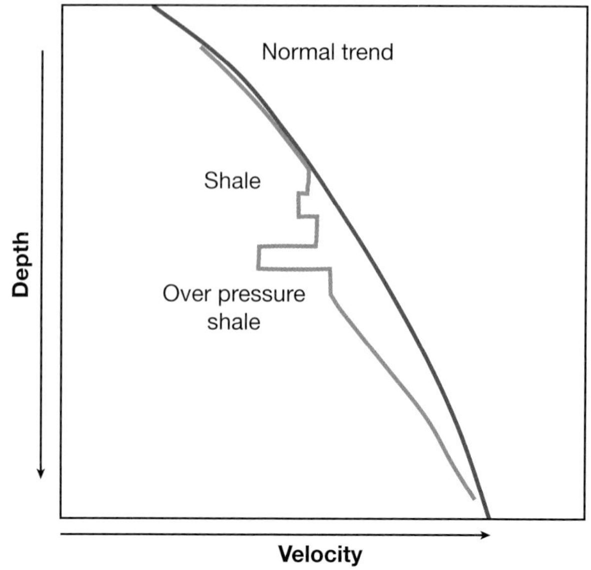

Fig. 8-7. Plot of shale velocity versus depth for average. The light gray line shows the effect of overpressure.

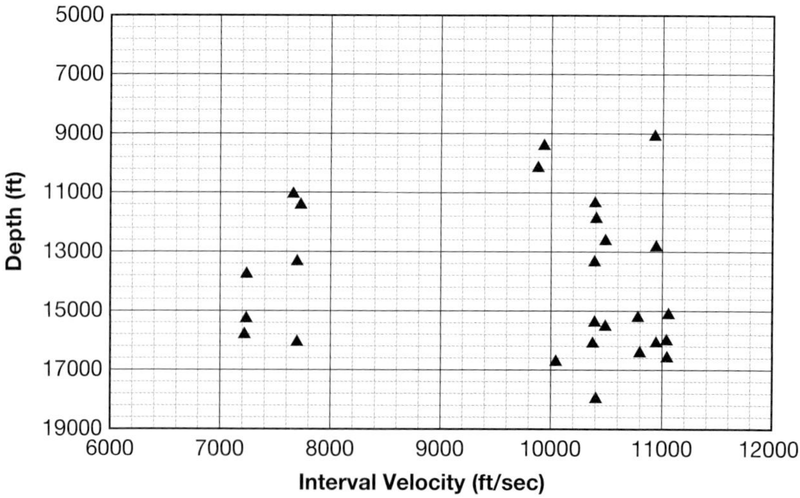

Fig. 8-8. Bakken shale interval velocities from various wells and locations, yielding a wide range of velocities irrespective of depth

FRACTURE GRADIENT

Hubbert-Willis frac gradient

The *fracture gradient* can be defined as the amount of pressure needed to induce fractures in a rock in the subsurface as a function of depth. Hubbert and Willis (1957) provided a simple fast approach to compute the fracture gradient (which serves as an initial step in a reservoir study). Theirs is one of the pioneering studies on fracking. Although this is the least accurate of the measuring techniques, it provides a quick approximation.

The effective stress, σ_{ef}, on a reservoir is the overburden pressure, σ_{ob}, minus the pore pressure, P_p:

$$\sigma_{ef} = \sigma_{ob} - P_p \tag{8.7}$$

However, from our previous discussion, the minimum horizontal stress would be ⅓ × σ_{ef} plus the static pore fluid; that is, the minimum stress is usually one-third of the effective overburden stress, and thus the minimum frac gradient must also be a function of one-third times the effective overburden plus the pore pressure (Hubbert and Willis, 1957). If the function is divided by the respective depth, the frac gradient can be estimated as follows (based on Hubbert and Willis):

$$\sigma_{min} = \tfrac{1}{3}(\sigma_{ob} - P_p) + P_p \tag{8.8}$$

where P_p is the pore pressure in psi. Equation (8.8) simplifies to

$$\sigma_{min} = \tfrac{1}{3}(\sigma_{ob} + 2P_p) \tag{8.9}$$

Dividing the result by depth, z, and noting that the lithostatic gradient is nearly 1 psi/ft and equal to σ_{ob}/z, where z is depth, yields minimum frac gradient, $F_{g\,min}$:

$$F_{g\,min} = \frac{1}{3}\left(\frac{1+2P_p}{z}\right) \tag{8.10}$$

The maximum frac gradient, $F_{g\,min}$, is estimated as follows:

$$F_{g\,max} = \frac{1}{2}\left(1+\frac{P_p}{z}\right) \tag{8.11}$$

Because of its simplicity and lack of detail, equation (8.11) should be used as a first approximation and can be easily mapped over the area of study.

Matthews-Kelly frac gradient

The Matthews and Kelly (1967) method is based on comparison of the normal pressure trends with the unique pressure at the reservoir. We will focus first on pressure relationships and then discuss other parameters as they apply, such as resistivity and velocity. The depth corresponding to the pressure gradient under normal conditions is designated as D, and the depth corresponding to where the overpressure zone is observed is designated as D_i; comparison of the two yields an estimate of the frac gradient. The difference between the two depths is a consequence of variations in the rock stress coefficient, K_{mk}. Often, the method is referred to as the *equivalent depth pressure determination*.

In essence, all of these factors are affected by normal compaction curves, and the presence of overpressure will impede further compaction and result in the deviation from the normal trend. In overpressure zones, the value of σ_{ob} at depth D is significantly affected by pore pressure and must be considered to establish the effective overburden pressure, σ_{ef}. K_{mk} represents the ratio between a specific attribute immediately overlying the target horizon and the characteristics of the target itself.

The Matthews-Kelly equation relates the overburden with the effective lithostatic gradient of Hubbert and Willis (1957) with adjustments. Returning to the effective pressure at depth, equation (8.7) can be rewritten as

$$\sigma_{ob} = \sigma_{ef} + P_p \tag{8.12}$$

A gradient can be arrived at by dividing both sides of equation (8.12) by depth (z). Furthermore, Hubbert and Willis (1957) noted that for every basin, a specific correction factor, matrix stress coefficient, K_{mk}, modifies the result as follows:

$$\frac{\sigma_{ob}}{z} = \left(\frac{\sigma_{ef}}{z} + \frac{P_p}{z}\right)K_{mk} \tag{8.13}$$

Hubbert and Willis (1957) suggested that, on the basis of their empirical data, the effective normal stress gradient is 0.535 psi/ft. Therefore, for the specific reservoir depth under normal conditions, the effective stress would be

$$0.535D_i = \sigma_{ef} \tag{8.14}$$

Hydrostatic pressure is the weight of a column of water at the target depth. The gradient ranges are as follows:

- Freshwater, 0.433 psi/ft
- Saltwater, 0.456 psi/ft
- Natural gas, 0.1 psi/ft

- Black oil, 0.32 psi/ft

The latter two affect capillary pressure, as discussed later in this chapter (see the section "Capillary Pressure, Wettabilty, and Pore Size").

The Matthews-Kelly matrix stress coefficients, K_{mk} (originally published as Ki) have been plotted for many parts of the world and are unique to each region. The value is obtained from the fracture gradient, F_g:

$$F_g = \frac{P_{p1}}{D} - \frac{K_{mk}\sigma_{ef}}{D} \tag{8.15}$$

where P_{p1} is the pore pressure at the point of interest, D is the depth at the point of interest (in ft), and σ_{ef} is the effective stress at the point of interest; $\sigma_{ef} = 0.535D_i$, as in equation (8.14).

The depth D_i is used only as a reference to properly define the K_{mk} correction factors. It is basically the reference depth at which the observed pressure should occur. Inputting the D_i value on the proper chart for the Gulf of Mexico gives the corresponding K_{mk} value for each depth D_i.

For a specific basin, the graph of K_{mk} values can be constructed by introducing values of K_{mk} at specific depths and subsequently varying the depths. The K_{mk} values can be confirmed by compiling leak-off tests in an area, discussed later in this chapter (see the section "Leak-Off Test (Minifrac) and Permeability").

As a function of depth, equation (8.15) can be rewritten as

$$K_{mk} = \frac{DF_g - P_p}{\sigma_{ob} - P_p} \tag{8.16}$$

where D is the depth, F_g is the frac gradient, P_p is the pore pressure, σ_{ob} is the overburden stress, and K_{mk} is the matrix coefficient.

Once the graph is generated (fig. 8–9), the following steps are taken to evaluate the fracture gradient:

- Determine the overburden pressure from equation (8.12)
- Obtain the normal pressure at the target depth, using hydrostatic if normal pressure and mud weight if higher pressure
- Determine the effective pressure as indicated in equation (8.7); if lithostatic gradient = 1 psi/ft, then $\sigma_{ob} = z \times 1$ psi/ft = z psi
- Calculate P_p using equation (8.6) for mud weight conversion
- Determine K_{mk} from the graph using $D_i = \sigma_e / 0.535$
- Calculate the fracture gradient F_g

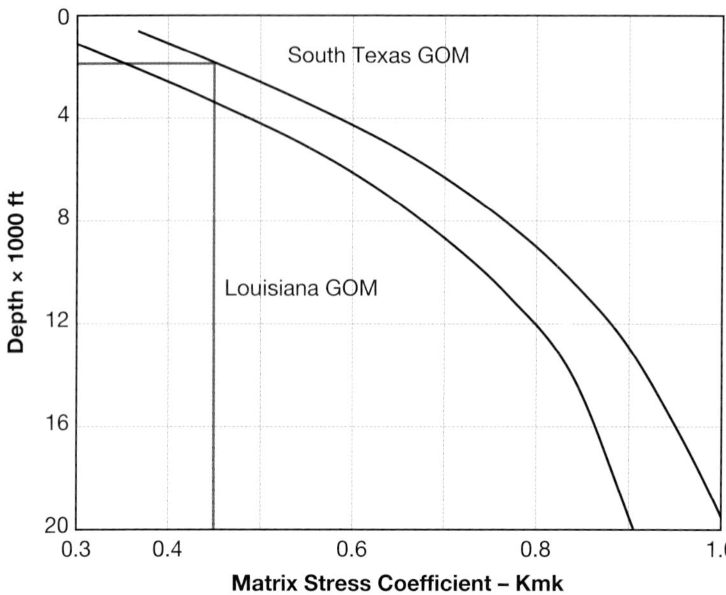

Fig. 8–9. Graph used to determine K_{mk} for the Gulf of Mexico (GOM) using the Matthews-Kelly method. Remember that the depth corresponds to the corrected Matthews-Kelly depth D_i. See equation (8.14) for the solution. (Graphs generated for various part of the world are available in Matthews and Kelly, 1967)

For example, a South Texas Eagle Ford well is selected where at a depth of 9,650 ft a mud weight of 14 ppg was used. If the lithostatic gradient is 0.9 psi/ft, then what is the Matthews-Kelly frac gradient?

A 14 ppg mud translates to a pore pressure at the reservoir (P_{pl}) of 7,000 psi. If the lithostatic gradient for the basin is used, then the overburden pressure (σ_{ob}) is 8,685 psi. The effective stress is 8,685 – 7,000 = 1,685 psi, according to the Matthews-Kelly equation, and the effective depth where the given pressure would be normal is $0.535D_i = 1{,}685$, or $D_i = 1{,}350$ psi. On the South Texas curve plot, the corresponding $K_{mk} = 0.45$. Using equation (8.15), Fg = (7,000/9.650) + [(0.45 × 1,685)/9,650)] = 0.8 psi/ft.

Matthews and Kelly (1967) went one step further. Using similar logic with resistivity and velocity data and applying the same *equivalent depth methodology*, they were able to predict formation pressures and fracture gradient. Eaton (1969) described this approach in more detail, and this will be expanded on in the next section. The concept is particularly important because the geophysicist can generate a velocity inversion profile that allows for pressure analysis based on seismic. Chopra and Huffman (2006) have also developed a technique to process the velocity data to obtain pore pressure data from seismic data. Matthews and Kelly (1967) used a similar approach with shale conductivity and shale velocities.

THE EATON METHOD AND THE DRILLING EXPONENT

For prediction of drilling problems in a region, aside from the mud weights per se, the drilling exponents (d_e) in adjacent wells are compared to the exponent at the well being drilled in real time. Because mud weights are reactive to the drilling conditions and somewhat subjective, the drilling exponent d_e is useful for ongoing analysis of drilling conditions as it compensates for the multiple parameters used in drilling. It is often adjusted in an area to fit the predicted values with the observed.

When drilling through a sediment, the greater the pressure is on the bit, the easier the drilling will be. In drilling, the pressure on the bit is known as the *weight on bit* and is registered on the drilling log (mud log) at each bit change. In addition, drilling is a function of the size of the bit. The restriction of the size of the bit, d_b, is the inner diameter of the last casing set. It is therefore variable and dependent on each well's drilling program. All modern drilling is accomplished with engine-driven rotary drills. The drill speed, in revolutions per minute (RPM), is variable and depends on the drive of the engine. The greater the rotation is, the more efficient the drilling will be; however, cooling the bit becomes more difficult.

The rate of penetration is primarily a function of the rock hardness. Drilling may be slower through very soft ductile material owing to sticking. Such formations can heave into the hole. This is where proper mud chemistry and mud weights play a key role in successfully drilling the well. For example, if drilling in a salt bed, oil-based muds should be used.

The drilling exponent, d_e, is calculated as follows:

$$d_e = \frac{\log(R/60N)}{\log[12W/(10^6 \times d_b)]} \times \frac{\rho_{normal}}{\rho_{observed}} \tag{8.17}$$

where R is the rate of penetration, W is the weight on bit, N is the number of RPM, d_b is the diameter of the bit, ρ_{normal} is the density of the mud above the overpressure, and $\rho_{observed}$ is the density of the mud in the overpressure zone.

Equation (8.17) is based on Bingham's model (1964). To normalize differences in drilling conditions, Bingham suggested the following solution:

$$R_T = K_{bi} N \left(\frac{W}{d_b}\right)^{d_e} \tag{8.18}$$

where K_{bi} is referred to as the *proportionality constant* and is a function of rock strength. Although there are no direct studies relating K_{bi} to the previously described K_{mk}, both describe similar relationships and must have an equivalency.

Note that the mud weight differences between expected and actual mud weights are not considered in equation (8.18). To correct this problem, Rehm and McClendon (1971) proposed a modified drilling exponent, d_{me}, defined as follows:

$$d_{me} = \frac{d_e \rho_n}{\rho_a} \quad (8.19)$$

where ρ_n is the normal or expected mud density at the corresponding level and ρ_a is the actual mud density in the current drilling (both in ppg).

The value of the drilling exponent, d_e, is lower as the drill bit begins to enter the overpressure zone; thus, one should drill with caution. Other indicators for overpressure are an increase in slowness (i.e., decrease in interval velocity) and an increase in the resistivity of the shales, but the bulk density continues to increase gradually. All three are combined and often included in MWD or LWD, not only to stay on target but also to assist in establishing drilling conditions.

The drilling exponent is applied to several equations that analyze overpressure. It is often a good idea to use redundant methods to ensure accuracy, such as the adaptation to the Matthews-Kelly relationship discussed in the previous section. Bowers (1995) observed the following stress relationship:

$$\sigma = \sigma_{normal}\left(\frac{V}{V_{normal}}\right)^n \quad (8.20)$$

where V is velocity and σ represents the expected normal and the observed values.

The Eaton (1975) method is commonly referenced and is considered the most accurate tool for frac gradient determination. Note that there are many variations to the basic relationship, including the similar approach developed by Bowers (1995). The Eaton method determines the expected velocity and stress under normal pressure. It then describes an exponent (n) that is a measure of the sensitivity of the velocity to stress. V is the observed velocity.

The common application of equation (8.21) is to begin with $n = 3$ and then calculate the pore pressure based on the interval velocities. Next, compare the results with the observed pore pressure in the well and adjust the exponent to match the existing well. Usually, n has regional characteristics. The reason most drillers consider the Eaton relationship the most accurate is that it is associated with

Poisson's ratio. As discussed before, the fracture gradient is close to or locally may exceed the lithostatic gradient. Once again, a new index is introduced, *n*, which is a correction based on the rock mechanics properties of the reservoir. It is logical that there is a relationship between the Eaton ***n*** factor, the Bingham factor K_{bi}, and K_{mk} in the Matthews-Kelly relationship. It remains critical to calibrate the results with existing wells.

Several adaptations have been made to allow other logs to be used to estimate P_p. The most common is the sonic log with *n* index correction:

$$P_p = \sigma_{ob} - \left(\sigma_{ob} - \sigma_{hyd}\right)\left(\frac{V_{observed}}{V_{normal}}\right)^n \tag{8.21}$$

where P_p is the pore pressure, σ_{ob} is the overburden stress, σ_{hyd} is the hydrostatic stress, $V_{observed}$ is the observed sonic (i.e., in the section of interest), and V_{normal} is the expected normal sonic (i.e., just above the problem area).

The basic expression of Eaton's equation is as follows:

$$\sigma_2 - P_p = \frac{\nu}{(1-\nu)(\sigma_{ob} - P_p)} \tag{8.22}$$

where ν is Poisson's ratio.

Eaton (1975) follows up on the Matthews-Kelly approach by preparing similar charts as in figure 8–9 for depth versus Poisson's ratio. The equation itself was derived by Eaton from equation (8.3) in the Matthews-Kelly study:

$$F_g = \left[\left(\frac{\sigma_{ob}}{z} - \frac{P_p}{z}\right)\frac{\nu}{1-\nu}\right] + \frac{P_p}{z} \tag{8.23}$$

Using equation (8.23) with z = 9,650 ft, P_p = 7,000 psi, σ_{ob} = 0.9 psi/ft × 9,650 = 8,685 psi, and Poisson's ratio of marl at 0.35, F_g = [(8.685/9,650) – (7,000/9,650)] [0.35/(1 – 0.35)] + (7,000/9,650) = (0.9 – 0.72)(0.54) + 0.72 = 0.73 psi/ft. Note that the Poisson's ratio reference in equation (8.23) is the confining stress ratio discussed in equation (8.3).

There are several offshoots from Eaton's (1975) method to determine pore pressure estimate. One commonly used invokes Poisson's ratio:

$$F_g = \frac{\sigma_{ob}}{z} - \left[\left(\frac{\sigma_{ob}}{z} - \frac{P_p}{z}\right)_n \left(\frac{d_m}{d_n}\right)^{b_e}\right] \tag{8.24}$$

where σ_{ob}/z is the observed overburden gradient, $[(\sigma_{ob}/z) - (P_p/z)]_n$ is the regional normal overburden gradient corresponding to that depth, d_m is the measured and d_n is the normal value of the drilling exponent for the region (Solano et al., 2007), and b_e, or E, is the Eaton exponent correction for sonic velocity. It ranges from 1.2 to 1.5 and is a function of overburden in the region.

Several other frac gradient equations have been generated. It is a good idea to test the various models and compare them to the observed values—and possibly adjust the indices to conform to the observed results. Each equation has a unique contribution that adds to the knowledge base in the region.

Another way of approaching the rock matrix stress coefficient is in terms of its relationship to brittleness. This topic was discussed previously (in chap. 5), but it is appropriate to reconsider the topic in the context of the above discussion. Eaton's equation relates stress to Poisson's ratio, and the rock matrix strength coefficient is also related to Poisson's ratio (ν). They provide the minimum value on the Hoek-Brown curve (figs. 8–4 and 8–5). Young's modulus (E) provides the correction to approach actual horizontal strength ratios. Furthermore, by definition, brittleness index is the average between the Young's modulus brittleness index and the Poisson's ratio brittleness index. The mathematical relationship between the Poisson's ratio and the Young's modulus is as follows:

$$E = 3K_b(1 - \nu) \tag{8.25}$$

where K_b is the bulk modulus, or compressibility modulus; a better term would be *incompressibility*, which is related to the Lamé parameters (see chap. 5).

In general, Poisson's ratio (ν) is more dependent on the lithology, whereas Young's modulus is more dependent on the porosity, as is the bulk modulus (Cho and Perez, 2014). Rickman et al. (2008) have provided a simple approach to the brittleness estimate (as described in chap. 6):

$$\frac{\mathrm{BI}_\nu + \mathrm{BI}_E}{2} \tag{8.26}$$

where BI_ν and BI_E are the brittleness indices using Poisson's ratio and Young's modulus, respectively.

A shortcoming of this simplistic stress model is a lack of integration of the anisotropy of the shales. As much as sands tend to be isotropic, shales are the exact opposite. (Shale anisotropy was discussed in chap. 6.)

In summary, the three calculations we provided yield the following results for the fracture gradient, F_g, for the same data set:

- Hubbert-Willis, 0.8
- Matthews-Kelly, 0.85
- Eaton, 0.73

The results require calibration according to experience in the region, and other models can also be used. One typical question asked in a project is whether Poisson's ratio can be rendered more precisely. If Poisson's ratio for the marl is 0.25, then by Eaton's equation, F_g = 0.84 psi/ft. If the petrophysicist and the geophysicist concur, then the correct frac gradient should be 0.84 psi/ft.

SEISMIC APPLICATIONS TO PORE PRESSURE STRESS ANALYSIS

Several studies have been carried out over the past decade on improving the integration of seismic response to reflect pore pressure and its relationship to frac gradients. Most have focused on velocity data. Two examples are described: The first approach uses seismic velocity and combines it with a Matthews-Kelly approach; the second approach is focused on rock physics. It is suggested that both be used.

Sayers et al. (2002) adapted the Eaton relationship to seismic. As discussed in chapter 5, there is a mathematical relationship between the petrophysically derived interval velocity and density (Gardner et al., 1974). The seismic interval velocity in an entire three-dimensional cube can therefore be converted to density by applying the same relationship. Chopra and Huffman (2006) have also related the velocity, emphasizing changes in velocity along a target horizon. Formento (2004) has summarized the steps involved in physically processing the seismic data to generate pseudo pore pressure data. The theory for the first approach (i.e., seismic) will be described in general terms, and the second approach (i.e., rock physics) will focus on computations to demonstrate how seismic can be used.

Prestack depth migration is recommended to obtain pore pressure data from seismic. On the seismic depth section, the normal velocity gradient for the region can be overlain and compared to the actual velocity data. Sayers et al. (2002) have used the concept of a *common image point* (CIP) to describe depth of a specific reflector, first using the actual velocity data and then using the normal velocity gradient in a depth migration to establish the depth difference. Sayers et al. used a depth differential equation similar to that used in the Matthews-Kelly equations; the difference is a function of pressure.

Goodway et al. (2010) have proposed a different modification to the Eaton equation. Because $\nu = \lambda/(2\lambda + 2\mu)$, they expanded the Poisson's ratio relationship in the equation (8.23) by incorporating the $\lambda\rho$ and $\mu\rho$ parameters (LMR or Lamé parameters; see chap. 6), as follows:

$$\frac{\nu}{1-\nu} = \frac{\sigma_2 - P_\mathrm{p}}{\sigma_\mathrm{ob} - P_\mathrm{p}} \tag{8.27}$$

Perez et al. (2011) and Goodway (2014) have described the following relationship as the *closure stress ratio* (CSR):

$$\frac{\lambda}{\lambda + 2\mu} = \frac{\nu}{1-\nu} \tag{8.28}$$

Given the relationship established in equation (8.28), the Eaton equation can be rewritten as follows:

$$\sigma_2 - P_\mathrm{p} = \frac{\lambda}{\lambda + 2\mu}(\sigma_\mathrm{ob} - P_\mathrm{p}) + \psi \tag{8.29}$$

where ψ is the correction needed because of the effect of Young's modulus on the relationship between σ_2 and σ_ob closure stress, shown previously in figure 8–5.

Once the LMR parameters are introduced, seismic inversion techniques can be applied. As noted in chapters 5 and 6, the LMR values can be inverted from the seismic:

$$\lambda\rho = I_\mathrm{p}^{\,2} - 2I_\mathrm{s}^{\,2} \tag{8.30}$$

$$\mu\rho = I_\mathrm{s}^{\,2} \tag{8.31}$$

Therefore,

$$\frac{\lambda}{\lambda + 2\mu} = \frac{I_\mathrm{p}^2 - 2I_\mathrm{p}^2}{I_\mathrm{p}^2} \tag{8.32}$$

where I_p is acoustic impedance and I_s is shear impedance.

By introduction of the overburden stress into the seismic data, combined with the estimated ψ value and pore pressure information from nearby wells, the maximum horizontal stress can be depicted on the seismic line. The ψ value can be calculated on the basis of the Lamé parameters (λ,μ), the bulk density (ρ), and each of the Thomsen (1986) anisotropy parameters (ε,δ,γ; see chap. 6). The γ Thomsen parameter is used for reference:

$$\psi = \frac{\lambda}{\lambda + 2\mu}\left(2\sqrt{\frac{\mu}{\rho}}\gamma\right) \tag{8.33}$$

The plot in figure 8–10 was developed by Perez et al. (2012) and describes Eaton's equation graphically. For the same rock type, the relationship of effective horizontal and vertical stress yields a constant slope, a function of $\lambda/(\lambda + 2\mu)$. From our previous discussion, in its simplest form, 0 is the y-intercept. This would correspond to the minimum values in figure 8–4. Obviously, the effective stress cannot be 0 because σ_1 cannot equal P_p. As the Young's modulus increases, so does the effective horizontal stress. To compensate for that difference, Perez et al. (2012) introduced the correction factor ψ. The plot in figure 8–10 can be used to estimate the values of the horizontal stress from the graph.

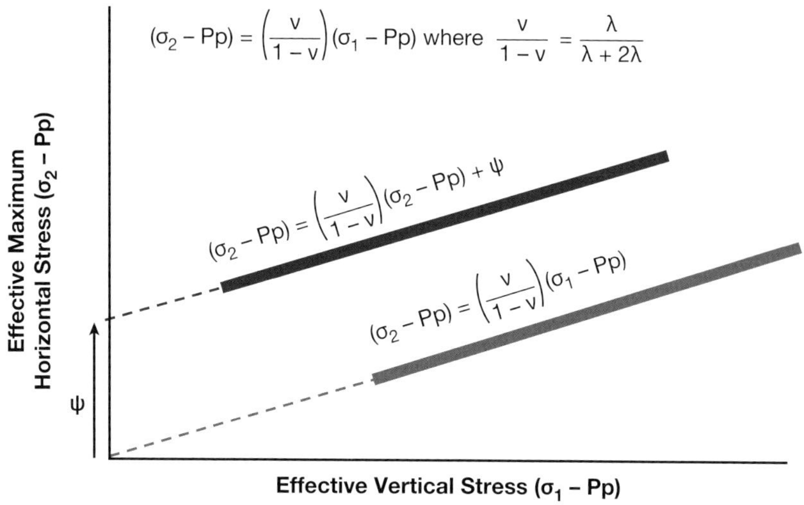

Fig. 8–10. Plot of effective vertical versus horizontal stress. Effective horizontal stress is $\sigma_2 - P_p$, and effective vertical stress is $\sigma_1 - P_p$. (Based on Perez et al., 2012)

It is evident that, because gas is highly compressible, the λ coefficient will be significantly affected by presence of gas and the pressure applied. Amplitude versus offset (AVO) analysis, discussed in chapter 6, is a proven indicator of gas and therefore has a direct application in shale resource assessment (Goodway et al., 1997). However, as noted in figure 6–10, both $\lambda\rho$ and $\mu\rho$ must yield low values, and such low values are not commonly observed in shales. Where present, potential AVO should be analyzed. More recently, Hunt et al. (2011) have incorporated curvature into the analysis, as will be discussed next. Any approach that yields a more accurate Poisson's ratio will be advantageous to the engineer planning a hydraulic fracture.

CURVATURE

It is self-evident that the amount of curvature a rock surface will influence the fracturing of the rock. Roberts (2001) has defined curvature as the rate of change in dip. This can be measured as the inverse of the radius of curvature that fits a surface at a particular point and is calculated by computing the rate of dip, trace to trace, on a seismic line, preferably in three-dimensional space. The minimum and maximum inverse radii are multiplied, and the resulting value can be mapped. The values are referred to as the *Gaussian curvature index* (GCI). The azimuth can be indicated by color-coding the maximum curvature direction (Marfurt, 2013).

Mapped curvature surfaces can also be used to establish the faulting pattern in the region. The effect of curvature is shown in figure 8–11, where an outcrop of the Madison Formation (350 Ma), which overlies the Bakken, contains open fractures. The detailed image demonstrates the function of a proppant in a fracture. In this case, the proppant is natural and is provided by the overlying Jurassic Sawtooth Formation (169 Ma), a sandstone filling the previously open fractures. As an aside, this evidence clearly dates the folding as Jurassic, forming part of the thin-skinned deformation of the Rocky Mountains associated with the Sevier orogeny (fig. 8–11).

The individual bed thickness also plays a large role in defining the intensity of folding and thus curvature. Currie et al. (1962) described the relationship between buckle folding of a thick single-layered unit versus folding of closely spaced multilayered units free to slip along bedding planes. Even though the formation in figure 8–12 was affected by equal compression where the bed thickness is millimeters thick, slippage allows for more intense folding.

It may be interesting to study rate of vertical curvature changes as a function of bed thickness on seismic data sets. Thin beds also are more strongly affected by anisotropy. As discussed earlier, anisotropy plays a vital role in the stress regime of thin-bedded shales.

Shales also fall into two other categories: hard shales and soft shales. Lonardelli et al. (2007) have described the degree of anisotropy in a shale on the basis of XRD. They used the term *orientation distribution* (OD) to assist in defining anisotropy. Physical properties of the two shales have significant bearing on the seismic response as well. Hard shales have higher densities (e.g., 2.5 vs. 2.2 g/cm^3 for soft shales). In the samples studied, both have the same illite/smectite ratio but have slightly higher quartz content. Porosities are very different: hard shales have 6% porosity, versus 26% for soft shales. The Thomsen anisotropy parameters are significantly different in the two types of shales studied by Lonardelli et al. (2007).

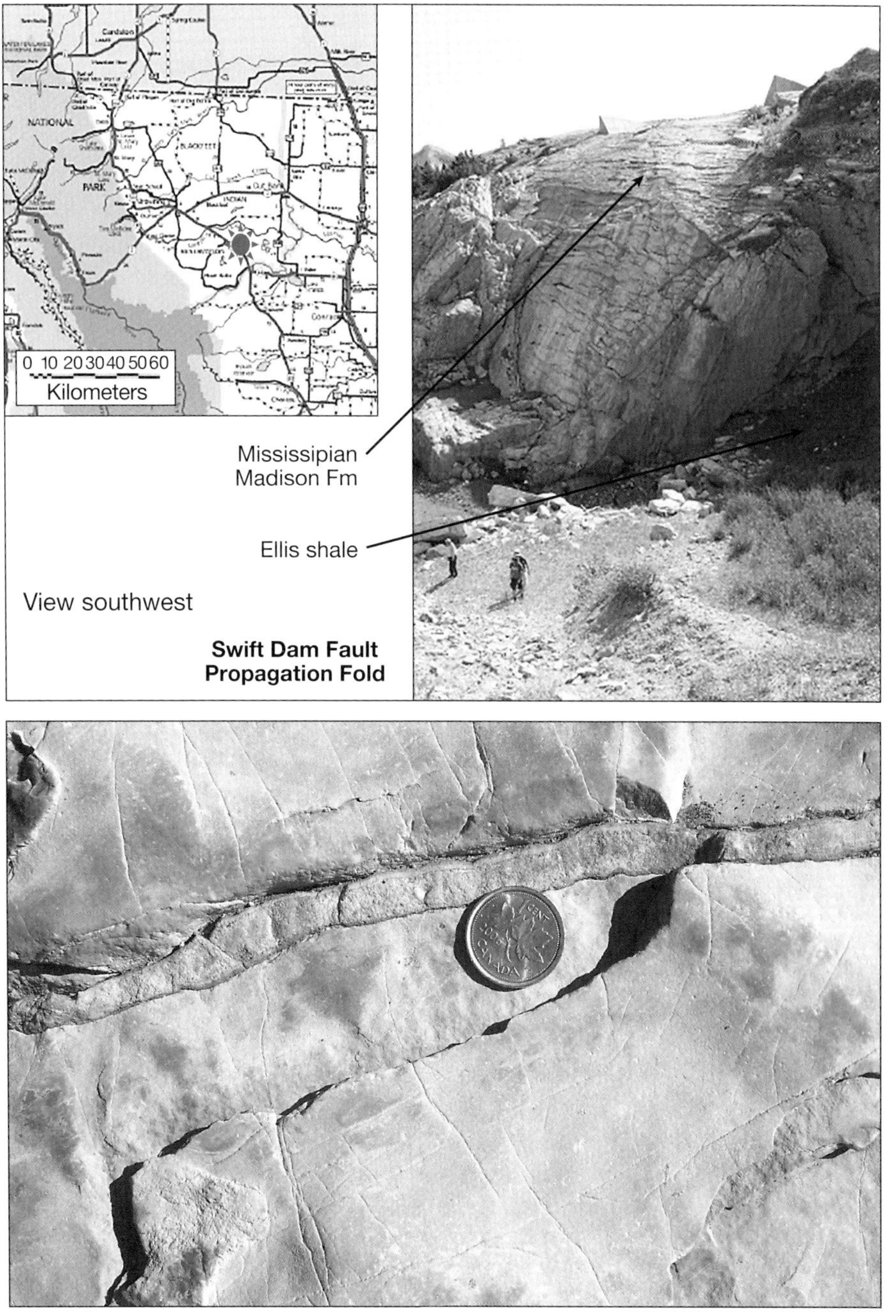

Fig. 8-11. Effects of curvature. A, Swift natural Dam, western Montana. (Photograph by the author) B, Fractured Carboniferous Madison Formation, showing natural proppant within the fracture. (Based on personal communication with Professor James Sears, University of Montana)

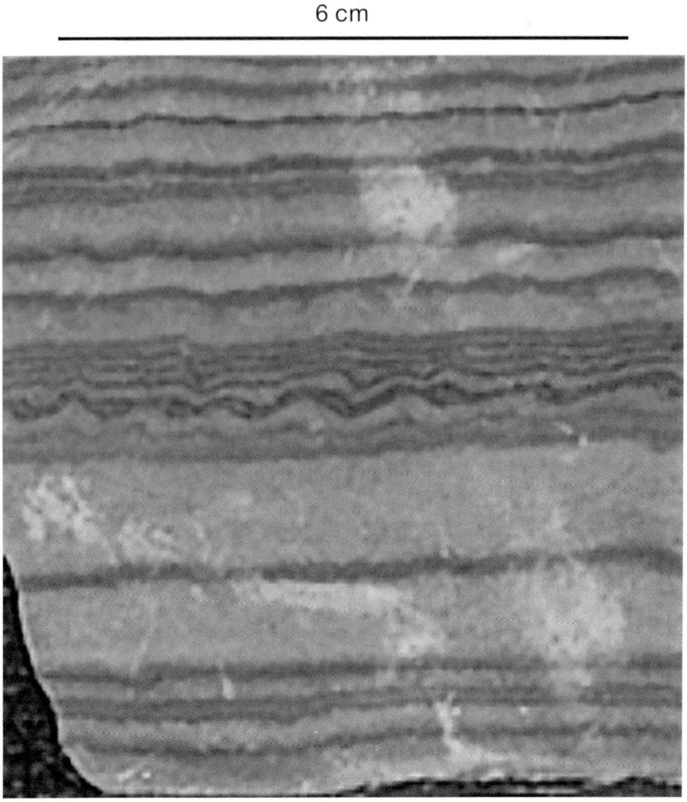

Fig. 8–12. Example of folding as a function of bed thickness, showing anhydrite from the Castile Formation, Permian Basin, West Texas (Photograph by the author)

GLOBAL AND REGIONAL STRESS MAPS

The ongoing 30-year investigation to produce a global tectonic stress map is being coordinated by a consortium of German universities, directed by the Heidelberg Academy of Sciences, in conjunction with government agencies, universities, and private oil companies (Heidbach et al., 2016). The results are series of continent-wide stress maps showing the dominant direction of the major horizontal stress field (fig. 8–13). Sibson (1977) stated that shear fractures should develop along the plane containing the intermediate axis (σ_2), also known as *greatest horizontal stress direction* in extension faulting terrains. The fractures form acute angles with the principal axis (σ_1). That is, if a horizontal well is drilled perpendicular to σ_2, then the hydraulically generated fractures will have wings horizontal and perpendicular to the well. If the well is drilled perpendicular to σ_3, then the fractures tend to be vertical and follow the wellbore (fig. 8–14).

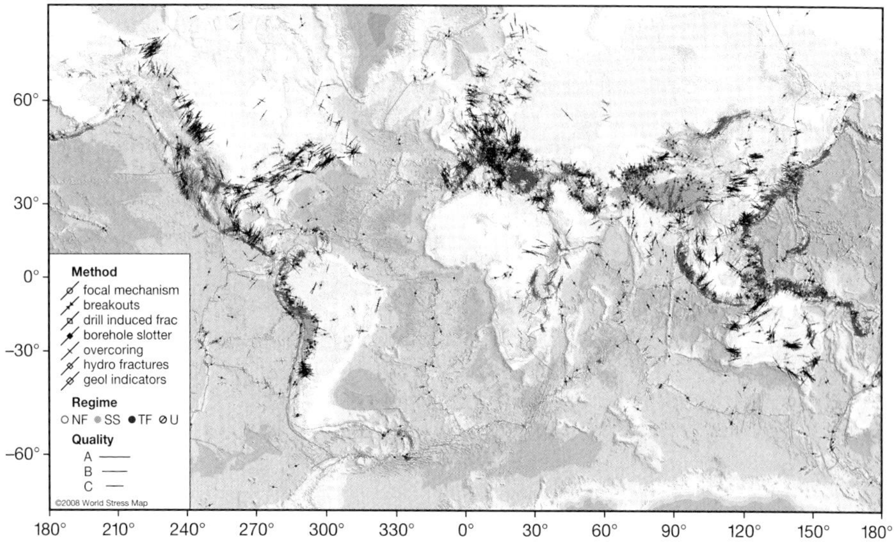

Fig. 8-13. Global and regional stress maps compiled by the Heidelberg Academy of Sciences, Germany (Reproduced with kind permission from Professor Oliver Heidbach; Heidbach et al., 2016)

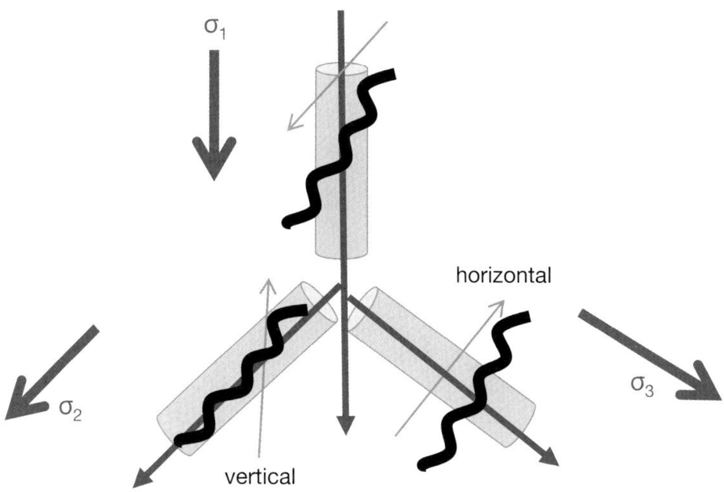

Fig. 8-14. Induced-fracture orientation given an east-to-west maximum horizontal stress as a function of well orientation (Based on data from Halliburton)

The acute angle that forms is the same as the computed angle of internal friction (ϕ) used in the Mohr circle (which will be discussed in the next section). Contractor companies provide extensive literature on the fracking process. The global tectonic stress maps indicate the direction of the maximum horizontal stress irrespective of the tectonic regime, as already mentioned. In regions of compressional tectonics, such as the Western United States and Western Canada, the maximum horizontal stress is in the σ_1 direction.

For most of eastern North America, the maximum horizontal stress (σ_2) direction is N 45 E. Therefore, fractures will tend to be directed northeast to southwest (fig. 8–15). The opposite is the case for western North America, except for the Western Canada basin. The overall stress field can be modified by local conditions, and either faulting or folding can modify the stress field. Once faulting occurs, the maximum horizontal stress is modified and must be taken into account in planning the direction of the horizontal well.

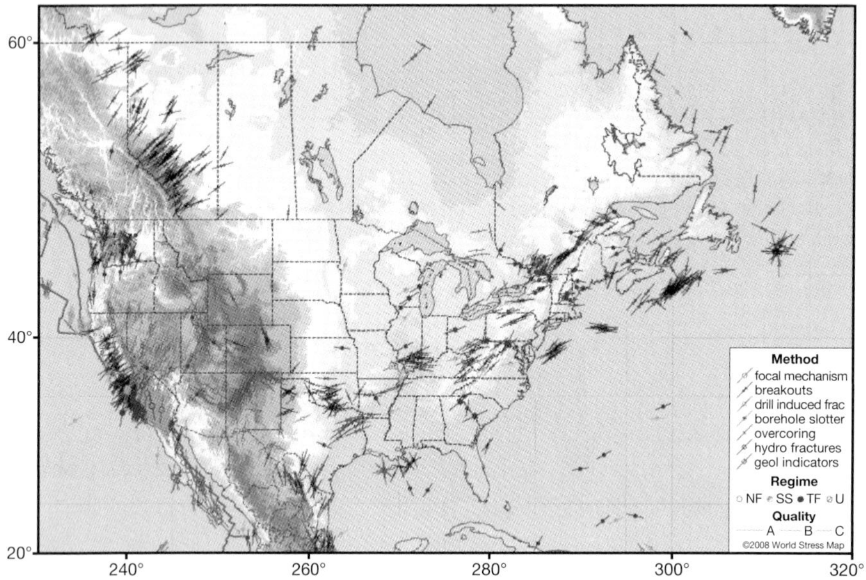

Fig. 8–15. Regional stress maps for North America. The maps show a maximum horizontal stress (σ_2) direction from northeast to southwest in the Eastern United States and from northwest to southeast in the western United States. The exception is Western Canada, where the Rocky Mountain trend is more intense. (Reproduced with permission from Professor Heidbach; Heidbach et al., 2016)

In McMullen County, South Texas (fig. 8–16), the effect of local tectonism is evident. In the lower center of the figure, the wells trend north-south whereas in the northern part of the county they are oriented northwest to southeast in order to cross the dominant regional natural fracture pattern. There is a significant fault between the two that modifies the stress regime resulting in an east-west dominant σ_2 direction to the south of the county. The fault corresponds to the underlying Lower Cretaceous Edwards reef trend.

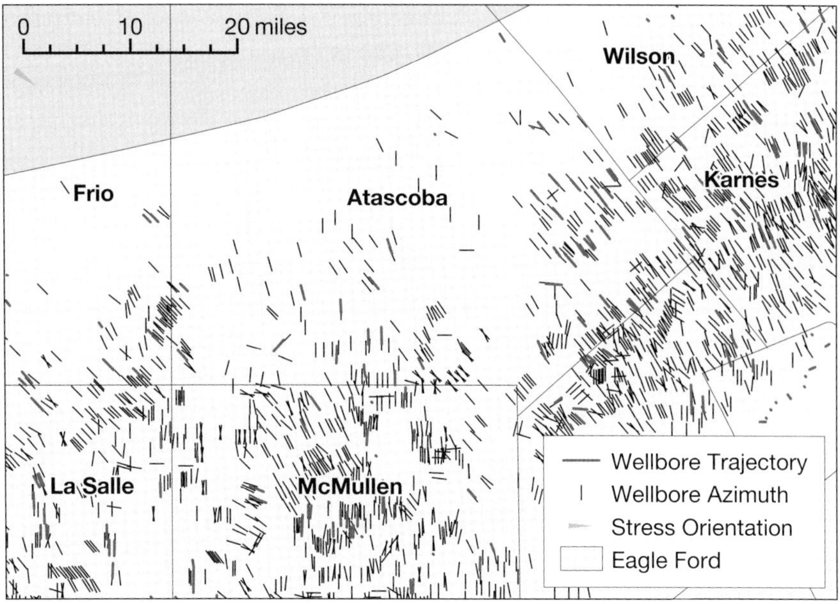

Fig. 8–16. Eagle Ford, South Texas, wellbore orientation. Most wells are drilled orthogonal to the northeast-to-south-west major stress direction, except in southern McMullen and La Salle counties. (Reproduced with kind permission from Drilling Info; Menchaca, 2014)

THE MOHR CIRCLE

Recall that the graph used to describe the Mohr circle is a representation of normal stress (σ_1 and σ_3) and its complementary shear stress (τ or σ_s). The Mohr circle allows for a more complete understanding of stress regimes and consequent fractures. The German civil engineer Christian Otto Mohr enhanced the application of the stress circle during the late 1880s, when he was tasked with calculating stresses on railroad bridges. He adapted and enhanced the existing stress circle to assist his analysis. His work led to the current Mohr circle, a technique applied in all construction, including the design of most civil engineering projects, and is also widely used in geoscience for structural analysis and fracture studies.

The Mohr circle represents stresses graphically, to predict the value at which structural failure occurs and establish the characteristics of the failure. The discussion below is hypothetical, and each case must be studied based on its own merits. In the model, the results were proven to be similar to the analysis generated by the uniaxial and triaxial compresses, as well as the computations using overburden stress.

When reviewing the literature on Mohr Circles, be aware of the convention to describe extension in the Mohr diagrams. The convention is opposite for civil engineering (where positive values are given along the *x*-axis) and geoscience (where negative values are given along the x-axis). For the purpose of this book, only the geoscience application is considered, although there are several other applications in structural geology; here, the objective is to determine the stresses required for fracturing to take place in the subsurface. Thus, we will first review the concept and then apply a real-Earth model to the results.

A core sample from the reservoir interval is analyzed in a triaxial compress. After application of an adequate vertical stress and a corresponding confining stress, brittle failure is observed. The sample is analyzed, and the angle of the fracture is calculated with respect to the vertical stress. By definition, the angle θ between the vertical stress, σ_1, and the plane of the fracture is equal to the angle between the normal (perpendicular) to the fracture, θ, and the horizontal stress, σ_3, because the major stresses are orthogonal (fig. 8–17).

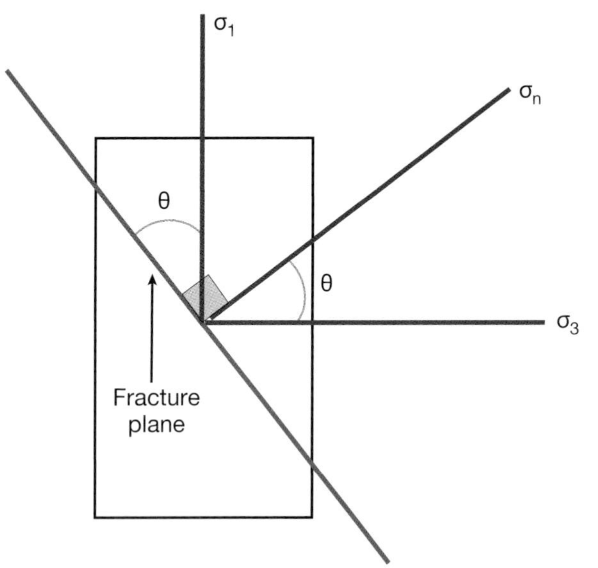

Fig. 8–17. Basic fractured block model used in Mohr circle analysis

From an evaluation of the stress regimes in the sample study area, it was determined that the σ_1 was 9,000 psi, and the minimum horizontal stress, σ_3, was 3,000 psi. For the sake of the model, incipient fracturing is observed. The two values are plotted along the *x*-axis in figure 8–18. A circle is drawn with a diameter of $\sigma_1 - \sigma_3$. An angle is drawn from the point with the coordinates (σ_1, 0) equal to the angle θ with the fracture plane determined in the triaxial test block (fig. 8–17). The angle line is extended to the circle, and at the point of intersection, a radius is drawn to center of the circle. By geometry, the angle between the new radius and the *x*-axis is 2θ. A tangent is drawn at the point of intersection with the circle and extrapolated to the *y*-axis. Because the failure is incipient, the line is described as the Mohr-Coulomb envelope.

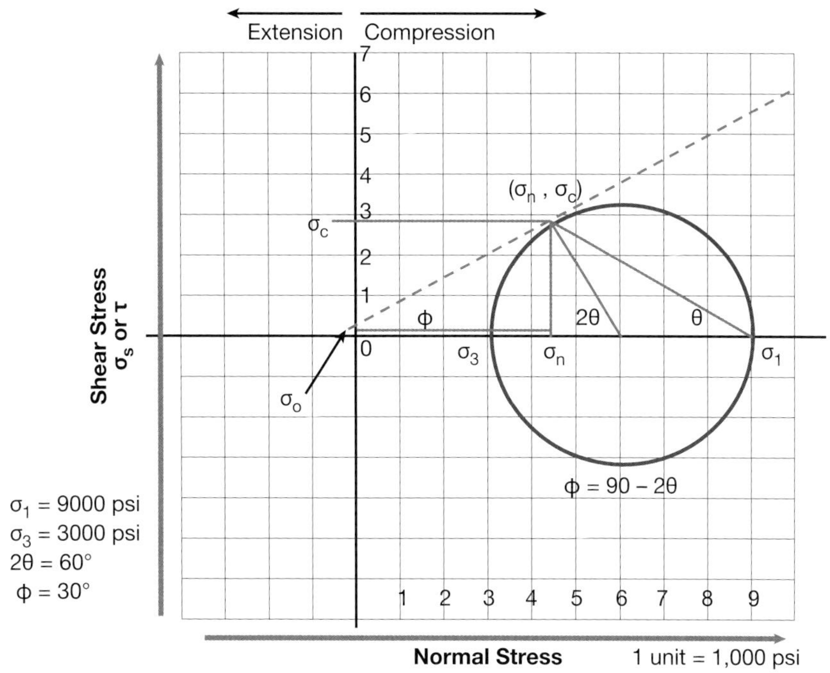

Fig. 8–18. Construction of the Mohr circle

To analyze this problem, let us assume that θ = 30°; thus, the angle 2θ = 60°. If σ_1 = 9,000 psi and σ_3 = 3,000 psi, by construction, angle ϕ = 90° – 2θ. The coordinates of the point of intersection of the tangent and the circle are (σ_n, σ_c). The slope of the line is tan ϕ. The angle of internal friction, ϕ, is a unique rock property and indicates the slope of the curve relating the strength of the rock to the increase in confining stress. As long as the same rock unit is tested and the stresses increase or decrease, the angle of friction should be approximately the same.

Several other observations can now be made on the Mohr circle. The y-intercept of the tangent line is $(\sigma_o, 0)$. The cohesive strength of the rock is the value corresponding to value of σ_o. Moreover, the purpose of carrying out uniaxial compression, or unconfined compression, is to obtain the true value of the cohesive strength, σ_o, which represents the point of failure during the unconfined stress test. As discussed previously, fractured rocks tend to have a cohesive strength of 0, meaning that the Mohr-Coulomb envelope for those rocks must pass through the origin. The coordinates for the point of intersection between the Mohr circle and its tangent translate to the normal stress value of the perpendicular to the fracture plane. For the model used, given σ_1 = 9,000 psi and σ_3 = 3,000 psi, σ_n = 4,500 psi. The sheer stress at which fracturing will occur is σ_c, corresponding to 2,900 psi. It should be pointed out that σ_c will increase as the vertical stress is increased.

Programs are currently available to compute the Mohr circles in both two and three dimensions. Several excellent website also discuss methods of constructing and analyzing the Mohr circle. The objective of the present discussion is to explain how the principle of the Mohr circle can be applied. As in the previous sections of this chapter, the value of σ_1 can be readily calculated from the overburden, and a reasonable estimate of σ_3 can be deduced from equation (8.3) because it uses the Poisson's ratio of the rock specimen. Three-dimensional Mohr diagrams simply plot the lower stress relationships inside the larger Mohr circle.

How are the values calculated? In a triaxial compress test, a reservoir sample is progressively subjected to vertical and confining stresses. Initially, a fixed confining pressure is maintained, and the vertical stress is increased until brittle failure is observed. A Mohr circle is drawn based on the observed values at the point of failure. A second test is run where the confining pressure is increased, and a new circle is drawn, again based on the point of failure (note that in a traditional triaxial compress, $\sigma_2 = \sigma_3$). In each case, the angle of the fracture to the σ_1 stress is recorded. As the σ_1 is increased and corresponding σ_3 is decreased, a family of circles and associated tangent lines, unique to the rock sample under investigation, are drawn that define the Mohr-Coulomb fracture envelope. Any stress beyond the Mohr-Coulomb envelope results in fractures.

One additional observation is that after fracturing, a dilation process is observed in the sample because fluid enters the cracks. Once unloading occurs, the fractures are affected by the closure index described previously (fig. 8–10). Any time the values of σ_c are exceeded, fracturing will occur. The circle is constructed on the basis of the relationships outlined above. The next logical step is to relate the Mohr circle data to the frac gradient.

The University of Toronto has been a center for extensive research on the topic of stress relationships because of its proximity to the Canadian mining industry. Initially under the direction of Eldon L. Brown, who was followed by Evert Hoek, research on the stress regime in mining began in the 1920s and culminated with the Hoek-Brown strength criteria for uniaxial and triaxial compression, and both investigators subsequently compiled a worldwide database of the relationship between σ_1 and σ_3 in rock mechanics (fig. 8–2). So how does the triaxial compress work?

As described earlier, a uniaxial compress is used to analyze fractures and, in particular, to obtain the unconfined tensile strength, σ_0, of the rock as the σ_1 is applied. The point of brittle failure was defined by Hoek and Brown as σ_0 and represents the cohesive strength of the rock. In the literature, the same value has been referred to as *c*, *C*, or *S*. As discussed earlier, the value of σ_0 is determined from the Mohr circle as well. It is the y-intercept of the plot of the Mohr-Coulomb envelope for a specific rock type and is derived directly from the point of fracture in a uniaxial (unconfined) stress field. The definition of Poisson's ratio is the internal deformation of the sample under stress and the corresponding shape change associated with the applied stress. The method used in a triaxial compression is to confine a sample and progressively increase the load until fracturing occurs. The relationship between σ_1 and σ_2 ($\sigma_2 = \sigma_3$) as a function of Poisson's ratio (ν) is given by equation (8.3).

Laboratory analyses have shown that the *stress concentration factor* (SCF) described by Hoek and Brown is very similar and closely related to the matrix stress coefficient, K_{mk}, used in the Eaton pore pressure determination. Thus, SCF can be expressed as follows:

$$\text{SCF} = K = \frac{\sigma_2}{\sigma_1} \tag{8.34}$$

Equation (8.34) is dependent on depth (see also fig. 8–3), the horizontal stress is the average between σ_2 and σ_3 and ν is Poisson's ratio. As discussed earlier, modern triaxial compresses allow σ_2 and σ_3 to vary.

Martin et al. (1999) have shown that the extension strain (ε_x) is then

$$\varepsilon_x = \frac{1}{E}[\sigma_3 - \nu(\sigma_1 + \sigma_2)] \tag{8.35}$$

where *E* is Young's modulus and ν is Poisson's ratio. The extension strain is defined as the change in length of the original sample versus the extended length after stress is applied. Note that in three-dimensional space, strain is described as ε_x, ε_y, and ε_z. It is a measure of the dilation that occurs after the fracture opens and fluids penetrate the fracture (Kwasniewski and Takahashi, 2010). Once the extension strain is calculated, the Young's modulus can be derived as shown in figure 8–19. In all of the analyses of stress/strain, it is assumed that only one sedimentary body is being affected. In a layered system, both the distribution of stress/strain in each unit and the effects of the boundary conditions need to be understood. For a more in-depth analysis of the problem, the reader is directed to Young and Budynas's (2007) book on stress/strain formulas.

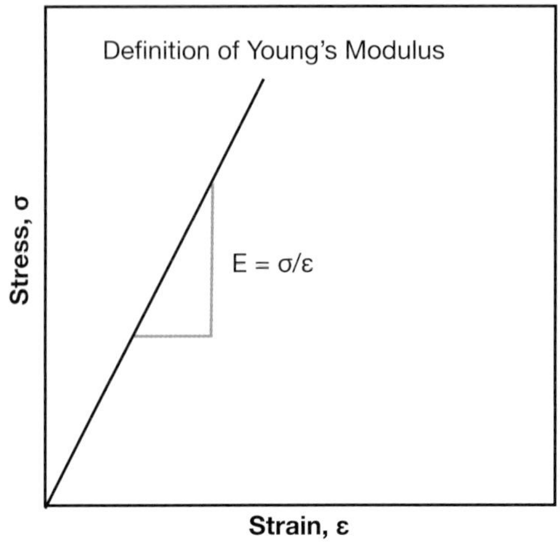

Fig. 8–19. Definition of Young's modulus

The basic principles of the Mohr circle are to define the points of brittle failure and determine the geometries of the consequent fractures. Given a particular fracture plane, the angle between the normal stress σ_n to that plane and the horizontal plane defines the failure criteria. Sibson (1977) described the shear stress (σc, often described as τ) as a function of the cohesive strength of the rock (σ_o is used in this book to ensure continuity), tan ϕ is the coefficient of internal friction and the slope of the tangent line ascribed to the Mohr circle, and P_p is the pore pressure:

$$\sigma_c = \sigma_o + \tan \phi(\sigma_n - P_p) \tag{8.36}$$

The value of σ_c corresponds to the *y*-value of the point of intersection of the circle with the Mohr-Coulomb envelope. The x-value of the same point is the value of σ_n and corresponds to the dimension of the normal stress to the fracture plane σ_n (fig. 8–18).

Once again, the Mohr circle is constructed using the relationships outlined above and progressively increasing the stress differential until the Mohr-Coulomb envelope can be generated.

Cohesive strength is a function of cementation. Sibson (1977) introduced a factor λ_f that can be applied to relate overburden pressure (σ_{ob}) and pore pressure (P_p) as follows:

$$\frac{P_p}{\sigma_{ob}} = \lambda_f \tag{8.37}$$

This λ_f value represents a practical approach to mapping pressure relationships in the subsurface, because it can be applied to individual formations rather than as a function of depth. As in the sample well from South Texas introduced in pore pressure discussion the well drilled to 6,728 ft (2,057 m). The lithostatic pressure (@1psi/ft) equals 6,728 psi and a pore pressure of 9.3 ppg yields 3,254 psi then $\lambda_f = 0.48$.

Fluids trapped in overpressure cells have an impact on the stress conditions in the subsurface. In essence, the fluids act to support the structure. Therefore, the effective stress due to pore pressure, P_{pe}, is the difference between the hydrostatic pressure P_h, or normal pressure, and the overpressure P_{po}:

$$P_{pe} = P_{po} - P_h \tag{8.38}$$

The frac pressure P_f is close to the overburden pressure. In an oil well, the value is obtained from a formation leak-off test as described in the next section of this chapter. Leak-off tests should be carried in several pilot wells to ensure the proper fracking procedure.

Fracking adds water laterally, diminishing the effects of σ_1 and σ_3 equally (fig. 8–20). In this manner, it simply shifts the Mohr circle to lower values until the fracture is achieved. In the real world, when the Mohr circle is first drawn, it is evident that no fractures are observed.

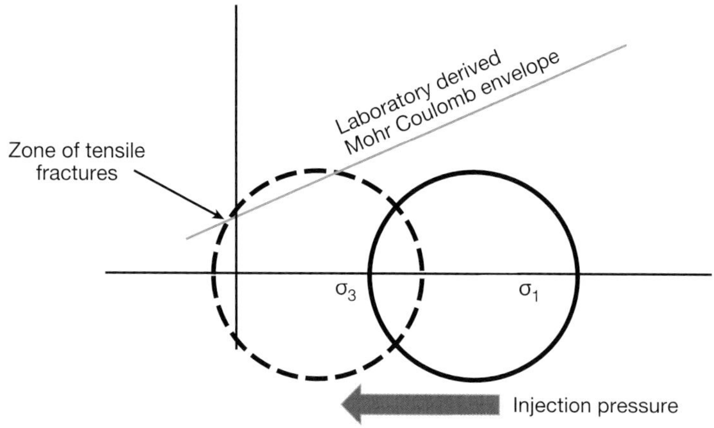

Fig. 8–20. Impact of fracking fluid injection on the Mohr circle. Mohr circle is forced to move to the left of the origin of the plot.

How much pressure must be introduced into the formation to cause fracking? The Mohr circle is constructed with the new parameters, and the Mohr-Coulomb envelope is drawn; by definition, any circle touching the Mohr-Coulomb envelope satisfies the condition of fracturing. Therefore, the pressure of injection should be such that the effective Mohr circle is moved to the left of the origin. Importantly, the effective pressure must be such that σ_3 is negative and, therefore, the Mohr-Coulomb envelope is on the tension side of the Mohr circle. Remember that fractures on the positive side of the circle will be compressional fractures.

How do we know if the induced hydraulic frac was extensional or compressional? Resolving the moments of displacement of the microseismic events provides the answer and will be discussed below. As will be discussed later (see the section "Capillary Pressure, Wettability, and Pore Size"), the limit of the fracking pressure is set by the seal capacity of the overlying unit.

LEAK-OFF TEST (MINIFRAC) AND PERMEABILITY

The leak-off test, or minifrac, is the most accurate determiner of the pressure required in order to produce fractures. Once the well is drilled above the target, a casing is set and the well continues drilling at least 30 ft (10 m) through the interval of interest. A bridge plug is set immediately above the zone of interest, and progressively mud is introduced under pressure. The mud loss is monitored, as well as the pressure at each stage. The pressure will increase until the formation fractures (fig. 8–21). The pressure buildup curve (i.e., the slope of the buildup) is a function of the permeability of the rock and can be computed from the standard pressure analysis Horner plot.

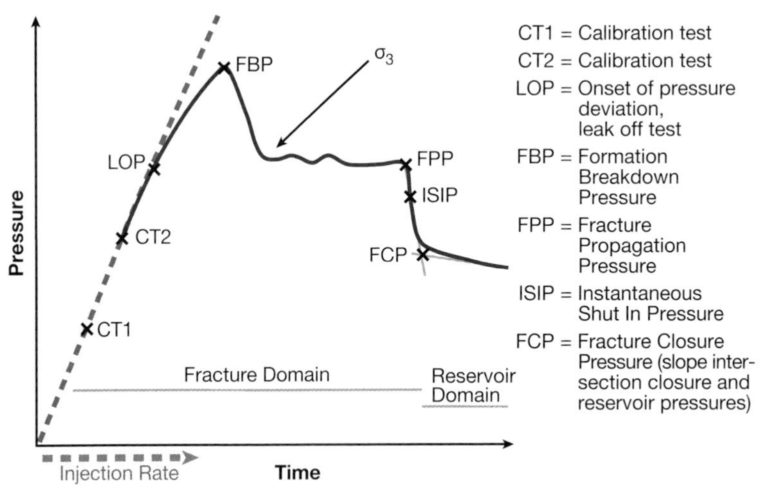

Fig. 8–21. Leak-off test plot, showing time and mud weight versus pressure

The leak-off test progressively builds up the pressure by increasing the mud weight until a pressure drop is observed (fig. 8–21). Currently, a wireline version is also available. At that point, the rock has fractured, and fluids are entering the newly created fractures. Up to the point of rupture, the mud added to the system is also recorded along the *x*-axis and continues to be added after fracturing the formation and until the *fracture breakdown pressure* (FBP) is reached. The FBP corresponds to the minimum horizontal stress (σ_3) and is one of the input data used to generate the global stress map (fig. 8–14 and fig. 8–15). Once the FBP is available the Mohr diagram should be redrawn. At that point, the well is shut in, and the pressure decline is monitored until it reaches a stabilized pressure known as the *fracture closure pressure* (FCP).

A leak-off test is fundamental for an understanding of the physical rock properties. Previous discussions in this book have focused on core analysis and log interpretations. By contrast, the leak-off test calculates the actual rock response. Based on all those parameters, the lateral variations are postulated and integrated with the seismic data to extrapolate the results laterally.

Permeability is the holy grail of a reservoir. If pores are not interconnected, they will not allow flow. In the preparation of thin sections for microscope petrography, the sample is injected with soft blue epoxy under vacuum conditions. All interconnected pores are filled with the bright blue epoxy; hence, interconnected porosity can be visibly seen and measured. However, it is often observed that some obvious pores are not filled with epoxy. These are isolated pores that do not participate in well production.

What is being sought is the effective porosity, as well as its relationship to the reservoir permeability. The most common formula for permeability is the Darcy equation, and in traditional laboratories permeability is determined with respect to *mercury injection capillary pressure* on a sample (as already mentioned in the contexts of core description of shales; see chaps. 2 and 5). Calibration of the measurements described above should lead to values similar to those from the leak-off test. For the leak-off test, the initial buildup curve is a function of the true permeability. The entry velocity of fluid into the formation is, by definition, its permeability.

APPLICATION OF MICROSEISMIC

Microseismic experiments are designed to measure emitted energy or vibrations associated with the hydraulic fracture and determine the extent of fracking (fig. 8–22). Adjacent pilot wells should have wireline geophones placed in their respective boreholes. Surface geophones are also placed on a semipermanent basis to record the individual and subsequent hydraulic fracs. Their bin size (described in chap. 6) should be appropriate for the target reservoir depth and sufficient to enable detailed location of the signal source—at subsurface source points known as *hypocenters*.

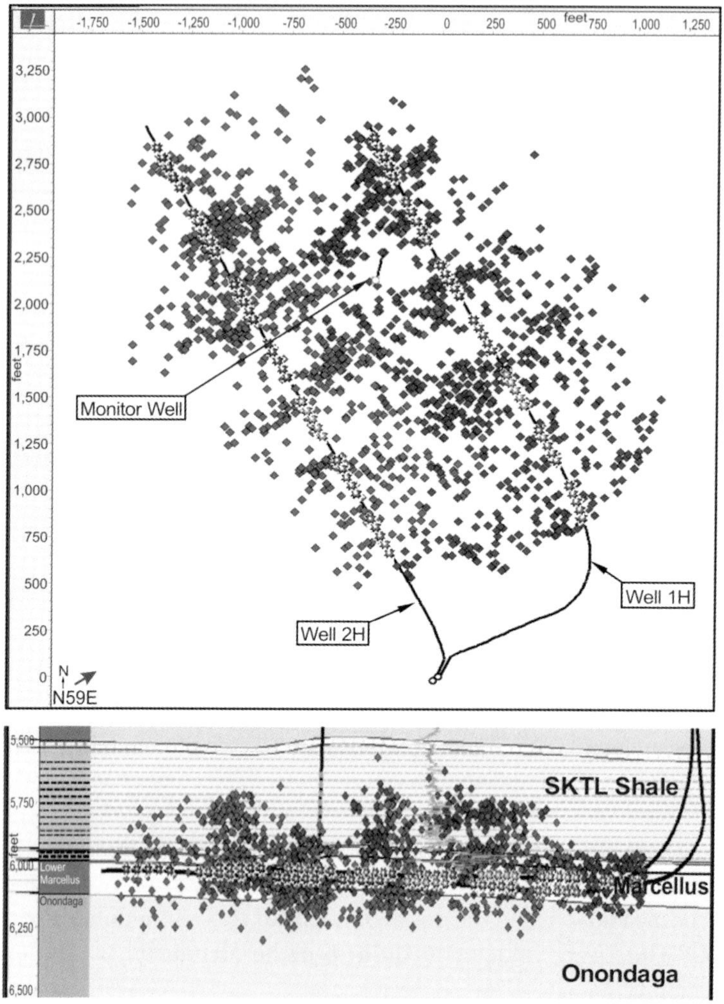

Fig. 8–22. Microfracture studies for the Marcellus shale (Reproduced from *The Leading Edge* with kind permission from the AGU; Warpinski, 2013)

Recording the arrival times at all geophones generates an accurate map, and a cross section can be determined for each hydraulic fracture event. Each event is usually color-coded for the purpose of differentiation (fig. 8–22). In addition, the acquired data can be analyzed to determine if the fracture behaves in an extension, compression, or sheer regime to determine the effectiveness of the hydraulic frac. This is done by analysis of focal mechanisms, and the results are shown on what is commonly referred to as a *beach-ball diagram* (fig. 8–23). A general description of the beach-ball diagram is available online (U.S. Geological Survey, 1969), and more in-depth discussion has been provided by Cronin (2004). The aspect most relevant to this discussion is that the displacement mechanism can be resolved and depicted on a map. Both the geologists and the geophysicist need to review the possible causes and means to predict trends prior to fracking.

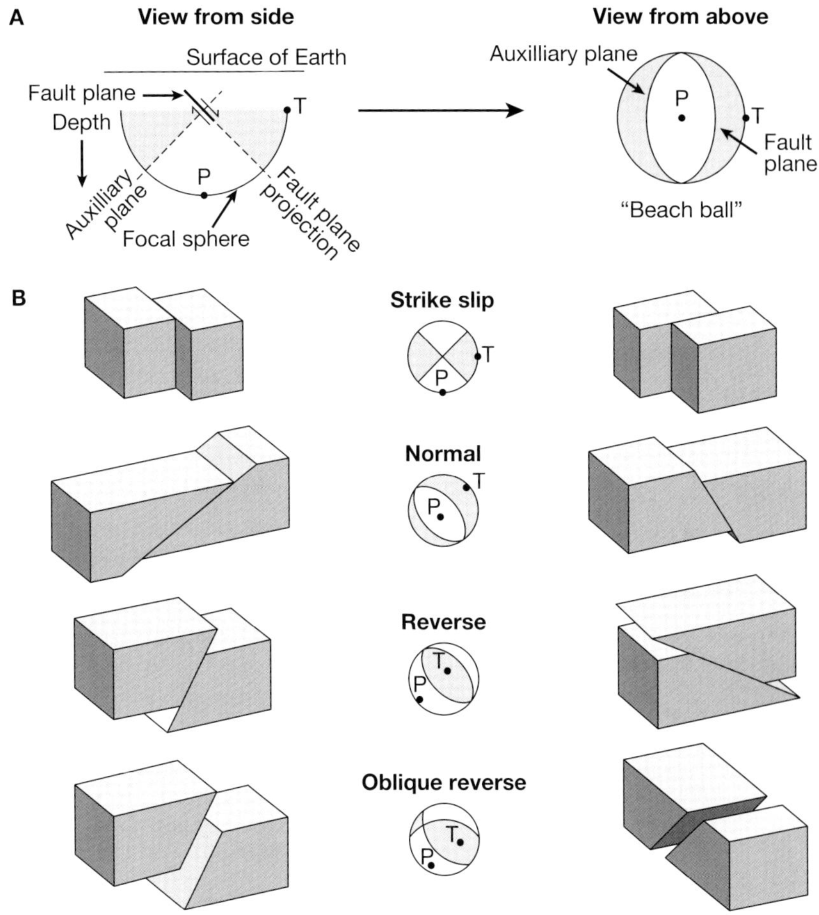

Fig. 8–23. Display of displacement of focal mechanism based on deformation style (Reproduced with kind permission from the U.S. Geological Survey, 1969)

Rich and Kahn (2017) have used an *R-T plot* to optimize description of the hydraulic fracture effect. In an R-T plot, a circle is drawn around the cloud of the frac wing, and the time is noted at the moment fracking stops and the seismic event ceases to record a relationship. Shapiro and Dinske (2009) have also pointed out that the lag time is a function of the enhanced permeability associated with fracking. The *r* value corresponds to the distance of the frac wing. The distance and the time are a function of the diffusivity of the hydraulic fracture.

The techniques used in processing microseismic data are similar to the velocity analysis and imaging techniques discussed earlier in this chapter (see the section "Seismic Applications to Pore Pressure Stress Analysis"). The hypocenters (subsurface source points) are determined using the same premise as for earthquake detection—namely, the differences in signal arrival times at the different geophones. Because of the distribution of geophones in various wells, it is also possible to determine the direction of fracturing (i.e., σ_2 direction). This is one source of information used for creation of a global stress map, as described above (see the section "Global and Regional Stress Maps"). The magnitude is provided by the actual test results.

In addition to the direction of fracture propagation during fracking of the horizontal well, the data also yield information on the height affected by the fracture. If there are preexisting faults at the reservoir level, then hydraulic fracking may not be appropriate (fig. 8–24). Several vertical wells should be used to determine the heights of frac fluid penetration. The goal is to frac inside the Middle Bakken and Three Forks formations. Any fracking fluids or proppant that penetrate the overlying Lodgepole and Mission Canyon formations (Madison Group) represent loss of proppant, because it cannot be produced; moreover, the fracture will act as a thief zone, allowing precious hydrocarbons to drain the new, unproducible upper interval.

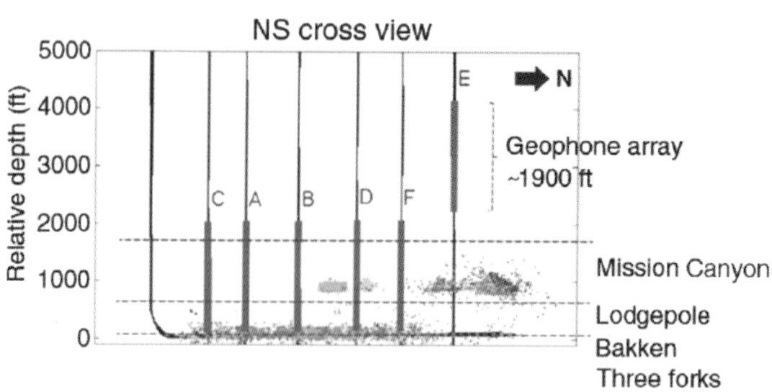

Fig. 8–24. Microseismic used to detect preexisting fractures and faults (Reproduced with kind permission from the SEG; Yang and Zoback, 2014)

FRACKING AND DRILLING PROCESS

In this section, the focus is on the application of geoscience methods to advanced exploration. Even though the mechanical aspects are not part of this discussion, geoscientists need to be familiar with the general procedures. Fracking horizontal wells requires multiple-stage hydraulic fracs. Currently, two techniques are used: sliding ball-and-sleeve system (fig. 8–25); and selected plug and perforations, or *plug and perf.*

Fig. 8–25. Staged fracking process showing sliding ball-and-sleeve system (Reproduced with kind permission from Drilling Info; Triepke, 2014)

Drilling the horizontal well is usually carried out with a drilling process known as coiled tubing, thus reducing drilling time. The drill string is continuous, allowing very efficient drilling. The sliding ball-and-sleeve method of fracking (fig. 8–25) is predesigned and uniform. The sleeves open sequentially, and balls are large at the toe and smaller toward the heel. As fracking takes place, the sleeves are closed, and balls are suctioned into position and then closed. For plug and perf, the perforations are selected on the basis of specific geological parameters. For example, in the context of microseismic (see the section "Application of Microseismic"), the sliding ball-and-sleeve system would not be able to avoid faulted intervals. Where there is little or no natural fracturing, the sliding ball-and-sleeve system is adequate.

Both methods are used in practice. One limitation of the ball-and-sleeve system is that the completion does not allow for refracking. This is becoming common practice in oil and gas field development, and there can be up to 40 frac stages in a well. It is required by state agencies and most countries that well schematics (fig. 8–26) be supplied to regulatory agencies to ensure that proper procedures are followed. It is incumbent on the geoscientists to review the drilling schematics and make recommendations based on their experience in the region. The well schematics should be compared to others in the area and updated to improve efficiency and reduce drilling time. We normally refer to the heel of the horizontal well nearest to the vertical section and the toe at its furthest reach. Similar completions and frac stages should be used in comparative studies.

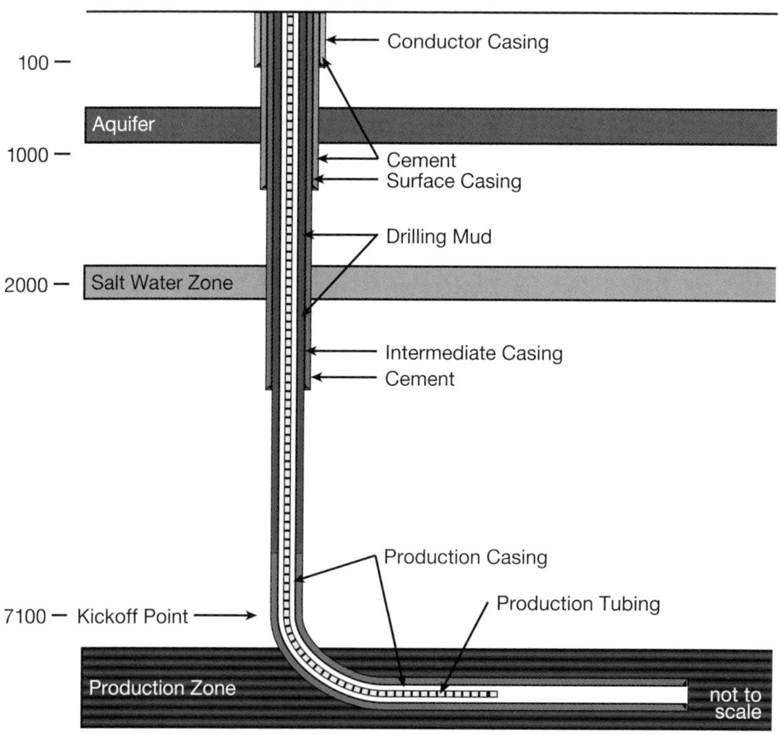

Fig. 8–26. Simplified drilling schematics for a horizontal well. More detailed plans are usually required, including the number of frac stages proposed. (Reproduced with kind permission from Drilling Info; Roach, 2014)

CAPILLARY PRESSURE, WETTABILTY, AND PORE SIZE

Capillary pressure, wettability, and pore throat size are discussed as a group because they play a significant role in producing hydrocarbons and are closely related. By definition, *capillary pressure* is the pressure differential observed in a reservoir between two immiscible fluids. Three combinations are possible: oil/water, gas/water, and gas/oil. In water-drive reservoirs, all three can be predicted and mapped on the basis of the distribution of three fluids and their pressure regimes. The simple water drive serves as a good model to understand the methods applied in this portion of the study. In a solution gas drive, common to most shale resource plays, determining the capillary pressures is of great importance in establishing the seal capacity of the overlying bed and the relative permeability of seal and reservoir, both of which are functions of pore throat size.

Wettability is the degree to which the solid in the pore space has an affinity to water (*hydrophilic*) or oil (*hydrophobic*). Because significant amounts of water coexist with the hydrocarbons in shales, the subject of water requires detailed examination. Moreover, as discussed earlier (in chaps. 2 and 5), water is either free flowing or bound by the clay and organic material. The wettability is directly related to the adsorption capacity of the reservoir, or the Langmuir effect (discussed in chap. 2).

Because of the effect of buoyancy, hydrocarbons tend to migrate upward into lower-pressure regimes. The smaller the pore throat is in the seal rock, the larger the hydrocarbon column will be that it can support before the seal is broken. Multiple small-diameter pipes will flow considerably less efficiently than one large-diameter pipe. For shales, there are two important considerations: first they have very small pore throats; and second, they are combined with organic materials that also have high wettability coefficients. As discussed previously, a water column has a hydrostatic gradient of 0.456 psi/ft, methane has a gradient of 0.1 psi/ft, and liquid oil has a gradient of 0.32 psi/ft. Therefore, their corresponding pressures can be calculated at the reservoir level.

Capillary pressure can be determined on a sample under laboratory conditions. A classic porous reservoir usually has a well-established oil/water contact and a gas/oil contact. The actual reservoir pressures detected in the well will be different for each of the three fluids owing to their density contrasts and the pressure gradients listed in the previous paragraph; by definition, those differences are the capillary pressures.

Three methods are commonly applied to evaluate capillary pressure: porous plate; mercury injection; and centrifuge. Other methods are being considered, but consistent, proven methodologies are preferred to enable better comparisons between various data sets. The objectives of the three methods are similar; therefore, only the first two are described in detail here.

The porous plate method is usually used for a gas/water system, and mercury injection is used for an oil/water system. In both cases, a sample core plug is cleaned and polished. For the first method, the polished end of the plug is placed against a porous plate with a known porosity and permeability. Water is injected into both the porous plate and the sample until fully saturated. Air is then injected under staged pressure changes until all the water is displaced. The reason the injection is staged is that the permeability of the sample is a function of the pore throat available and the consequent delay in the transmissibility of the evacuated water. Initial low air pressure results in little or no displacement. The velocity of transmissibility is a function of permeability by definition. Once the surface tension can be overcome, air will begin to enter the sample. This critical entry pressure is related to pore throat size. As equilibrium is reached at each stage, the corresponding water saturation is noted. The pressure is increased until the air is detected at the porous plate. In most reservoirs, this occurs when the water saturation is 10%–20%. The same process should be repeated several times to ensure accuracy. The lag time in the equilibration is used to determine the relative permeability.

Mercury is a good substitute to represent an oil/water system because it has similar wettability characteristics and hence can be translated into its oil equivalency. A hysteresis plot is generated for a mercury injection test (fig. 8–27). In essence, a water-soaked sample is immersed in mercury and sealed in a container with a long neck to allow measuring the mercury column (although dry samples may also be used). After cleaning the sample of hydrocarbons, brine should be introduced to simulate the original reservoir conditions. An error is also introduced in cleaning and preparation stages because some of the native organic material in the sample may also be removed during cleaning. As mercury is introduced, under staged pressure, the entry pressure can be determined by the ease of entry at each stage. (The theory and calculations for this method have been summarized in an online resource by Professor Paul Glover, 2017).

For the North Sea Chalk in figure 8–27, the entry pressure (P_e) of 10–25 psi (50–100 kPa) can be sustained as the water saturation decreases. As the mercury enters the rock pore space and replaces the water, the level gauge indicates the relative water saturation (stage A-B in fig. 8–27); this is known as the *drainage stage* (Tweheyo et al., 2001). The injection is continued until the level of irreducible water is reached (point B in fig. 8–27). As the pressure is reduced, the result is plotted as the imbibition curve. Once the water pressure exceeds the oil pressure, the capillary pressure yields negative values.

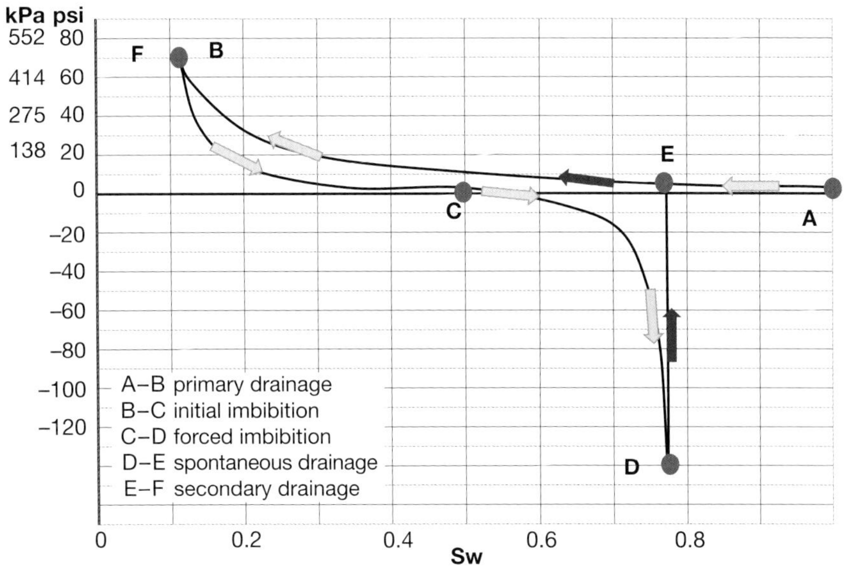

Fig. 8–27. Typical hysteresis plot for a porous media using example from a North Sea chalk (Adapted from Tweheyo et al., 2001)

Using the data set from which the relative permeability was obtained, Tweheyo et al. (2001) also demonstrated how to determine the Archie resistivity index, I (as discussed in chap. 5). The resistivity index can be obtained because in their centrifuge analysis they used a hydrocarbon and brine. By placing electrodes and a charge into the solution, Tweheyo et al. were able to determine the resistivity index. The theory can be described by a simple model of a capillary tube of a given diameter placed in a dish of brine water; the container is then filled with oil, and as the oil volume is increased, the height of the water in the capillary tube rises as a function of the density change between the two fluids multiplied by gravity, the diameter of the capillary tube, and the surface tension of the brine. The difference in nature is that in place of density times gravity (a measure of pressure), actual pressure differentials are used; hence,

$$P_c = hg(\rho_w - \rho_o) \tag{8.39}$$

where P_c is the capillary pressure, h is the height of the column, ρ_w is the density of water, and ρ_o is the density of oil.

We also know that the capillary pressure is a function of the wetting angle, φ, and the radius of the pore throat, r_c:

$$p_c = p_o - p_w = \frac{2\tau_w \cos\varphi}{r_c} \times A \tag{8.40}$$

where p_c is the capillary pressure, p_o is the oil pressure in contact with water, p_w is the water (wetting fluid) pressure, τ_w is the interfacial tension, φ is the contact angle, r_c is the pore throat radius, and the constant $A = 0.145$ converts the result to psi (Glover, 2017).

The third method applies the centrifugal force as an injecting pressure agent (Tweheyo et al., 2001; Glover, 2017.). One common approach follows the U.S. Bureau of Mines (USBM) wettability standard (Donaldson et al., 1969). The sample is desorbed, saturated with wetting fluid, and placed in a centrifuge to separate the movable from the adsorbed fluids; often hydrocarbons are used. However, this method is costly and time consuming. Thyne (2013) has proposed different approaches, once again, using techniques from other disciplines—namely, nuclear magnetic resonance (NMR). In chapter 5, it was noted that NMR is able to distinguish between bound water and free water and provide a measure of pore throat size, and the same procedure can be applied on a sample to determine the wettability and correlates well with the USBM method.

Porosities also need to be studied. Either crushed-sample methods (described in chap. 2) or less destructive methods can be applied (Kuila et al., 2014). The key relationship for a reservoir is to tie the porosity to the permeability.

WETTABILITY

Once again, it is easier to understand wettability when related to a simple system. Consider a metal plate laid flat and measure its wetting properties. Wetting characteristics are a function of two fluid interfaces and their response to the wetting surface (fig. 8–28). A drop of oil on the plate will bead, and there will be a measurable obtuse angle between the substrate surface and the droplet. Water on the same surface will spread, and an acute angle will form between the medium and the water drop. A particular surface can have an affinity to either water (hydrophilic) or oil (hydrophobic).

How is the contact angle measured? Special microscopes have been designed to do just that because measuring the contact angle is also an industrial issue. In addition, several other techniques are used, and the best known and easiest to understand of these is the *Wilhelmy plate* method (fig. 8–28). In this method, a thin roughened platinum plate, is used as an adsorbing surface. It is suspended in the liquid to be tested. As the liquid rises on the plate, the contact angle can be measured. The Wilhelmy plate experiment serves two functions: First, while the plate is suspended in the fluid, the force applied is the weight of the plate, and the angle, φ, can be measured; second, from the angle, the surface tension, τ_w, can be calculated (expressed in dynes/cm^2). Other studies have measured the weight of a drop and determined the relationships associated with the surface tension,

τ_w, of the drop. By repetition of the Wilhelmy plate analysis with brines of different salinity (i.e., changes in density), the contact angle for different oils with various viscosities and densities can be determined.

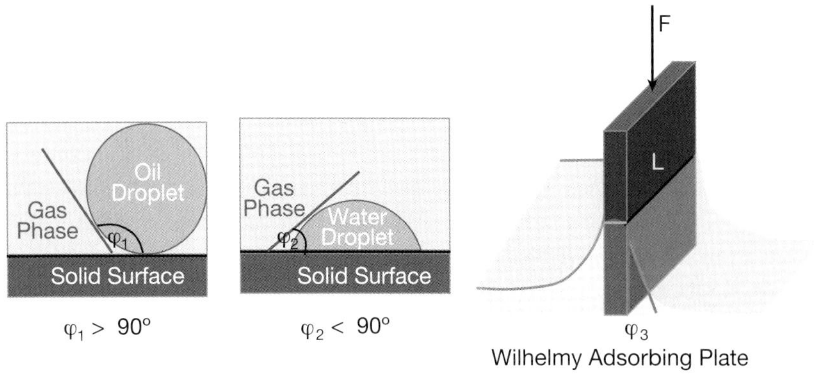

Fig. 8–28. Wetting phase. An oil droplet on a plate in the presence of gas (air) and a water droplet on the same plate that has an interface with the same gas will have different wetting angles (φ). For measurement of the wetting angle, a Wilhelmy plate is suspended in the liquid, and the angle is measured. Force applied to the plate is related to the surface tension.

An additional issue to consider is the roughness of the surface. Organic material tends to have rough surfaces that modify the contact angle. As the force F is applied to the plate, the contact wetting angle will increase, and the interfacial tension, τ_w, can be calculated as follows:

$$\tau_w = \frac{F}{L\cos\varphi} \tag{8.41}$$

where L is the length of the wetting surface along the plate.

Certain materials are more hydrophilic than others. For example, common soap can be compared to an organic system present in source rocks: an oxalic acid root attached to a fatty acid (paraffin) tail (fig. 8–29). The oxalic acid belongs to a family of carboxylic acids that attach themselves to several normal parraffins (Eglinton et al., 1987). The clear similarity between soap and the oxalic complex and improves our understanding of how the latter interacts with the clay mineral.

The basis for soap chemistry is a mixture of a normal alkane (fatty acid) and an organic salt (sodium-oxygen compound). Recall the basic formula for the normal alkanes, C_nH_{2n+2}; each carbon has two hydrogen atoms, and the carbons form a straight chain, with the orientation of hydrogen atoms being a systematic rotation of the unit cells. Because water is a polar molecule, it will be attracted to any surface that has polarity or relative charge. The tail of the soap is polar neutral and hence repels water (i.e., hydrophobic). The head is polar owing to the presence of oxygen and sodium and hence attracts water (hydrophilic) and would also be adsorbed on the charged clay surface (fig. 8–30). Therefore, a portion of the soap will attach to water, and the remainder will collect the oils. As an aside, this is the function of surfactants added to the fracking fluids. In a simple case, the same applies to the oxalic acid/paraffin complex. While the oxalic head is attached to the substrate, the paraffin tail repels water and is oil wet. The polar water molecule and the carried cations are also adsorbed as shown in figure 8–30.

So, are resource shales hydrophilic or hydrophobic? Because resource shales contain abundant organic materials and the clay minerals have either a T-O or T-O-T crystal structure with either oxygen atoms or hydroxyl (OH) groups facing outward, shales should by definition be hydrophilic. Unfortunately, it is not that simple, for several reasons:

- First, OH-dominant surfaces provide hydrophobic surfaces; however, in kaolinite, the tetrahedral portion is not OH dominant, but the octahedral layer is. Therefore, kaolinite is more hydrophobic than smectite and illite. The external surface of the smectite tends to be more hydrophobic, and the internal space tends to be more hydrophilic—hence, the bound water.
- Second, there is an abundance of organic material. Many organic compounds but not all of them are hydrophobic.
- Returning to wettability, a third factor that requires study is the rugosity of the clay minerals. The higher the rugosity is, the greater will be the adsorption capacity.
- Fourth, smectite can be dioctahedral or trioctahedral. Dioctahedral smectite contains trivalent cations (Al^{3+}, Fe^{3+}) in an octahedral layer, whereas trioctahedral smectite contains divalent cations (Ca^{+2}, Mg^{2+}, Fe^{+2}). The difference is important because the surface charges are different; hence, the wettability is also different. As noted in chapter 2, smectite favors larger ,smoother platelets (fig. 2–21) whereas illite has a much more rugose crystal habit (fig. 2–20). Little has been published on this topic, and its final effect on wettability is not clear. In general, both the organic-rich Bakken Formation of the Williston Basin and the Eagle Ford of South Texas tend to be more oil wet. Discussions usually focus on the end members, oil wet or water wet; however, in practice, wettability in resource shales range between the two.

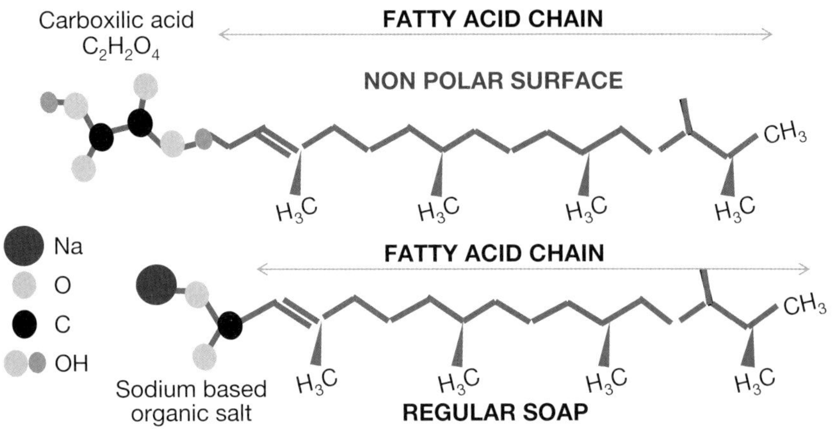

Fig. 8-29. Comparison of soap molecule to oxalic acid/paraffin complex to evaluate wetting behavior

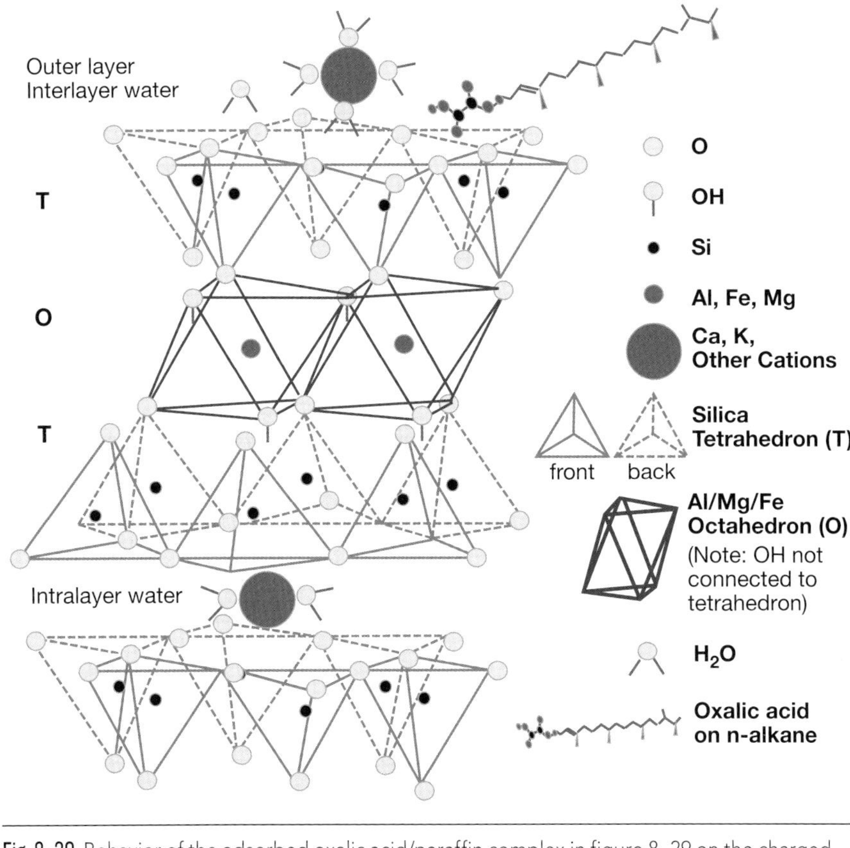

Fig. 8-30. Behavior of the adsorbed oxalic acid/paraffin complex in figure 8–29 on the charged clay surface

PORE THROAT SIZE AND SEAL DEFINITION

The third and fourth component of the study can now be determined—namely, the pore throat size and the relative permeability. Solving equation (8.40) gives a range of pore throat values. Two independent confirmations of these results should be made: NMR estimates pore throat size, and, even more accurately, ion beam milling (IBM) with SEM and the focused ion beam (FIB; see chap. 2) provide details of the pore geometry and flow characteristics. As pressure is applied and the injecting mercury is forced into the sample, the mercury will first enter the larger pore throats and reach equilibrium. The time involved for entry is directly proportional to the pore throat size. From the previous discussion, note that as a force (i.e., pressure) is applied to a wetting angle, the angle increases; hence, the surface tension decreases. During fracking, additional pressure is applied. As the surface tension is reduced, are there any noticeable changes in the oil chemistry and, as a consequence, in the adsorption capacity? Testing of the oils before and after fracking may answer this question.

Now let us return to shale problem. As can be seen from all the previous discussions, clean treated samples are used. Shales have many characteristics, but their surfaces are rarely clean. Furthermore, for the most part, the analyses listed were not designed to measure nanopores. To better understand this problem, Josh et al. (2012) have taken a fully integrated approach to the analysis of shales. By combining all the tools described in chapter 2, they extrapolated their observations to the capillary pressure problem in shales.

Chalmers and Bustin (2010) have raised another important issue. They studied the problem of wetness in resource shales and noted that pore space fill is a function not only of the surface available inside the pore space but also of how the fluid migrated into the pore space. They proposed three conditions to fill available pore space. We will focus on the principal two: First, is the smallest pore throat hydrophobic (quartz/carbonate), thus allowing hydrocarbons and water to enter the pore and attach to hydrophilic clays and aliphatic kerogens (paraffins) and hydrophobic bitumen-coated clays as well as aromatic-rich kerogens? Second, is the smallest pore throat hydrophilic, thus reducing the entry capacity of hydrocarbons?

Low-permeability media give rise to additional problems. Akin and Kovscek (1999) used specially designed equipment to test imbibition in low-permeability material. To test for gas/water interaction, they used air and water in their study. More recently, to compare how capillary pressure affects injected fluids in various resource plays, Donnelly et al. (2016) carried out extensive investigations, following the Brooks-Corey equation:

$$\theta_v = \theta_r + (\phi - \theta_r)\left(\frac{P_e}{P_c}\right)^{\lambda} \tag{8.42}$$

where θ_v is the volumetric wetting; θ_r is the residual wetting fluid content; ϕ is the porosity, as measured by helium gas injection; λ is a function of pore size; P_e is the entry pressure, the capillary pressure at which pores can drain; and P_c is the capillary pressure across the curved interface between the two immiscible fluids. Also refer to the work of Szymkiewicz (2013) for additional information on this topic and background on the theoretical aspects of wettability.

The residual water content is a function of water saturation and porosity (Szymkiewicz, 2013). For $P_c > P_e$, equation (8.42) is directly proportional to water saturation. Further,

$$P_c = \frac{2\tau_w \cos\varphi}{r} \tag{8.43}$$

where τ_w is the interfacial tension, φ is the wetness angle, and r is the pore radius.

The entry pressure is defined as the critical pressure at which the injected material is able to enter the host rock. In other words, as mercury is injected into the sample, the small pore throat impedes entry until the threshold is overcome. At the same pressure, additional mercury is able to further saturate the sample (Pittman, 1992). The entry pressure (P_e) is directly proportional to the pore throat size:

$$P_e = \ln\left(\frac{p}{p_o}\right)\frac{RT}{V_m} \tag{8.44}$$

where p is the vapor pressure within the porous media, p_o is the vapor pressure of pure water at the same temperature, p/p_o is the humidity factor, R is the universal gas constant, T is the temperature (in K), and V_m is the molar volume of water.

References to p/p_o ratios for clay minerals are common in clay mineral literature. For example, Moore and Lockner (2007) have indicated that montmorillonite (smectite) with p/p_o ranges from 0.5 to 0.6 for intralayer humidity tend to have one or two layers of water. At p/p_o ranges greater than 0.95, it has three layers. On the external surface, a p/p_o of between 0.1 and 0.25 is monolayer, whereas there are two layers starting at $p/p_o = 0.6$, three layers at $p/p_o = 0.85$, and six layers at saturation. Studies have shown that a simple water/air interface on a substrate is a function of the surface tension of the water and the radius of curvature of the pore lining surface. This is expressed by the Laplace equation (Szymkiewicz, 2013):

$$\Delta p = p_a - p_w = \sigma_{aw}\left(\frac{1}{r_{c1}} + \frac{1}{r_{c2}}\right) \tag{8.45}$$

where p_a is the pressure of air, p_w = pressure of water, and $[(1/r_{c1}) + (1/r_{c2})]$ is the facing-grain radius of curvature.

As mentioned previously, the velocity of filling the pores in the imbibition study was an indication of relative permeability:

$$V = \frac{1}{\alpha} \times \frac{g}{\mu} \times \frac{\Delta p}{L} \tag{8.46}$$

where V is the velocity, $1/\alpha$ is the permeability factor, g is gravity, μ is the viscosity, Δp is the pressure differential, and L is the pore length. One of the key studies on this topic was by Swanson (1981).

A very detailed review of the methodologies available to analyze relative permeability was undertaken by Li and Horne (2006). Once again, independent determinations of permeabilities should be compared. The best comparison is with the leak-off test described earlier in this chapter. Naturally, all the results have to be reintroduced into the petrophysical analysis.

The mercury injection pressure is directly proportional to pore throat. As pressure is increased, there is little entry until the critical threshold pressure is reached that allows penetration into the tight rock. The controlling factor on seal capacity is the buoyancy differential. Bolger and Reifenstuhl (2008) have provided the Sneider seal classification shown. Due to the effect of buoyancy, a column of oil can exert an entry pressure in excess of 6000 psi. If the fracking pressure combined with the reservoir pressure exceeds the required entry pressure minimum for the seal integrity, then the seal will be broken.

Table 8–1. Sneider seal classification of critical threshold pressure (*P*)

Class	Seal entry pressure psi	Column held ft
A+	>6,868	>5,000
A	1,373–6,868	1,000–5,000
B	687–1,373	500–1,000
C	137–687	100–500
D	69–137	50–100
E	14–69	<50
F	<14	Waste zone

It is important to consider the observed behavior of water and hydrocarbon in the shale play and to attempt to understand differences in water and hydrocarbon production as a function of the many variables discussed in the capillary pressure analysis. As the problems present themselves in predicting production efficiencies, more detailed analyses are recommended.

POSTDRILL REVIEW

A common practice in many oil companies is to have a detailed review of the following:

- What was known prior to drilling
- What was learned during drilling
- What questions need answers
- What immediate recommendations to make

The postdrill review is commonly done in two stages. The first meeting is a short discussion with upper management, bringing them up to speed on the results of the well. The geologists present the proposed depositional model, along with the expected physical and chemical conditions proposed for the well, including diagenesis of sediments, pore pressure, and temperature issues. The geologists summarize what has changed and how well the prediction model functioned. The geophysicist follows up with a similar review focused on what the seismic data proposed at the target zone and how accurately they predicted the results. Is there anything new that needs to be done to improve future outcomes? The petrophysicist reviews the predictions and the results and makes suggestions. The drilling engineer reviews the proposed drilling time and schematics, discusses the results, and suggests modifications. The reservoir engineer discusses the potential and the results of the well and summarizes the cost of the well. Either the reservoir engineer or, preferably, the environmental engineer should review the anticipated and actual impact of the well on the environment. At these meetings, the manager in charge decides on the way forward and approves the time and expenditure for follow-up studies.

As soon as possible, a follow-up meeting is called where the highlights of the previous meeting are reviewed and improvements are proposed. Before the management meeting, the team discusses the recommendations and decides on the financial impact of the changes and modifications based on each member's recommendations. The key aspect of both meetings is teamwork. Each member plays a vital role, and management must hear one voice. Once the recommendations are made, they should be either approved or denied, and the process is then repeated for the next well.

PRODUCED WATER

Water disposal following fracking and production is often considered of secondary importance. However, recent studies have shown that earthquakes (Boak, 2017) and potable water contamination can be associated with water injection (Osborne, et al., 2011). Therefore, a study of injector wells is of the utmost importance to correct any errors. The Society of Exploration Geophysicists (SEG) dedicated a special issue of *The Leading Edge* to this topic in 2015. Still, the answers are not always clear.

Oklahoma, for example, has several major producing basins for the Woodford shale and Mississippian lime, of which the Anadarko and Arkoma basins (fig. 8–31) are two large producers. Area "E" in figure 8–31 is flanked by the Nemaha Uplift Fault (west) and the Wilzetta Fault (east) and produces from the Mississippian lime (Boak, 2017). Because this area produces from an interval that has a high water saturation, it also produces significant wastewater. Reinjection of the produced water plus any excess fracking fluid has generated earthquakes. The Oklahoma Geological Survey is trying to mitigate the effect of these earthquakes by reducing the rate of injection.

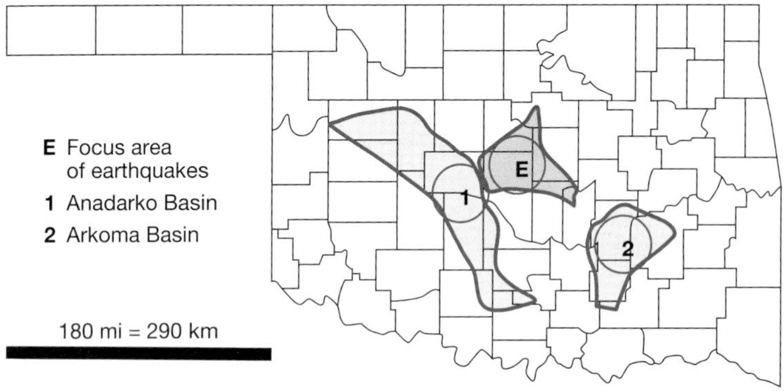

Fig. 8–31. Area where fracking is taking place (1 and 2) and principal areas of earthquake activity ("E") (Data from U.S. Geological Survey Earthquake Hazards Program website)

South Texas, on the other hand, has reported few seismic events; there are no significant earthquakes, and tremors appear to be small scale. To avoid even the possibility of connectivity and lubrication of major faults it is recommended to inject wastewater into reservoirs that are not connected or near any the major faults. Microseismic monitoring may provide valuable information in establishing a connection between the two. Because the injection amounts have been reduced in recent years owing to the industry downturn, so have the earthquakes. Therefore, the relationship between the two is evident (Boak, 2017).

AQUIFER CONTAMINATION

As in any geological problem, the first step is to analyze the process involving aquifers. Second, study the specific area in which you are working. Third, if anomalies are detected, determine the significance and the incorporate corrective measures. The problem with water injection problems is that they are generally detected after major investments have been made. How best can problems be avoided? Study the aquifers before injecting water.

An area of significant research—both currently and in the future—is the optimization of recycling of formation water and fracking fluids by removal of caustic compounds and mixing with additional freshwater. Today, most of the water used in fracking is freshwater, with small concentration of added chemicals. The mixture is injected under pressure. As mentioned earlier, only 30% of the water injected is recovered. However, there are several formations that have a high water saturation and hence also produce formation water. Because the produced water is brine, currently most of the water is reinjected into nonreservoir intervals as waste.

The basic controls on aquifers are provided by observation wells. All wells require a gamma-ray log and a resistivity log by law. Most states require clear definition of the characteristics of the aquifer and the *aquiclude* (confining bed).

Aquifers by their nature fluctuate. By definition, an aquifer must have a recharge zone and an outcrop section where rainwater and surface runoff filters into the aquifer. Great care must be taken to protect the recharge zone against spills and any contaminant. Most states and countries require detailed reporting on the water well, and the parameters reported can serve as basic criteria of the aquifer. The following suggested recording system for aquifer wells follows the guidelines of the Indiana Department of Environmental Management Agency:

- Well design and location of the well
- Use of water (residential, industrial, nonfarming irrigation)
- Type of pumps used
- How well was tested, bailer or pump
- Estimation of piezometric head (application of Bernoulli's equation)
- Test of rate and length of time
- Casing design and screens

- Inside diameter of the casing
- Total depth of the well and aquifer (well should be drilled at least 25 feet below the base of the aquifer)
- Static level of the water (over a prolonged period)
- Information on the driller
- Geological description of the aquifer and aquiclude

Contamination of potable water is a concern. In any area where oil and gas development will take place, pilot water wells should be drilled as early as possible. Their number is dependent on the problems encountered. They should be completed in such a way that, upon completion of the testing period, they could be utilized as producing water wells. The depth should be adequate to test all potable aquifers and penetrate several candidate water injection horizons.

These water wells serve two important functions. First, detailed analysis of the water provides a baseline for issues regarding contamination. In gas-prone basins, it is not uncommon for groundwater to contain some gas. The study should be carried out by an independent entity selected by both the communities served by the aquifer and the company, to ensure impartiality. Second, the piezometric surface is defined as accurately as possible (i.e., it is a computed number and not a real horizon). If there is incursion of injected water, then the piezometric surface will also rise. Among the best studied aquifers in the world are the Chicot, Evangeline, and Jasper aquifers of southeast Texas and southwest Louisiana. Of particular interest is the Jasper aquifer (the deepest of the three) because it is the only one of these three aquifers that has a well-defined aquiclude. Several studies of recharge rates have been performed in the area by the U.S. Geological Survey, and the recharge rate is estimated at 5–18 mm/year. The water in the Jasper aquifer has been dated using ^{14}C, and the ages range from 8,000 to 12,000 years before the present. From the standpoint of contamination from drilling fluids, the aquifer can be assumed to be static. Both the Jasper and Evangeline aquifers contain natural methane contamination and require separation.

Determining the presence of potable water is made easier by the resistivity log. Because the log was designed to test for brines and hydrocarbons, the response is a negative deflection on the log in the presence of freshwater. Any interval with freshwater must be carefully mapped, and any faulting affecting the aquifer should be scrutinized with care. Water injectors should target deep brine aquifers and must avoid any faults. Because the objective of injecting water into the subsurface is to use a highly porous conduit, communication with younger beds may create difficulties. The problem is that in contrast to the slow recharge rate, injection rates can be very high.

It is necessary to monitor both injected water and the water tested in the monitor wells. Characteristics that should be monitored include dissolved hydrocarbon and chemicals used in drilling, pH, salinity, and temperature. As discussed already, heated brines can result in diagenesis of minerals encountered and alter clays. The ultimate goal is to confine the injected fluids in the subsurface, as well as to avoid inducing new fractures by monitoring microseismic events that are external to the injection process. Major faults in the region should be avoided because they can act as conduits or have fractures zones that extend a significant distance from the main fault. Once filtration into a fracture zone is detected, remedial action must be taken to first reduce inflow rates and then suspend injection. The presence of newly opened fractures can be detected by a pressure drop even as flow rates are increased.

Before injection of fluids into an aquifer, there must be a clear understanding as to who are the primary and secondary controlling agencies. The operator must comply with all their requirements. The U.S. Environmental Protection Agency provides certain control, but in most cases, it is superseded by state regulatory agencies. Often, detailed maps of the aquifer and cross sections are required. In particular, it is important to note the distance to the point sources where water is used for human consumption.

Groundwater hydraulics are well understood. The velocity of the flow is a function of the permeability of the sand. Darcy in 1856 developed the basic formula used today:

$$V = \frac{PI}{\phi} \tag{8.47}$$

where V is the velocity (in ft/day), P is the coefficient of permeability, I is the hydraulic gradient (in ft/100 ft), and ϕ is the porosity of the medium.

A second important equation determines the fluid discharge:

$$Q = PIA \tag{8.48}$$

where Q is the discharge is based on the permeability and hydraulic gradient, I is the hydraulic gradient, and A is the cross-sectional area of the aquifer.

Can similar calculations be used in reverse?

SUMMARY

Application of the static model is used to improve geosteering and provide engineers with models to quantify the potential of an area. It serves as a basis for future dynamic models.

By analysis of overburden pressures and pore pressure data, estimates can be generated of fracture gradients. Three methods are described:

- Hubbert-Willis—simple but not very precise, thus used only as a starting point
- Matthews-Kelly—provides a good estimate, using pressure data and well logs
- Eaton—the most commonly used, focusing on Poisson's ratio, drilling conditions, and well logs

All three should be used to ensure reasonable values. Other models can also be applied.

Seismic data can also be applied to fracture prediction studies. Curvature and global stress maps assist in predicting fracture patterns and in defining well orientation.

Uniaxial and triaxial compress analyses of the core data allow for calibration of fracture parameters and preparation of Mohr circles. Mohr circles guide analysis of the normal stress field to evaluate the shear stress required for fracturing.

Leak-off tests (minifracs) on pilot wells are vital to determine fracking parameters. Microseismic evaluates the efficiency of the hydraulic fractures. Proper design of the wellbore and drilling and fracking design are important to attain a positive outcome.

The postdrill review enables improvements on follow-up wells. Water disposal is a significant problem. Potential aquifer contamination also requires attention.

REFERENCES

Aadnøy, Bernt S., and Reza Looyeh. 2011. *Petroleum Rock Mechanics: Drilling Operations and Well Design*. Amsterdam: Gulf Professional Publishing.

Akin, S., and A. R. Kovscek. 1999. "Imbibition studies of low-permeability porous media." Paper presented at SPE Western Regional Meeting, Anchorage, AK, May 26–27.

Bingham, M. G. 1964. "How rock properties are related to drilling." *Oil & Gas Journal*. 62: 94.

Boak, Jeremy. 2017. "Earthquakes in Oklahoma: Trends in injection induced seismicity and regulatory responses." Paper presented at Houston Geological Society Applied Geoscience Conference, The Woodlands, TX, March 7–8.

Bolger, George W., and Rocky R. Reifenstuhl. 2008. "Mercury injection capillary pressure and reservoir seal capacity of 26 outcrop samples, Miocene to Triassic Age." *Alaska Division of Geological & Geophysical Surveys, Report of Investigation 1D.*

Bommer, Paul Michael. 2008. *A Primer of Oilwell Drilling.* Austin, TX: University of Texas at Austin.

Bowers, Glenn L. 1995. "Pore pressure estimation from velocity data: Accounting for overpressure besides undercompaction." *SPE Drilling & Completion*. 10: 89–95.

Burke, Lauri A., Scott A. Kinney, and Temidayo B. Kola-Kehinde. 2011. *Digital Archive of Drilling Mud Weight Pressures and Wellbore Temperatures from 49 Regional Cross Sections of 967 Well Logs in Louisiana and Texas, onshore Gulf of Mexico Basin: Open-File Report 2011–1266*. Reston, VA: U.S. Geological Survey.

Chalmers, Gareth, and R. Marc Bustin. 2010. "The effects and distribution of moisture in gas shale reservoir systems." Paper presented at the AAPG Annual Convention and Exhibition, New Orleans, April 11–14. Search and Discovery article 80113. www.searchanddiscovery.com/documents/2010/80113chalmers/ndx_chalmers.pdf.

Cho, David, and Marco Perez. 2014. "Brittleness revisited." Paper presented at GeoConvention 2014, Calgary, AB, May 12–16.

Chopra, Satinder, and Alan R. Huffman. 2006. "Velocity determination for pore pressure prediction." *CSEG Recorder*. 31. csegrecorder.com/articles/view/velocity-determination-for-pore-pressure-prediction.

Cronin, Vince. 2004. "A draft primer on focal mechanism solutions for geologists." www.ocean.washington.edu/courses/oc410/reading/Focal_mechanism_primer.pdf.

Currie, J. B., H. W. Patnode, and R. P. Trump. 1962. "Development of folds in sedimentary strata." *Geological Society of America Bulletin*. 73: 655–673.

Deere, Don U., and Don W. Deere. 1989. *Rock Quality Designation (RQD) after Twenty Years: Contract Report GL-89-1*. Longmont, CO: Rocky Mountain Consultants, Inc. and U.S. Department of the Army.

Donaldson, Erle C., Rex D. Thomas, and Philip B. Lorenz. 1969. "Wettability determination and its effect on recovery efficiency." *Society of Petroleum Engineers Journal*. 9: 13–20.

Donnelly, B., Edmund Perfect, L. D. McKay, P. J. Lemiszki, V. H. DiStefano, L. M. Anovitz, J. McFarlane, R. E. Hale, and C.-L. Cheng. 2016. "Capillary pressure–saturation relationships for gas shales measured using a water activity meter." *Journal of Natural Gas Science and Engineering*. 33: 1342–1352.

Eaton, Ben, A. 1969 "Fracture gradient prediction and its application in oilfield operations." *Journal of Petroleum Technology*: 21.

———. 1975. "The equation for geopressure prediction from well logs." Paper presented at the Fall Meeting of the Society of Petroleum Engineers of AIME, Dallas, September 28–October 1.

Eglinton, T. I., C. D. Curtis, and S. J. Rowland. 1987. "Generation of water-soluble organic acids from kerogen during hydrous pyrolysis: Implications for porosity development." *Mineralogical Magazine*. 51: 495–503.

Formento, Jean-Luc. 2004. "Seismic pore pressure prediction." Paper presented at the European Association of Geoscientists and Engineers Conference, Utrecht, The Netherlands, September 6–9. www.cgg.com/technical document/cggv_0000001681.pdf.

Gardner, G. H. F., L. W. Gardner, and A. R. Gregory. 1974. "Formation velocity and density—the diagnostic basics for stratigraphic traps." *Geophysics*. 39: 770–780.

Glover, Paul. 2017. "Capillary pressure." Chapter 8, in *Formation Evaluation MSc Course Notes*. Leeds, UK: University of Leeds. homepages.see.leeds.ac.uk/~earpwjg/PG_EN/CD%20Contents/Formation%20Evaluation%20English/Chapter%208.PDF.

Goodway, Bill. 2014. "The magic of Lamé." Interview by Satinder Chopra. *CSEG Recorder.* 36: 10–17. csegrecorder.com/interviews/view/interview-with-bill-goodway.

Goodway, Bill, Marco Perez, John Varsek, and Christian Abaco. 2010. "Seismic petrophysics and isotropic-anisotropic AVO methods for unconventional gas exploration." *Leading Edge*. 29: 1500–1508.

Goodway, Bill, Taiwen Chen, and Jon Downton. 1997. "Improved AVO fluid detection and lithology discrimination using Lamé petrophysical parameters; 'λρ', 'μρ', & 'λ/μ fluid stack', from P and S inversions." In *SEG Technical Program Expanded Abstracts 1997.* Society of Exploration Geophysicists, 183–186.

Heidbach, Oliver, Mojtaba Rajabi, Karsten Reiter, and Moritz Ziegler; WSM Team. 2016. "World stress map: Database release 2016." doi.org/10.5880/WSM.2016.001.

Hoek, Evert. 1988. "The Hoek-Brown failure criterion—a 1988 update." In *Proceedings of the 15th Canadian Rock Mechanics Symposium*. Toronto: Department of Civil Engineering, 31–38.

Hubbert, M. King, and David G. Willis. 1957. "Mechanics of hydraulic fracturing." *Transactions of the AIME*. 210: 153–163.

Hunt, Lee, Scott Reynolds, Scott Hadley, Jon Downton, and Satinder Chopra. 2011. "Causal fracture prediction: Curvature, stress, and geomechanics." *Leading Edge*. 30: 1274–1286.

Ikelle, Luc T., and Lasse Amundsen. 2005. *Introduction to Petroleum Seismology.* Tulsa: Society of Exploration Geophysicists.

Josh, M., L. Esteban, C. Delle Piane, J. Sarout, D. N. Dewhurst, and M. B. Clennell. 2012. "Laboratory characterisation of shale properties." *Journal of Petroleum Science and Engineering.* 88: 107–124.

Kuila, Utpalendu, Douglas K. McCarty, Arkadiusz Derkowski, Timothy B. Fischer, and Manika Prasad. 2014. "Total porosity measurement in gas shales by water immersion porosimetry (WIP) method." *Fuel.* 117: 1115–1129.

Kwasniewski, M., and M. Takahashi. 2010. "Strain-based failure criteria for rocks: State of the art and recent advances." In *Rock Mechanics in Civil and Environmental Engineering*, Jian Zhao, Vincent Labiouse, Jean-Paul Dudt, and Jean-François Mathier (eds.). Boca Raton, FL: CRC Press, 45–56.

Li, Kewen, and Roland N. Horne. 2006. "Comparison of methods to calculate relative permeability from capillary pressure in consolidated water-wet porous media." *Water Resources Research*. 42: W06405.

Lonardelli, Ivan, Hans-Rudolf Wenk, and Y. Ren. 2007. "Preferred orientation and elastic anisotropy in shales." *Geophysics*. 72: D33–D40.

Marfurt, Kurt. 2013. Seismic attribute course. SEG, Convention Short Course, San Antonio.

Martin, C. D., P. K. Kaiser, and D. R. McCreath. 1999. "Hoek-Brown parameters for predicting the depth of brittle failure around tunnels." *Canadian Geotechnical Journal*. 36: 136–151.

Matthews, W. R., and John Kelly. 1967. "How to predict formation pressure and fracture gradient." *Oil & Gas Journal*. 65: 92–106.

Menchaca, Matt. 2014. "Stressing out on the Eagle Ford: Shale play geology." *Drillinginfo* (blog). May 24. https://info.drillinginfo.com/stress-eagle-ford-shale-play-geology/.

Moore, Diane E., and David A. Lockner. 2007. "Friction of the smectite clay montmorillonite." In *The Seismogenic Zone of Subduction Thrust Faults*, Timothy H. Dixon and J. Casey Moore (eds.). New York: Columbia University Press, 317–345.

Nikravesh, M., F. Aminzadeh, and L. A. Zadeh (eds.). 2003. *Soft Computing and Intelligent Data Analysis in Oil Exploration*. Developments in Petroleum Science, vol. 51. Elsevier.

Osborne, Stephen G., Avner Vengosh, Nathaniel R. Warner, and Robert B. Jackson. 2011. "Methane contamination of drinking water accompanying gas well and hydraulic fracturing." *PNAS*. 108: 8172–8176.

Perez, Marco, David Close, Bill Goodway, and Greg Purdue. 2011. "Developing templates for integrating quantitative geophysics and hydraulic fracture completions data: Part I—principles and theory." In *SEG Technical Program Expanded Abstracts 2011*. Society of Exploration Geophysicists, 1794–1798.

Perez, Marco, Bill Goodway, and David Close. 2012. "Geomechanics, anisotropy and LMR." Paper presented at GeoConvention 2012, Calgary, AB, May 14–18.

Pittman, Edward D. 1992. "Relationship of porosity and permeability to various parameters derived from mercury injection-capillary pressure curves for sandstone." *AAPG Bulletin*. 76: 191–198.

Rehm, Bill, and Ray McClendon. 1971. "Measurement of formation pressure from drilling data." Paper presented at the Fall Meeting of the Society of Petroleum Engineers of AIME, New Orleans, October 3–6.

Rich, Jamie, and Dan Kahn. 2017. "Exploiting microseismic event characterization to optimize completion strategies." Paper presented at Houston Geological Society Applied Geoscience Conference, The Woodlands, TX, March 7–8.

Rickman, Rick, Michael J. Mullen, James Erik Petre, William Vincent Grieser, and Donald Kundert. 2008. "A practical use of shale petrophysics for stimulation design optimization: All shale plays are not clones of the Barnett Shale." Paper SPE-115258-MS, presented at SPE Annual Technical Conference and Exhibition, Denver, September 21–24.

Roach, Eric. 2014. "Well completion 101 part 2: Well perforation." *Drillinginfo* (blog). October 7. https://info.drillinginfo.com/well-completion-101-part-2-well-perforation/.

Roberts, Andy. 2001. "Curvature attributes and their application to 3D interpreted horizons." *First Break*. 19: 85–100.

Sayers, C. M., G. M. Johnson, and G. Denyer. 2002. "Predrill pore-pressure prediction using seismic data." *Geophysics*. 67: 1286–1292.

Shapiro, S. A., and C. Dinske. 2009. "Fluid-induced seismicity: Pressure diffusion and hydraulic fracturing." *Geophysical Prospecting*. 57: 301–310.

Sheorey, P. R. 1994. "A theory for in situ stresses in isotropic and transversely isotropic rock." *International Journal of Rock Mechanics and Mining Sciences & Geomechanics Abstracts*. 31: 23–34.

Sibson, R. H. 1977. "Fault rocks and fault mechanisms." *Journal of the Geological Society.* 133: 191–213.

Solano, Yully P., Rodolfo Uribe, Marcelo Frydman, Nestor Saavedra, and Zuly H. Calderon. 2007. "A modified approach to predict pore pressure using the d exponent method: An example from the Carbonera Formation, Colombia." *Ciencia Tecnologia y Futuro.* 3: 103–112.

Swanson, B. F. 1981. "A simple correlation between permeabilities and mercury capillary pressures." *Journal of Petroleum Technology.* 33: 2498–2504.

Szymkiewicz, Adam. 2012. *Modelling Water Flow in Unsaturated Porous Media: Accounting for Nonlinear Permeability and Material Heterogeneity.* Berlin: Springer-Verlag.

Thomsen, L. 1986. "Weak elastic anisotropy." *Geophysics.* 51: 1954–1966.

Thyne, G. D. 2013. "A review of the measurement of wettability in shales." Technical report for Sandia National Laboratory, September 16.

Triepke, Joseph. 2014. "Well completion 101 part 3: Well stimulation." *Drillinginfo* (blog). October 30. https://info.drillinginfo.com/well-completion-well-stimulation/.

Tweheyo, M. T., M. S. Talukdar, and O. Torsæter. 2001. "Hysteresis effects in capillary pressure, relative permeability and resistivity index of north sea chalk." Paper SCA 2001-65, presented at the International Symposium of the Society of Core Analysts, Edinburgh, UK, September 2001.

U.S. Geological Survey. 1969. "Focal mechanisms." U.S. Geological Survey Earthquake Hazards Program website. earthquake.usgs.gov/learn/topics/beachball.php.

Vachaparampil, Alex M. 2017. "The influence of the intermediate principal stress on the strength of the Mancos shale." Paper presented at the Houston Geological Society Applied Geoscience Conference, Houston, March 7–8.

Warpinski, Norman R. 2013. "Understanding hydraulic fracture growth, effectiveness, and safety through microseismic monitoring." Paper presented at the ISRM International Conference for Effective and Sustainable Hydraulic Fracturing, Brisbane, Australia, May 20–22.

Xiong, Yi, Phil Winterfeld, Cong Wang, Zhaoqin Huang, and Yu-Shu Wu. 2015. "Effect of large capillary pressure on fluid flow and transport in stress-sensitive tight oil reservoirs." Paper SPE-175074-MS, presented at the SPE Annual Technical Conference and Exhibition, Houston, September 28–30. inside.mines.edu/~yxiong/SPE-175074-MS.pdf.

Yang, Yi, and Mark D. Zoback. 2014. "The role of preexisting fractures and faults during multistage hydraulic fracturing in the Bakken Formation." *AAPG Interpretation.* 2: SG25–SG39. dx.doi.org/10.1190/INT-2013-0158.1.

Young, Warren Clarence, and Richard Gordon Budynas. 2007. *Roark's Formulas for Stress and Strain*, 7th ed. New York: McGraw-Hill.

9

BUSINESS AND RISK

TAKEAWAY

- Keep it simple (KIS)
- Reserves—volumetric, material balance, production decline curve
- Basic cost/benefit estimate
- Risk assessment—GDE and common risk segment (CRS) analysis
- Cash flow and payout time

The goals of this chapter are to develop and apply meaningful data to determine the commercial feasibility of a project, as well as to perform cost-benefit analysis and calculate rate of return on investment. Two related aspects influence the economics of a project: the estimate of reserves; and assessment of the risk of the venture on the basis of available data. Once these have been established, the next step is to carry out a financial analysis using standard investment analysis techniques.

The most appropriate reserve estimate in shales is based on material balance, but other techniques are also discussed. The most important consideration is that the estimate for the economic value of the project be reasonable. Geostatistical tools should be applied to determine the risk in a project, and two rules need to be followed: use industry standards; and apply acceptable geological models. The process must be consistent because, in most cases, outside consultants must certify your reserve estimates, and the government regulates the dissemination of information to third parties. Cash-flow analyses and establishing payback periods are fundamental for ranking projects and determining the economic viability of each project.

THE BUSINESS SIDE OF UNCONVENTIONAL SHALES

During a morning assembly meeting held at a major oil company for young professionals, one of the senior managers began his speech with a question, "What is the purpose of the company?" The answers ranged from finding oil, to refining oil, to marketing gasoline. Only one person in the back gave the correct answer: "Make a profit." If the venture is not profitable, the science is irrelevant.

How is the profit forecast? Two axioms should be observed: first, as a general principle, keep it simple (KIS); and second, use realistic data for analysis. All states and countries require annual certified corporate accounting, financial/budget forecasting, and reserves reports, which are the means by which taxes are levied and investors are informed. All must follow uniform standards to ensure comparison; however, given their nature as public information, they represent only a snapshot of the company's activities.

Four parts comprise the official report:

- Balance sheet
- Income statement
- Projected income statement
- Hydrocarbon reserves

The first of these reflects historical data, and the second category offers a glimpse into the future of the company. Both data sets should be used internally for planning purposes and parsed for individual projects and subprojects.

This chapter will focus on analysis of individual projects and only the first two parts of the report. It will also be assumed that the venture is not the first high-risk project in a basin. Usually, petroleum engineers carry out the economic evaluation in conjunction with finance staff and geoscientists.

Before embarking on large investments, corporations base their forecast on the answers to the following questions:

- How much oil/gas/condensate can be produced in the lifetime of the field? To answer this, the recoverable reserves are estimated.
- How quickly can the company produce the hydrocarbons? To answer this, historical production decline curves in the area are analyzed, and a cash-flow prognosis is based on high- and low-price scenarios.
- What is the risk component or exposure for the venture? The answer is based on geological and political risks (the latter category also encompassing environmental risks).

- What are the site-to-market requirements for the project? The answer takes into account surface facilities, pipelines, and so forth.
- What discount rate will be used? The answer depends both on optimistic and pessimistic ranges and on how to apply *discount cash flow versus rate of return analyses* (DCFROR).
- What impact will taxes, fees, and royalty rates have?
- What would trigger a back-out point? Criteria therefore should consider what constitutes a failure—and specifically whether the concept is wrong, as opposed to a technical failure.

This chapter will summarize the first three geotechnical aspects of the analysis and include a brief discussion on environmental risk, followed by a review of the financial aspects of the investment. Values for the other parameters, including political risks, can be assigned. However, in the real world, they must be estimated correctly, and ranges must be set.

RESERVE ESTIMATE

The first purpose of economic analysis for any project is to present the estimate of recoverable reserves. This must be formalized and requires certification by a third party before the reserves can be made public. In other words, it is not sufficient merely to state a number; instead, an independent certified appraiser must be convinced that the number provided is quantifiable.

The U.S. Securities and Exchange Commission (SEC) has published specific criteria that must be met. The SEC guidelines ensure that the company stock price is not overinflated by falsely adding reserves and that everyone abides by similar guidelines. The SEC rules apply only to corporations registered in the United States and are considered more restrictive than those in other countries. Nevertheless, many countries follow similar rules.

Before the unconventional shale boom, the procedures for estimating reserves followed a straightforward method. For the purpose of the economic analysis in this book, only oil will be considered. The amount of gas can be used as barrel oil equivalent (BOE) for reserve estimates, with the conversion factor of 5,800 ft^3 gas per BOE; for simplicity, this value is often rounded up, to 6.000 ft^3 per BOE.

There are three classes of accessible reserves:

- *Proven*—also known as *P90* or *1P*
- *Probable*—also known as *P50* or *2P*
- *Possible*—also known as *P10* or *3P*

As the names above suggest, P90 has a 90% chance that the reserve estimate is correct; P50 has a 50% chance the estimate is correct; and P10 has a 10% chance that the estimate is correct. These estimates apply only to technically producible hydrocarbons. There is an additional category known as *contingent reserves* that are inaccessible given present technology, environmental concerns, or cost, and these are beyond the scope of the present text.

Proven reserves are further subdivided into *proven developed* (PD) and *proven undeveloped* (PUD). PDs are based on drilled wells where test results confirm the estimated reserves. PUDs are based on the experience of the area or if appraisal wells have been drilled reserves can be extrapolated beyond the initial wells with a 90% level of confidence—hence, the P90 designation. PDs are further subdivided into *producing* and *shut-in* wells.

Several techniques are used in the estimation of reserves:

- Volumetric—areal distribution, porosity, thickness and fluids
- Material balance—thickness, geochemistry of the rocks, maturity, density, thickness, size, pressure, temperature
- Decline-curve analysis—based on past performance of existing wells, lacking geological input
- Other, more sophisticated numerical models.

The two most commonly accepted techniques are decline-curve analysis and material balance. For traditional prospects, the simple volumetric computation shown in equation (9.1) provides an adequate result. The standard formula for an oil-filled structure in the English system is as follows:

$$N = \frac{7{,}758A(1-S_w)h\phi}{\beta} \tag{9.1}$$

where N is the number of barrels, or stock tank original oil in place STOOIP; A is the mapped surface area of contour closure considered economically recoverable, limited to the area of interest; $1 - S_w$ is the oil saturation (S_o), which is the complement of water saturation; h is the height of the oil column; ϕ is the porosity; 7,758 is a conversion factor for acre-feet to barrels; and β is the liquid compressibility factor (usually near 1.0 for oil).

A common map used in the economic analysis of an area is to contour a portion of equation (9.1), the *SoPhiH* (i.e., $S_o\phi h$). It is important to understand the limitations of this map as well. As Brinkerhoff et al. (2016) have suggested, the significance of the SoPhiH map can be borne out with production statistics. However, only similar sedimentary environments and physical conditions should be compared.

Because in the metric system, the volume of oil is given in cubic meters, equation (9.1) does not require a conversion factor. Also note that the estimate is for the oil in place. The reserve is dependent on the historical recovery factor in the area, which may range from 10% to 30% or higher. Moreover, the structure map used to define maximum closure is based on the accuracy of the seismic and geological interpretation and thus will change as more data is introduced. The porosity and oil saturation are based on petrophysics, core studies, and engineering testing of the well. The water saturation can be calculated using imbibition studies as described in chapter 8.

The conventional gas reserve formula calculates the *gas in place* (GIP) as follows:

$$\text{GIP} = \frac{43{,}560A(1-S_w)\phi h}{\beta_g} \tag{9.2}$$

where 43,560 is the conversion factor of acre-feet to cubic feet of gas and $b_g = 0.00289ZT/P$, giving the total compressibility of gas (T is the reservoir temperature, P is the reservoir pressure, and Z is the compressibility, which is dependent on the molecular composition of the gas and the universal gas law).

For shale gas, GIP is slightly different than for oil because gas is highly compressible and can be either free gas or adsorbed gas. A similar equation exists for traditional gas reservoirs in the metric system and is simply stated in cubic meters. The graph for the Z factor of gas compressibility is used because gas does not behave as an ideal gas under reservoir conditions but must be expressed as a function of the true gas in pore versus the ideal gas in the pores. At low pressures, gases behave as ideal gases. As the pressure is increased, the gas molecules interfere with each other and deviate from ideal gases. The temperatures and pressures are the ratios between the observed and critical temperature and pressure of the ideal gas. (This topic is discussed further in all introductory petroleum engineering texts.)

The chart in figure 9–1 represents a simplified example of the behavior of methane. For the actual values to use in technical evaluation, refer to McCain's (1990) chapter on the *equations of state*.

Tables for Z factors, as a function of temperature and pressure, are available for each gas type, and weighted averages provide the overall Z factor for a mixture. A pure gas reservoir can have, as a first approximation, a reserve estimate calculated by simple volumetrics in accordance with the universal gas law, the available pore space, and the dimensions of the reservoir storage tank. If there is a liquid fraction, when subjected to surface pressure conditions, the gas is considered a wet gas. Reserve calculations should be performed separately for the liquid (condensate) and gas phases and under reservoir conditions.

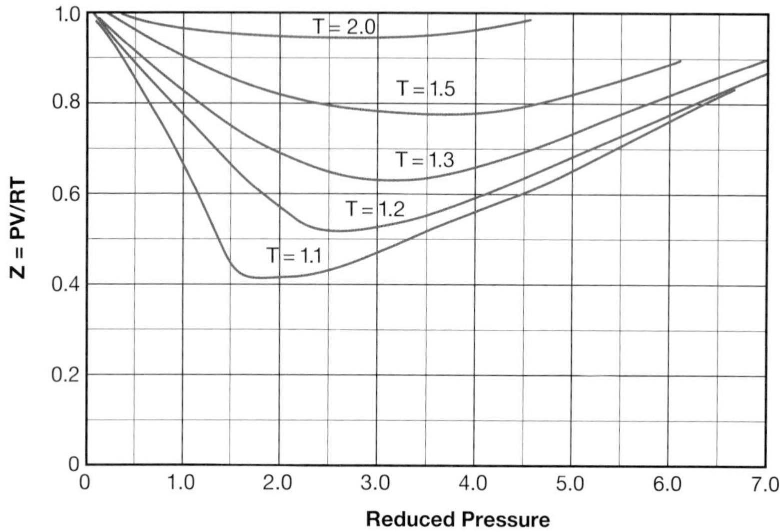

Fig. 9–1. Typical sample compression factor (*Z* factor) for methane gas as a function of reduced pressure and reduced temperature. *Z* factors are slightly different for various gases. Reduced pressure and temperatures are ratios of critical and ideal *T-P* conditions.

A second problem with gas production is the pressure balance in the reservoir. As the pressure in the reservoir is reduced, the phase diagram for the particular molecular composition can change from its original gas state to a mixed, liquid-and-gas phase. The pressure decline is due to production, and the point of phase change is known as either the *bubble point*, for an oil system, or the *dew point*, for a gas system. The phase change can occur only when condensate liquids are present in the original mixture. Once the dew point is reached, it is more difficult to produce the gas owing to the small pore throat available and the liquid droplets or oil film plugging the pores.

The bubble point is determined in the *PVT* (pressure, volume, temperature) analysis performed after drilling of the well. Importantly, the phase diagram applies only to hydrocarbons with the same composition and must be calculated individually for the specific hydrocarbon under investigation (fig. 9–2). In laboratory tests, the pressure is decreased while the temperature is kept constant. The sample is assayed by the purchasing refinery to determine the true value of the hydrocarbon in PVT analysis. This is commonly performed on gas and liquid samples acquired during well testing to determine the value of the discovery (McCain, 1990).

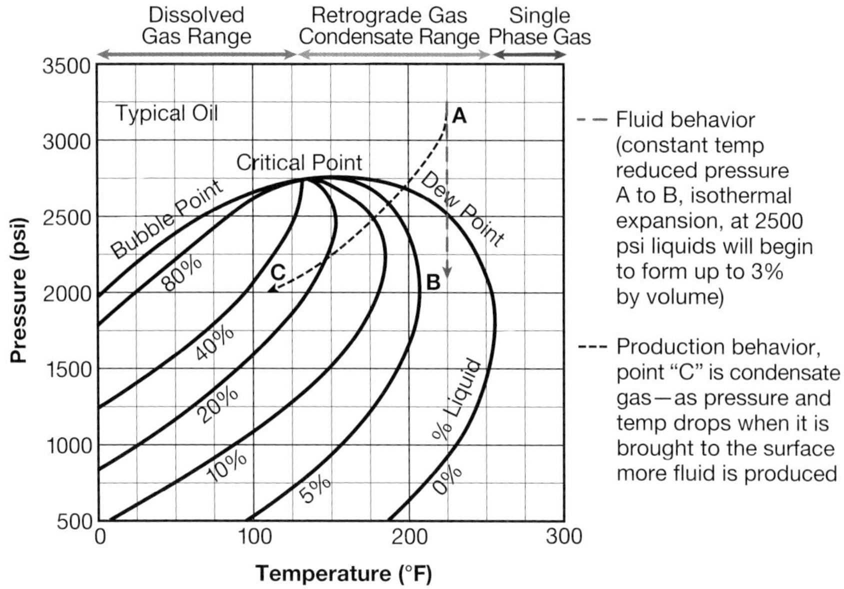

Fig. 9–2. Sample phase diagram based on a PVT analysis (Modified from Terry and Rogers, 2015)

Three tasks are key in the well fluid analysis:

- Describe the original composition of the fluids
- Define the pressure/temperature regime in the well, and use it to determine the dew point of the gas
- Determine the compressibility factor of the gas

For the reservoir engineer, the trap is a complex storage tank of porous material filled with hydrocarbon and water for which the engineer must estimate reserves prior to drilling by applying a material balance approach. The engineer needs to ensure that the tank has the following characteristics:

- Constant pore volume
- Constant temperature
- Uniform pressure
- Uniform composition and physical properties
- Equilibrium between the oil and gas phase
- Uniform total organic carbon (TOC) distribution
- Uniform brittleness at the fracked level
- Uniform maturation
- Uniform initial and actual hydrogen index (HI)

Usually these parameters are calculated after testing wells and include dynamic aspects of flow such as transmissibility of fluids and pore throat size, etc. The following equation from figure 9–3 is an approach at material balance prior to drilling:

$$Gp = \text{area} \times \text{thickness} \times \text{bulk density} \times \text{TOC} \times (\text{original HI} - \text{observed HI}) \quad (9.3)$$

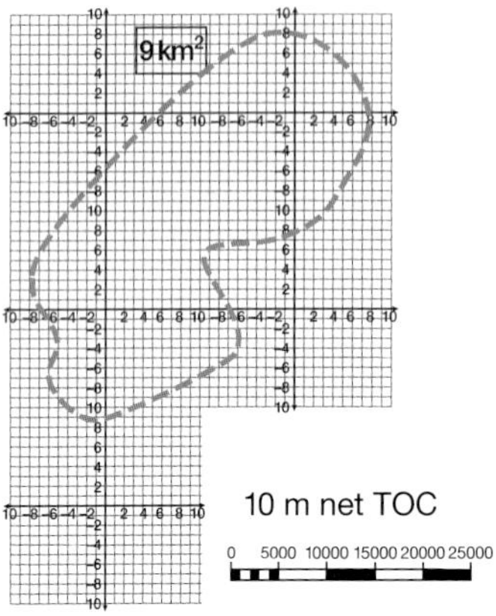

210 sq km if you assume TOC 18%

area	210,000,000 m^2
thickness	10 m
rho	2250 kg/m^3
TOC	.18
wt TOC kg	8.5E+11 kg TOC
wt TOC g	8.5E+14 g TOC
diff. HI	400 mg/g TOC
wt oil potential	3.4E+17 mg oil
kg	3.4E+11 kg
metric ton	340,200,000 T
Bbl (N)	2,493,666,000 BBLs
N × PI × RF =	22.5 MM BBLs

Estimation of oil generating capacity (Gp):
Gp = area × thickness × rho (bulk) × TOC × (HI original – HI current)

(HI original – 800, HI current – 400)
Diff HI = 400
Where HI is S2*100/TOC
N = 210000000 sq m * 10m * 2250 kg/m^3 * .18 (TOC) * 400 mg oil/gm TOC × 10^6 = N mg oil
Production Index (PI) = (S1/S1+S2) = 0.15 for Bakken
Recovery factor (RF) of 6%
Conversion is 7.33 bbl/metric ton

Fig. 9–3. Material balance approach to STOOIP calculation for a resource shale. Values are given for HI, production index (PI), and TOC (see chap. 7). Density values are based on core data and petrophysics. (Adapted from Kuchinskiy et al., 2010)

The resource estimation shown in figure 9–3 is representative of a large typical lease block in North Dakota. All calculations should be performed using the metric system. The net thickness of organic-rich shale is provided by a detailed analysis of several well logs in the area using the Passey equation (see chap. 7) and corroborated by Rock-Eval analysis. The TOC contents are given as ranges in the resultant plot but still provide reasonable averages and should be compared to literature estimates. If the plot shows a significant gradation, then a weighted average can be computed giving more weight to the rich interval. Once the average TOC content is estimated, the volume of rock that has this characteristic (sweet spot) can be calculated.

Rock-Eval analysis satisfies the following series of conditions included in the calculation (values used in the calculation):

- Determine the surface area of the sweet spot or lease (210 km^2)
- Provide a complete analysis of the pilot well or wells in the region
- Determine the average density of the high TOC interval based on the FDC/CNL log (2,250 kg/m^3)
- Determine the current HI at the reservoir (400)
- By analysis of an immature section of the source, establish what the original HI might have been (800)
- Determine if other, younger reservoirs received hydrocarbons from the source rock being studied
- Based on Rock-Eval, estimate the transformation ratio (0.15), higher if the shale is more mature
- Estimate the current average TOC content and provide a best estimate for the original TOC content in the organic-rich interval (0.18)
- Analyze all potentially significant parameters that may assist in reducing risk and generate a radar plot of the parameters

Wells studied in the immature section of the Williston Basin provide an indication of the potential original HI of the Lower Bakken shale. Along the eastern portion of basin, an average HI of 800 has been observed (Pollastro et al., 2013). The average HI in the producing trend is 100. In the above example, which is more marginal, an HI of 400 is used. The average density for the Lower Bakken is 2,250 kg/m^3 (Kohlruss and Stamatinos, 2014), and the thickness of the main source interval is 10 m. Laboratory calculations have yielded a transformation ratio or production index (i.e., S1/(S1 + S2) in Rock-Eval; see chap. 7) of 0.15 and TOC content of 18% (Jin et al., 2016). The recoverable unrisked reserves are 22.4 million barrels of oil (MMBO) for the specific area of study. In resource plays, it is not uncommon to assume 70% chance of meeting the original target. Therefore, the risked reserves become 15.7 MMBO, or 75,000 BO/km^2.

A regular vertical gas well can usually drain 160 acres. Horizontal wells are significantly better. With good completions, 1,000-acre (4 km^2) spacing is not unreasonable. To drain the outlined block in figure 9–3 with given the risked reserves of 15.7 MMBO would require 20–30 wells. With an initial production of 400 barrels of oil per day (BOPD), dropping to 150 BOPD after 1 year, and so on (fig. 9–4), it is not unreasonable to assume a recovery of over 200 MBO per well over the lifetime of the well.

The reserves can be estimated from the area under the curve, using the production history curve. If each well costs \$6 MM and the oil price is \$60/BO or \$40/BO, the investment calculations can be computed as follows. Let us assume that this block was available early in the Bakken exploration; 210 km^2 = 51,870 acres. At a price of $1,000 per acre, the block will cost $52 MM to acquire. Lease rentals in the early exploration days would have been $5.2 MM ($100/acre/year). Royalties commonly range up to 18% of gross production. Should my company invest in the block? Before acquisition of the block, an estimate must be made of the resource, followed by an economic evaluation to determine its commerciality:

Cost of the block:	$52,000,000
Cost of drilling 20 wells ($6 MM/well):	$120,000,000
Miscellaneous costs:	$10,000,000
Total cost:	$182,000,000

Thus, assuming that lifetime production from 20 wells (200 MB/well) is 4,000,000 BO,

Price/BO	Value	Gross Revenue (less $182 MM cost)
$60	$240,000,000	$58,000,000
$40	$160,000,000	($22,000,000)

At first glance, this investment is marginal. For staff in a corporation, it is of the utmost importance to consistently work to improve drilling conditions and lower costs. However, as a word of caution, if too many parameters are modified in simultaneously for a given drilling program, it becomes difficult to discern which approach yields the more significant cost and time savings. Can production be increased? Can drilling cost less? Define all the critical aspects that have an impact on cost. Review area production history. For example, the North Dakota Department of Mineral Resources publishes monthly summaries of wells and production historical data. Figure 9–4 demonstrates a typical decline on a good Bakken well.

Assume that, of the 770,000 km^2 of the basin, 45,000 km^2 (Pollastro et al., 2013) are considered as the main productive assessment unit and the net optimum area is 26,000 km^2 and the same U.S. Geological Survey (USGS) report estimates the recoverable reserves to be 3.65 billon barrels of oil (BBO) (Pollastro et al., 2013; Demas, 2013), equivalent to 140,000 BBO/km^2. Thus, For the example calculation in figure 9–3, for an area of 210 km^2, the reserves should be 29.5 MMBO. These values are close to the material balance amount and suggest that the initial calculation is reasonable.

As stated previously, the published reserve estimate must adhere to the logic of the KIS principle. Therefore, for a shale resource play, the area of closure is defined as the sweet spot, or limit of economically recoverable hydrocarbons. However, internal company approaches need to be more robust. As more data are introduced into the model, the statistical validity of the first calculation must be corroborated or rectified as required.

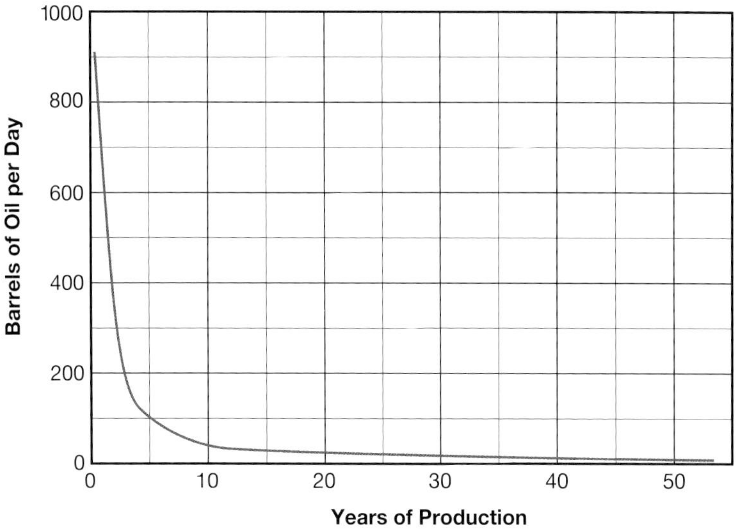

Fig. 9–4. Typical recent Bakken production decline curve. The initial curve has a hyperbolic decline, and the tail has an exponential decline. (Based on data from the North Dakota Department of Mineral Resources)

Defining the surface area of the sweet spot requires in-depth review of the seismic cube over the area of investigation. Field size distribution plots for the basin should also be generated, with maps prepared of production (per acre or km^2). In shale gas reservoirs, there is an additional problem that requires redress. For gas, the formulas of equations (9.1) and (9.2) give the total gas generated. However, gas can either be free gas or adsorbed gas. Only the free gas can be produced without enhancement. Thus,

$$G_c = \frac{V_L P}{P_L + P} \tag{9.4}$$

where G_c is the adsorbed gas volume, V_L is the Langmuir volume, P is the reservoir pressure, and P_L is the Langmuir pressure. (Langmuir analyses were described in chap. 2.)

Several problems are evident when studying generation potential. The efficiency of transformation, expressed as the *transformation ratio*, can be determined from Rock-Eval. For example, the transformation ratio in the Bakken is 15%. Such a low value is common for early generation of oil. Once the gas window is reached, the transformation ratio may be as high as 40%–50%. If the area has undergone mature generation, it is possible that, of the oil generated, 30%–40% can leave the source rock and migrate to younger units; the remaining 60%–70% of the oil in the shale is still considered the target of the shale resource assessment, but the migrated oil should be considered in the overall material balance calculation.

For example, we can work the equation for material balance (9.3) in reverse. North Dakota has produced from conventional trapped Bakken oil since the 1950s and continues to do so. The cumulative production is 1 BBO, according to the North Dakota Geological Survey. Therefore, based on these numbers, what volume of hydrocarbon is generated? For the Bakken examples of North Dakota:

- Produced to date: 1,000,000,000 BO
- Potential conventional reserves: 500,000,000 (rough estimate)
- Total conventional oil: 1,500,000,000 (equivalent to 15% recovery factor)
- If 15% recovery factor is assumed, then 100% = 10,000,000,000 for trapped oil, which represents only 30% of oil migrated
- For 100% of oil generated = 34,000,000,000 (rounded up for ease of calculation)
- With 70% as the target objective, 24,000,000,000 is the target STOOIP for the region
- With 6% as the maximum recoverable, 1,400,000,000 is the recoverable estimate from the shale

Thus, by the above estimation, the total recoverable, from conventional and unconventional resources, is 2.9×10^9 barrels.

The best areas of the Bakken shale comprise approximately 6,000,000 acres (24,000 km^2). The more marginal area is nearly equal, but only half of the area is applied. The total area used in the USGS estimate was 9,000,000 acres (Pollastro et al., 2013).

Given the general form of the original equation, the preceding estimate is reasonable. The computed value refers only to what is remaining in the shales. In other words, it corresponds to the value 1.4×10^9 barrels, as calculated above. The two numbers appear to be consistent. The estimate based on production and that based on material balance provide very similar results.

Using volumetric calculations and applying equation (9.1),

- $S_w = 0.2$, and $S_o = (1 - 0.2) = 0.8$
- Porosity = 8%
- Net thickness = 10 m = 34 ft
- Area = 9,000,000 acres

Using the classic formula and the same area and thickness as used in the model, noting that 1 km^2 = 247 acres and 1 m = 3.37 ft, as well as a shale porosity of 8% and an oil saturation of 80%,

- $N = 7{,}758 \times 9{,}000{,}000 \times 34 \times [(0.08 \times 0.8)/1] = 1.5 \times 10^{11}$ BO
- Only 15% transformed, 6% recovery = 1.4×10^{9} BO (i.e., if only 15% is generated and 6% can be recovered of the N value)
- The value would yield 152 BO/acre. At 52,000 acres (210 sq km) yields 7.9×10^{6} BO

This value assumes a layer-cake distribution with optimum conditions across the entire study area. It is not valid for all assumptions but provides an estimate that can and should be adjusted as new information is introduced.

Subsequently, a more precise estimate can be obtained by focusing on a smaller area and introducing more accurate parameters. The area can be further constrained by introducing the seismic anomalies. This will provide a higher degree of confidence on a reserve estimate.

Furthermore, the recovery factor is applicable only to the available free gas. A high recovery factor in a shale is 6% of STOOIP. The average is 4%–5%, and a poor recovery factor is less than 3% (International Energy Agency, 2013). The statistics per well are variable owing to the presence of faults, misfired fracs, and variations and inefficiencies in the well completions. Corrections on the production histories also have to compensate for different frac lengths, different methods, and well spacing. What is important to learn from the various estimates of reserves is the need for multiple approaches to ensure that the posted values are within the range of realistic results.

PRODUCTION FORECAST

Once production begins, the same material balance calculation with introduction of the following additional parameters:

- Free gas expansion
- Free oil expansion
- Free water expansion
- Pore volume contraction
- Water influx
- Net gas withdrawal
- Net oil withdrawal
- Net water withdrawal

Also, in areas where there is historical production data in the form of tables and curves, detailed analyses are required. Decline curves are generated for all wells to determine the life stage of each well and determine what remedial work can be done to improve the general decline in production. Note that basins behave differently, and both flow rates and decline curves can vary dramatically from basin to basin.

Usually, plots are provided that indicate not only the oil and gas produced but also the water produced and the gas/oil ratio—all of which vary as a function of time. Care must be taken when reviewing the data. There are basically two drive mechanisms that can work on a reservoir: *water drive*; and *gas expansion*, which is solution gas drive with no gas cap (compaction also plays a role). Water drive is very efficient because it sweeps the hydrocarbons in the formation. Because most shale resources are in low-dip areas, gravity plays a minimal role. Water drives can have recovery factors of 15%–45%. By contrast, gas expansion can have a recovery factor of 10%–20%. Because shale reservoirs require fracking and therefore are less efficient, the recovery factor is 5%–6%. The principal drive mechanism for shales is gas expansion.

The common form of the production decline in shales is a rapid decline following a hyperbolic decline-curve equation, and a long tail follows an exponential decline (fig. 9–4). Given a particular decline-curve equation, the tail end can be extrapolated with confidence. The only caveat is that the pressure decline must be plotted as well. Once the dew point is reached, production becomes uneconomical for most fields. Hence, the total volume of hydrocarbons that the well produces is the area under the curve up to the dew point.

Thus far, we can estimate the reserves on a block, and the production of individual wells in the basin can be analyzed. How can the data be extrapolated with confidence?

GEOLOGICAL RISK ASSESSMENT

A necessary approach in the interpretation of reserves is to demonstrate to an external auditor that the model generated allows for extrapolation with a particular level of confidence. The very nature of risk assessment is dependent on statistics. Therefore, a few basic statistical concepts warrant review before applying them to reserve estimation. One of the most powerful statistical tools applied in geoscience is known as the *semi-variogram* and is used to analyze the data and, subsequently, by applying kriging algorithms, to map the results. Several others techniques exist, but their objectives are similar, and kriging is commonly accepted.

Probability theory has been a discipline in mathematics for over a century but was rapidly enhanced with the advent of supercomputers. The more data become available, the more mathematically intense the computation becomes. Probability theory also has a wide variety of applications in many disciplines.

In most cases statistics focus on observational data only; however, in geostatistics, the geographic position of the data also plays a key role. Furthermore, what interests the geoscientist is the superposition of multiple independent variables that, when combined, can increase the probability of success. A visual display of how risk reduction can be achieved is to take a thin glass plate as a map and indicate a single dot where that specific parameter has a favorable description. Repeat the process with other parameters and overlay the glass plates. Where the points have the darkest density the probability of success is enhanced. Unfortunately, visual results are insufficient. A mathematical sound and quantifiable probability analysis is required. (For more detailed analyses of geostatistics, see Davis, 1973; McCammon, 1975; and Dubrule, 2003.)

Statistical analysis serves two important roles in any major investment: first, the data are analyzed to determine their significance; second, the probability for predicting results is established, in terms of degrees of probability. Only a few aspects of statistics will be reviewed as they pertain to geoscience. Statistics can be descriptive or inferred from numerical data. An example of descriptive statistics is defining cutoffs in reservoir porosity. The porosity in a reservoir may be above or below a certain cutoff, and the result is either a yes ("Y") or a no ("N"). The probability of success should be predictable. If you drill a discovery well containing a reservoir above the minimum cutoff, what is the likelihood that the next well will also encounter porosity above the cutoff? If the presence or absence of adequate porosity is random and totally independent in the next well, then there are two possible outcomes, Y or N. The probability is 50% chance of success. What are my chances when drilling three wells? If the first offset well was ½, the second offset well was ½ × ½, or 25%, then three consecutive successes will occur ⅛ times, as indicated in the following sample list of occurrences below: YNN, YYN, YNY, YYY, NNN, NYN, NNY, NYY.

The second type of statistics relates to the analysis of numbers. The simplest pure data analysis tools are averages, variances, and standard deviations. Two simple data sets are analyzed in table 9–1 and figure 9–5 to demonstrate the problem.

Table 9–1. Two random lists of numbers are used to demonstrate the effect of averages. The total for each column is 180, and the average (mean) for both is 18; however, the scatter in column A is much greater than that in B. How can I describe the scatter?

A ▲	B ●
29	21
3	17
15	17
6	23
10	15
52	18
6	14
11	21
30	15
18	19

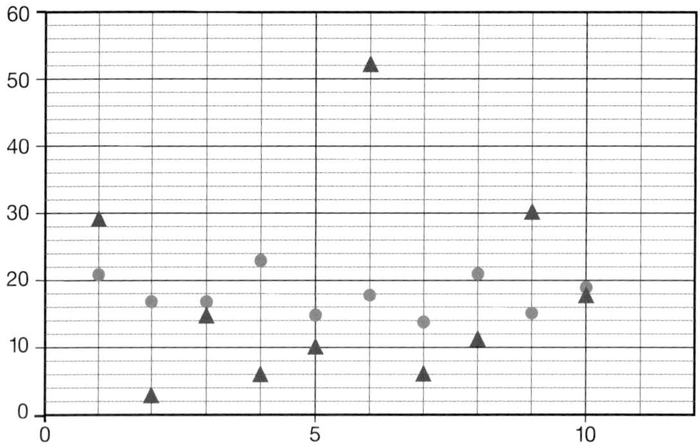

Fig. 9–5. Scatterplot of the data from table 9–1. The sequence is plotted along the x-axis.

All data sets used in the study should follow the same rigorous statistical approach. It is common in any data set to have spurious data that must be analyzed, corrected, or discarded. Examples of critical values for an interval that require an average calculation are matrix density, TOC, and porosity.

A simple plot of the data will visually depict the scatter. Statistics will quantify it. First, the variance in each data set must be calculated (table 9–2). The second step is to analyze the standard deviation, σ_v. The variance, ϖ, is defined as follows:

$$\varpi = \frac{\Sigma(x-\bar{x})^2}{n-1} \tag{9.5}$$

Table 9–2. Computation of averages, variances, and standard deviations using same data as in table 9–1

Sequence	Value x	Mean xm	Diff x–xm	Diff² (x–xm)²	Value Y	Mean ym	Diff y–ym	Diff² (y–ym)²
1	29	18	11	121	21	18	3	9
2	3	18	-15	225	17	18	-1	1
3	15	18	-3	9	17	18	-1	1
4	6	18	-12	144	23	18	5	25
5	10	18	-8	64	15	18	-3	9
6	52	18	34	1156	18	18	0	0
7	6	18	-12	144	14	18	-4	16
8	11	18	-7	49	21	18	3	9
9	30	18	12	144	15	18	-3	9
10	18	18	0	0	19	18	1	1
TOT	180			2056	180			80

VARIANCE (ϖ) = (sum((x-xm))/(n-1) = 2056/9 =		228.4
sum((y-ym)/(n-1)) = 80/9 =		8.9
STANDARD DEVIATION = SQRT(ϖ) =	x values	15.1
	y values	3.0

The variance is the difference between the observed value and the mean squared divided by the degrees of freedom, $n - 1$. The reason it is divided by $n - 1$ is because when you compare two numbers, you automatically lose one degree of freedom. The reason it is squared is because the computation searches for the difference and the direction of the difference. The standard deviation (σ_v) is the square root of the variance (ϖ):

$$\sigma_v = \sqrt{\frac{\Sigma(x-\bar{x})^2}{n-1}} \tag{9.6}$$

The meaning of the standard deviation can also be elaborated. In a normal Gaussian distribution, 68% of the data will lie within one standard deviation of the control point, and 95% of the data will lie within two standard deviations.

Several statistical tests can be run on the data to evaluate significance (e.g., *t* test, chi-squared; Davis, 1973). Two of the more common tests in geostatistics are R (coefficient of correlation) and R^2 (coefficient of determination). The R factor compares two variables and relates their dependence or independence. The closer the value is to one, the greater is the dependency. It may also be a simple multiple. The R^2 value allows analysis of the data for a single set of numbers referred to as a *quantile* (reference number) and reflects the *fit* of the data to the equation of the trend line used.

Excel automatically quantifies the R^2 value. For example, in figure 9–6, triangles yield $R^2 = 0.0185$, and circles yield $R^2 = 0.049$. Thus, circles have a better correlation, or fit, to the line function.

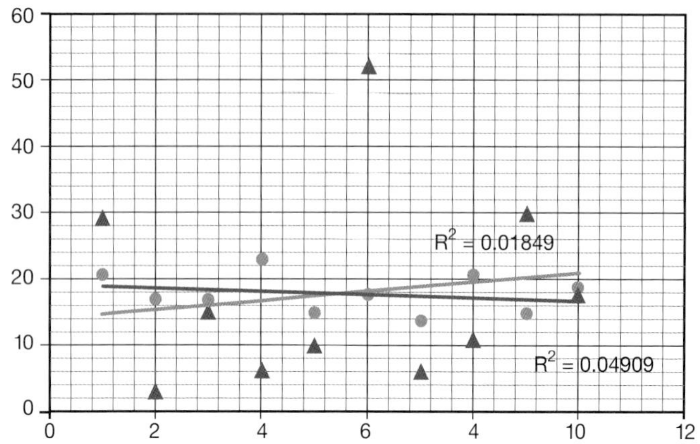

Fig. 9–6. Scatterplot showing the coefficient of determination, R^2 (i.e., relationship of the data to the trend line)

Each data set has a variance and standard deviation. If the two data sets are related, they can be compared by *covariance* as follows:

$$\mathrm{Cov}(x,y) = \frac{\Sigma(x-\bar{x})(y-\bar{y})}{n-1} \tag{9.7}$$

Covariance is the sum of the product of the data deviation from the mean divided by the degrees of freedom:

$$R = \frac{\mathrm{Cov}(x,y)}{\sqrt{\mathrm{var}(x)}\sqrt{\mathrm{var}(y)}} \tag{9.8}$$

The R factor described above is a function of the covariance given by equation (9.8). So far, the data themselves have been analyzed. The next step is to consider the data in the context of spatial distribution. The key question to ask is, What influence does a data point exert on a nearby point—and is there a distance factor?

The spatial data is evaluated using the *variogram*. The easiest way to define the variogram is in terms of what it does. A variogram collects data, and by measuring the distance between the points and the covariance between neighboring points, it assigns a function that defines the sphere of influence of each point in the data set with its neighbor.

The example variogram in figure 9–7 was first designed by Krige (1951) as a mathematical model to describe the procedure required when following up a discovery of a nugget of gold in the mines of South Africa. Once the nugget is located in a mine, how far can one excavate to find additional nuggets? The terminology used in all variograms follows the mining model.

Table 9–3. Establishing covariance in a data set using equation (9.6) in Excel

Sequence	Value x	Mean xm	Diff x–xm (A)	Value Y	Mean ym	Diff y–ym (B)	Covariance (A)*(B)
1	29	18	11	21	18	3	33
2	3	18	-15	17	18	-1	15
3	15	18	-3	17	18	-1	3
4	6	18	-12	23	18	5	-60
5	10	18	-8	15	18	-3	24
6	52	18	34	18	18	0	0
7	6	18	-12	14	18	-4	48
8	11	18	-7	21	18	3	-21
9	30	18	12	15	18	-3	-36
10	18	18	0	19	18	1	0
TOT	180			180			6

COVARIANCE (x,y) = (sum((x-xm)(y-ym))/(n-1) = 6/9 = 0.67

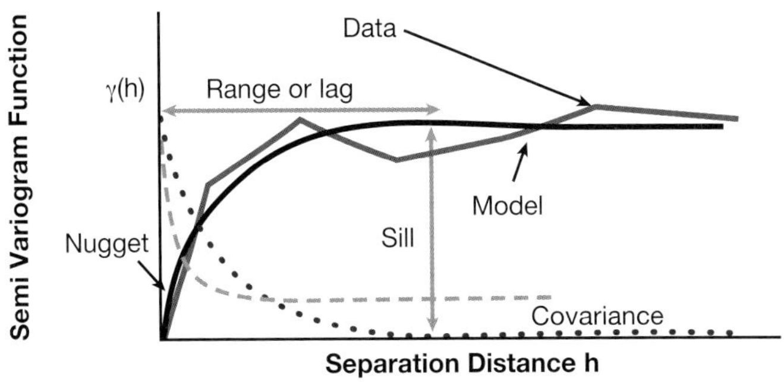

Fig. 9–7. Definition of the variogram and relation to covariance. The dashed line represents the actual data, and the model is shown in black. The black dots correspond to the model covariance.

Several terms require clarification in figure 9–7. The x-axis is the separation distance between the point starting with the nugget and the next target. By definition, the nugget is referred to as having a covariance of h, C(h), or C(0). The y-axis is the variogram function defined below. As the reference points become more distant from the nugget, the variogram function, $\gamma(h)$, increases; h corresponds to the difference between Δx and Δy directions. The procedure is to measure the distance between two values and calculate half of the difference in distance. To yield the variogram function, then multiply by the difference in values squared:

$$\gamma(h) = \frac{1}{2N(h)} \Sigma(z_i - z_j)^2 \tag{9.9}$$

Once the difference attains the maxima, there is no additional increase in $\gamma(h)$, and equation (9.9) levels off. The $\gamma(h)$ value at which this event occurs is known as the *sill*, and the distance, h, corresponding to the distance at which the sill first appears is known as the *lag* or *range*. All points to left of the lag point have some degree of correlation; all points to the right of the lag lack correlation. The lag represents the critical distance to which the nugget exerts an influence and defines the sill. The covariance of (h) yields $\gamma(h)$ at each point with a lag vector h.

Covariance is the inverse of the variogram function. By definition, where the variogram is 0, the point is called the *nugget*. Various mathematical equations are used to fit the data. They include circular, spherical, exponential, Gaussian, and linear. These equations allow prediction of trends and permit comparisons with other variograms in the project to determine consistency. Once the fit is confirmed and the sill can be established, values with a separation vector beyond the sill value have no validity, and the maximum range can be set to generate a map using the variogram data (for additional information on variograms, see Dubrule, 2003).

The initial map generated is known a *variogram map*. The variogram is rotated to meet the various points of control to map the effect, and the angle, or azimuth, as well as the distance and the value of $\gamma(h)$ are recorded. A follow-up map that is generated is the *kriged map*, where a weight factor is applied to the distance between the points and a clear relationship can be established as to the sphere of influence of the control points. The computation itself is time consuming, but many workstations and statistical packages have incorporated variogram programs and kriging mapping functions. Kriging can be used to analyze any variable. Many of the geological, petrophysical, geochemical, geophysical, and engineering parameters can and should be mapped using this technique. Of particular importance are the characteristics that define reserves (porosity, thickness, TOC, HI, Poisson's ratio, Young's modulus, etc.).

The kriging process compares parameters from various control points (seismic or well data) for which either covariance or variogram has been calculated and place the data in a conformable matrix. A vector matrix, known as the *Lagrange multiplier*, is applied to ensure that the proper limits are applied to data that are not spatially equidistant. The product of the two matrices yields the kriged map. (For a more in-depth discussion on this topic refer to chap. 3 in Dubrule, 2003.) The importance of the kriged map is that it has statistical significance. Where multiple favorable parameters coexist, the sweet spot can readily be identified. Krivoruchko (2012) has provided samples of ideal work flows for kriging data using ArcGIS Geostatistical Analyst (Esri). The ideal work flow includes the following steps:

- Analyze the data distribution, using histograms
- Generate quantile-quantile plots, or *Q-Q plots*, which compare actual data to expected normal data or two separate data sets to visually detect relationships (Ford, 2015)
- Explore spatial relationship with semivariograms and trend surface analysis
- Generate a kriging map

An important consideration in geoscience is to understand the true meaning of statistical analyses. For example, What is the true meaning of an average? Consider an average TOC if the thickness is disregarded. If the average TOC in an interval is 3% and there are two blocks, one containing a 100 m unit and a second containing a 200 m thick interval, are they comparable in importance? If the same example is modified such that, in the first model, the lower half has a TOC content of 4% and the upper half has a TOC content 2%, then the average remains 3%; however, the exploration would focus only on the lower portion of the first example.

It is imperative that equal conditions be compared. One must also view the distribution of the data. If there is a pattern such as a bimodal distribution, the resulting average will once again be misleading. This is why a histogram of the data must be prepared. Serious problems can arise if the data points between two wells, one containing shales dominated by smectites and the other by illites, are averaged to determine the efficiency and accuracy of the logging tool.

As a best practice, select a pilot area with at least four wells and three-dimensional seismic coverage. Of the four wells select three and fit the data to the results of the kriged map then test the fourth well. Rotate the three control wells and use a fourth well as your blind test. Once an adequate unit cell size is established, extend the search radius to additional surrounding wells and test the unit cell size. The concept of geocellular modeling was described in chapter 8 (see fig. 8–1). By necessity, the maximum cell cannot exceed the lag distance on the variogram. Establish whether the same cell size can be applied to other physical parameters. Then select an intermediate cell size that best fits all the basic parameters. The minimum cell size is also dictated by your earth model capacity to resolve.

The kriged map of the critical reserve estimators and seal definitions can be applied to establish a best estimator of risk factors. In particular, a Bayesian approach allows establishing *risk segment analyses* (Allen and Allen, 2013). Bayes, an English statistician, in 1770 developed a theory of probability that is widely used to this day. Although the underlying mathematics behind the theorem are complex, it is sufficient to understand how to apply the concept to the exploration effort. There are basically two approaches to statistics: *frequentist* and *probabilistic*. The frequentist states that either an event will or will not occur, as in the Y/N porosity threshold discussed earlier in this section. The probabilistic approach, by contrast, analyzes data to determine trends and assign probabilities.

Bayesian statistics ascribes probabilities and can be applied to kriging. The basis for Bayesian method is the Venn diagram. Where two favorable conditions intersect, the union (fig. 9–8) will have a higher probability of success.

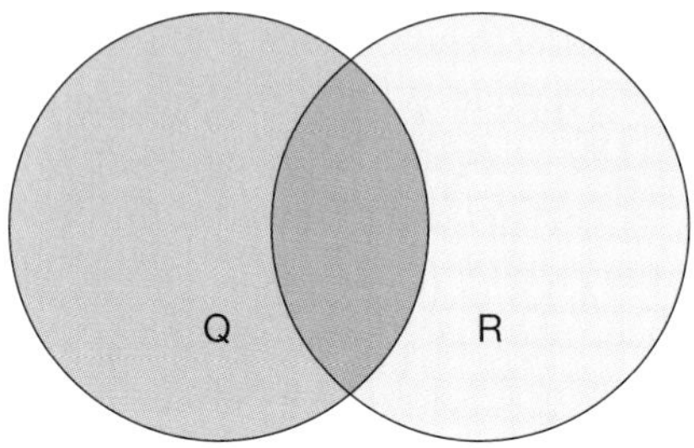

Fig. 9–8. Venn diagram showing intersection (union) of two parameters

Therefore, in figure 9–8, we can describe the areas of circles Q, R and the union of Q and R. The union is written as follows:

$$Q \cap R \tag{9.10}$$

Thus, the probability (P) that condition Q intersects with condition R is written,

$$P(Q|R) = \frac{P(Q\cap)R}{P(Q)} \tag{9.11}$$

In other words, if Q and R are present, they intersect at $Q \cap R$. In the shale example, if Q represents porosity and R represents TOC content, where the two intersect will be a better target than where they do not intersect. In the same manner, where three parameters intersect, the target will be even better. The result of the parameters can then be mapped to define a *common risk segment* (CRS), as will be discussed in the next section. When the Bayesian approach is combined with kriging, the final output is known as *empirical Bayesian kriging*.

Common risk segment (CRS) analysis

If the original map used in the reserve estimate (fig. 9–3) can be separated into two sectors, the potential reserves for each can be calculated. When it is assumed that the northern and southern half each have 50% of the unrisked reserves, each has 8 MMBO *unrisked* reserves; using the respective multipliers for success from fig. 9–3, the risked recoverable reserves are calculated as follows:

- Northern half: 8 MMBO × 0.54 = 4.3 MMBO
- Southern half: 8 MMBO × 0.42 = 3.4 MMBO

The total reserves thus are 16 MMBO unrisked and 7.7 MMBO risked, and the latter value should be used in all economic analyses.

Where do these numbers come from? By definition, CRS analysis associates similar conditions in a reservoir and assigns a probability of success for that segment (fig. 9–9). The CRS analysis is the product of the principal risks involved in the project.

CRS analysis is commonly combined with *gross depositional environments* (GDE) mapping. For example, GDE maps for offshore Nova Scotia have been drawn by Wilson et al. (2011). GDE maps are used to improve visualization of the CRS analysis. All data should be input into standard Monte Carlo simulation packages and determine the probabilities of the recoverable reserves (fig. 9–10). In addition to the Monte Carlo simulation, field size distribution curves should be generated to validate the simulation. Where new concepts or strategies can be invoked, the existing field size distribution may underestimate the potential. For example, underestimation occurred in the Bakken shale, where initial estimates computed the net worth of the Bakken interval only. However, it was later discovered that the same oil system extended to the Three Forks Formations, which significantly increased the earlier estimate of reserves.

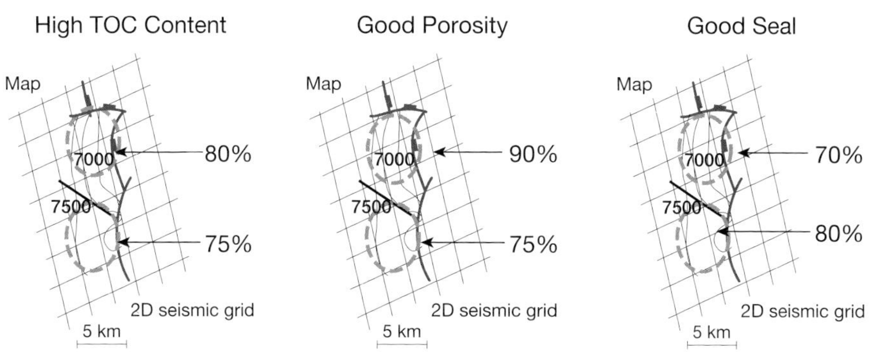

Sector	TOC		Porosity		Seal		CRS
North	0.8	×	0.9	×	0.75	=	0.54
South	0.75	×	0.7	×	0.8	=	0.42

Fig. 9–9. CRS analysis. The major risk components are quantified and multiplied to provide the CRS for the area in question.

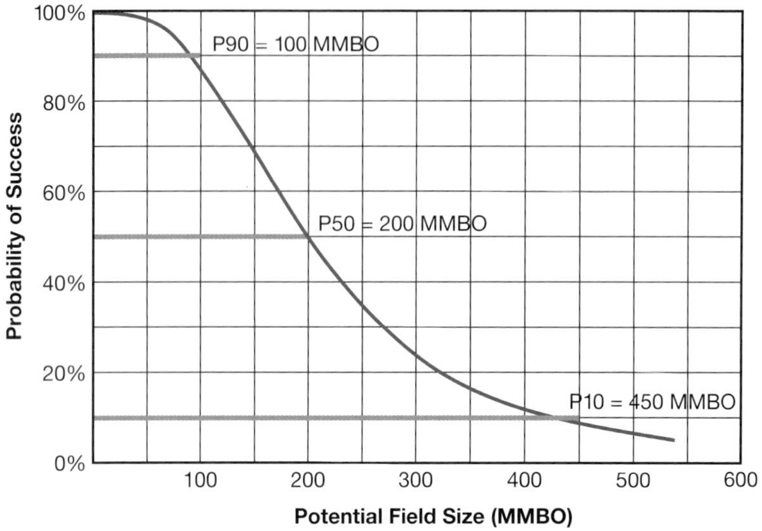

Fig. 9–10. Monte Carlo simulation to define probability for reserve estimate

POLITICAL AND ENVIRONMENTAL RISK

Political risk is beyond the scope of this review. However, in many cases, it overlaps with environmental risks, which is the focus of this section. As discussed in chapter 8, the industry has great need for environmental engineers. They should be in charge of three basic studies:

- Carry out predrilling baseline studies. Many water wells in a region need to be analyzed in detail. If not available, pilot water wells should be drilled and tested. Test potable water in the nearby communities. If there is the risk of earthquakes in the region, carefully study predrill earthquakes epicenters, to determine their causes and focal depths. Evaluate any aquifer that might be used for water injection. Determine if there is any communication with shallower aquifers.
- During the drilling and water injection phase, monitor the same wells used in the baseline study. Before drilling begins, ensure that all water issues have been discussed with the various communities. For example, in a drought, how will freshwater use be prioritized?
- For several years after production has ceased, the same water well will require monitoring. Other environmental hazards (noise, gas leaks, etc.) must also be investigated.

PROJECTED CASH FLOW

Before corporate management can approve a project, an estimate of cost and benefits must be established. For simplicity, only an oil case will be considered. In the real world, the gas and liquids have to be dealt with separately because of the different price structure. The initial vehicle is a cash-flow forecast.

For a specific project, the cost involves three phases:

- The initial study costs include data purchase and study costs
- The second phase involves land acquisition and any additional seismic that must be shot
- The third phase comprises drilling

In all three, overhead expense must be added. As discussed before, each individual element of the project should be self-sufficient. This topic is addressed routinely in most petroleum engineering texts (e.g., see chap. 12 of *Principles of Oil Well Production*, by Nind [1981] and *The Business of Petroleum Exploration*, edited by Steinmetz [1992]).

The sweet spot—that is, the key exploration target—requires a certain number of wells to produce the reserves efficiently. To do so, the engineers must run a series of scenarios based on existing wells or local knowledge. These include field wide and individual-well decline curves. Generic *approvals for expenditure* (AFEs) are generated to predict potential problems in the cost estimation. AFEs are divided into dry-well cost and completed-well cost. A review of existing wells in a basin provides a reasonable guideline as to the number of either dry holes or wells not completed owing to mechanical problems.

The following are the highest-cost items on the AFE (values are given in million U.S. dollars):

- Drilling rig, $6.0
- Location and road costs, $0.3
- Drilling muds and chemicals, $0.2
- Cements, $0.2
- Supervisory staff, $0.15
- Land fees and damages, $0.2
- Tubulars, $0.5

The remaining costs are related to miscellaneous items. The information should be based on historical data or actual cost estimates obtained from contractors. As a rule, an experience factor allows for improvement in drilling rates for wells as drilling progresses. Competitor analyses are also important. But beware: drilling faster can damage formations.

Are those that are drilled more efficiently better producers? What was done on the wells to improve efficiency? On the basis of the results and by use of government instructions on design, an appropriate spacing pattern is established.

INCOME STATEMENT FUTURE PROJECTION

The income statement is built on the estimated cash flow. There is always a *sunk cost* in any project. People are paid to carry out an initial investigation, and data have to be acquired. Today, that is likely to include two-dimensional seismic, some three-dimensional seismic, and well data. Combined with salaries, the sunk cost may exceed US$6 MM. The original fund for the venture is loaned by the corporation or an external bank and must be returned with interest. In the United States, one initial cost is for leasing. Outside the United States, projects require bid offers and signing bonuses equivalent to the lease costs.

For the sake of the model project, it will assumed that the area of study has 210 km^2. The cost per acre may vary from US$600 to $2,000 per acre. Therefore, the leasing cost can be in excess of $10,000,000. In the model case, the lease cost was $1,000 per acre. The above area translates to 52,000 acres. Thus, the cost would be $52,000,000.

Conventional engineering economic analyses cover two basic aspects: *payback period* and *rate of return on investment*. The procedure commonly followed is known as *discount cash flow versus rate of return* (DCFROR), which can be done before (BFIT) or after (AFIT) federal income taxes. At the same time, it is necessary to ensure a proper work flow by applying *process optimization*. Del Castillo (2007) describes this as an analysis of a frequently carried out process, involving testing or experimentation to improve the results. This is particularly applicable where the final results are not well understood and the performance cannot be guaranteed.

The following rules of thumb can be used in evaluating a project (Jones, 1992):

- A common hurdle is that the investment you choose should be able to yield 30–40 times what you invested over the life of the investment.
- Estimate an appropriate rate of return, which is a function highly dependent on risk. The higher the risk is, the higher will be the rate of return needed: high, 30–40%; medium, 20–30%; low, 15–20%.
- Determine payout time.

- Estimate two cost elements: the finding cost; and the cost per barrel produced.
- The first wells should pay for all initial costs, including land costs. Jones refers to this as the *first-well rule*.

A simple model is chosen for descriptive purposes. The field has nine wells drilled; eight were successful, and the producing wells flowed at initial rates of 400 BOPD with an exponential decline (fig. 9–3). An eight-year life span for the field is projected. Each year, at most, two wells are drilled. Each well cost US$6 MM, and the predrill cost plus leases amounted to $52 MM. The tax and royalty is 12%, and the cash discount rate (r) is 5%. What is the annualized rate of return on the investment? What is the payout time for the investment?

The calculation of the discounted net present value (NPV) is as follows:

$$\text{Discounted NPV} = \frac{CF_1}{(1+r)^1} + \cdots \frac{CF_n}{(1+r)^{n-1}} \tag{9.12}$$

where n is the time involved in the investment study and "CF" denotes the cash flow at each time point. The NPV is the value of the investment over time (multiply the revenue by the discount factor). The purpose of this exercise is to discern whether to invest in the field or to invest in another opportunity (e.g., see table 9–4).

The example given in table 9–4 follows the guidelines discussed earlier in this section. Two wells are drilled per year at $6 MM each, and the production is staggered following the sample decline curve from figure 9–4. The discounted rate of return determined for the exercise (40%) is reasonable for a high-risk venture. Several questions should be answered:

- Are the leases more expensive?
- Can more wells be drilled sooner?
- Is the price of oil going to drop?

For simplicity, assume the entire investment is set aside at the outset of the project. The future value (FV) is calculated as follows:

$$FV = P(1+i)^n \tag{9.13}$$

where P is the initial cash principal, i is the annual interest rate, and n is the time period. If the funds were invested at 6% and the inflation rate was 4% and the net interest was close to 2%, then, by use of equation (9.13), FV = $53,000,000 $(1.02)^8$ = $62,000,000.

Table 9–4. Example of a projected cash-flow analysis

Year	Expenses	Lease Rentals	Total Expenses	Prod	at $60/B	Royalty plus tax 12%	Net revenue	Discount rate	Discounted cash flow
Pre drill cost 52000000			52000000				-5200000	1.0000	-52000000
1	14000000	52000000	19200000	896000	53,760,000	2,304,000	37,456,000	1.0000	37456000
2	14000000	52000000	19200000	1472000	88,320,000	4,147,200	70,172,800	1.1025	63648798
3	14000000	52000000	19200000	1236000	74,160,000	5,068,800	55,091,200	1.1576	47589850
4	14000000	52000000	19200000	460000	27,600,000	5,529,600	8,070,400	1.2155	6639538
5	14000000	52000000	19200000	208000	12,480,000	5,587,200	(7,107,200)	1.2763	-5568677
6	3000000	52000000	8200000	84000	5,040,000	3,312,000	(1,272,000)	1.3401	-949186
7	3000000	52000000	8200000	24000	1,440,000	1,497,600	(3,057,600)	1.4071	-2172979
8	3000000	52000000	8200000	20000	1,200,000	604,800	(2,404,800)	1.4775	-1627663
9	3000000	52000000	8200000	20000	1,200,000	172,800	(1,972,800)		
10	3000000	52000000	8200000	20000	1,200,000	144,000	(1,944,000)		
11	3000000	52000000	8200000	20000	1,200,000	144,000	(1,944,000)		
12	3000000	52000000	8200000	20000	1,200,000	144,000	(1,944,000)		
13	3000000	52000000	8200000	20000	1,200,000	144,000	(1,944,000)		
14	3000000	52000000	8200000	20000	1,200,000	144,000	(1,944,000)		
15	3000000	52000000	8200000	20000	1,200,000	144,000	(1,944,000)		
				4540000	Undiscounted rev		104,948,800	NPV	93,015,680.46
Total inv		Total expenses	230000000			Discounted cash flow ROR			40%

The discounted annualized rate of return is nearly 40%, and the payout is 1 year. Would you invest in this project? And if so, when would you sell it? The cumulative cash flow analysis provides an answer to one of the key questions asked in any project: "How quickly will my project pay out?" For figure 9–11, the payout is in one year. This is as usually expected, although payout time may extend to two years for some projects.

Important differences between various basins must be considered when calculating economic benefits. For example, if your company is investing in the North Dakota Bakken shale play and the South Texas Eagle Ford shale, the production decline in the two are very different (figs. 9–12 and 9–13). The Eagle Ford wells have a much longer production life. By contrast, the Bakken wells have a higher initial production (IP) rate. Which is better depends on your finances.

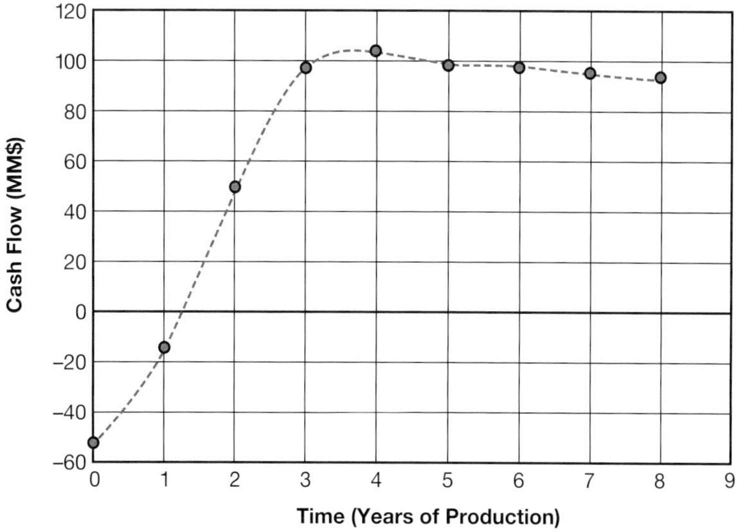

Fig. 9–11. Cash-flow plot using cumulative discounted cash flow from table 9–4. Note payout in the first year.

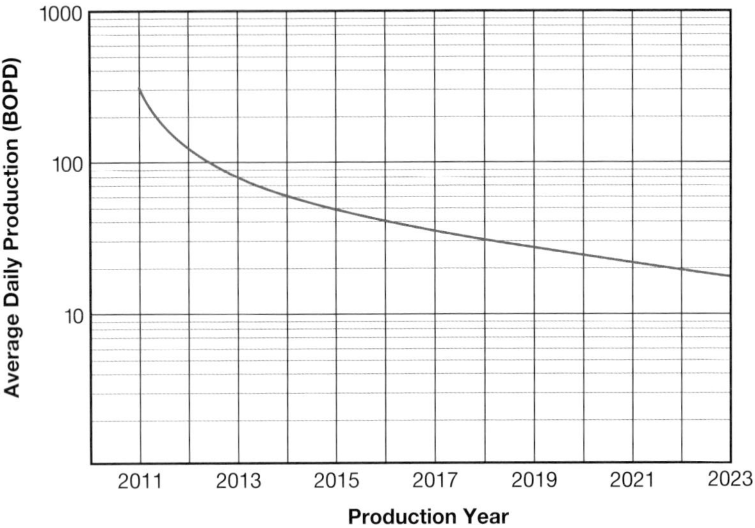

Fig. 9-12. Estimated per-well production forecast for the Eagle Ford shale. (The primary source is the Texas Railroad Commission. Note that these charts are very coarse and are used here only for descriptive purposes. They are not intended for financial planning purposes.)

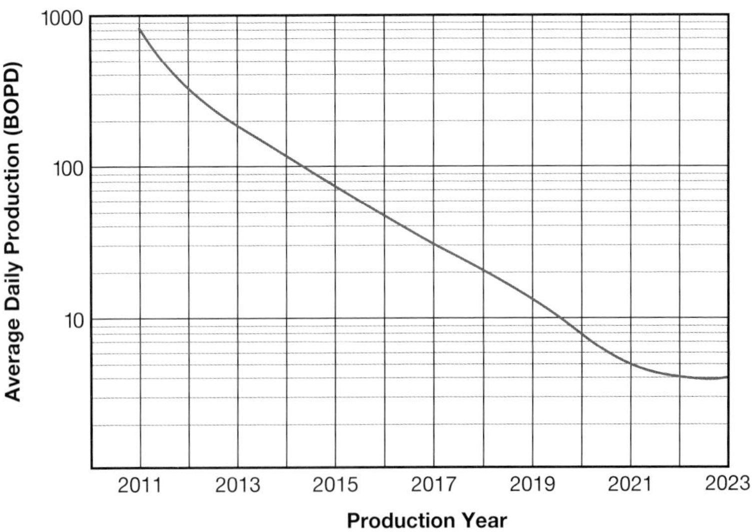

Fig. 9-13. Estimated average well production prognosis for Bakken wells in North Dakota. (The primary source is the North Dakota Geological Survey. Note that these charts are very coarse and are used here only for descriptive purposes. They are not intended for financial planning purposes.)

SUMMARY

Economic analysis meets three specific needs: government regulatory requirements, investor information, and internal company requirements. Analyses are based on historical information and future projections. All three require an estimate of reserves, a production forecast, and a cash-flow analysis. The rule is KIS (keep it simple).

Reserves are classified as technically producible or contingent. Contingent reserves are not accessible with present technology. Technically producible reserves are either P90, P50, or P10. P90 reserves are proven producing (PD) or proven undeveloped (PUD), and PD are further divided as producing or shut-in reserves.

Reserves are given by volumetric calculations, decline-curve analysis, and/or material balance. All three are recommended to act as controls. Gas and oil require different formulas owing to the behavior of gas. Gases are compressible and under pressure do not behave as ideal gases. Gases with condensate must follow their composition phase diagrams. Gases can be either free gas or adsorbed gas (Langmuir isotherm).

Material balance requires significant geochemical input and usually include physical oil properties. Production history in the basin provides the basic input into the cash-flow analysis.

Geostatistics provide the basic tools to assess risk of the reserves. Only risked reserves can be input into the cash-flow analysis. Geostatistics include variograms and kriging, and the results are plotted as common risk segment (CRS) maps to compute the overall risk.

Projected cash flow, DCFROR, is based on projected costs, including sunk costs and projected revenues. The cumulative cash flow is plotted versus time, and the rate of return is calculated as well as the payout time.

REFERENCES

Allen, Philip A., and John R. Allen. 2013. *Basin Analysis: Principles and Application to Petroleum Play Assessment.* 3rd ed. Chichester, UK: Wiley-Blackwell.

Brinkerhoff, Riley, Sarah Edwards, and Mark Millard. 2016. "Pitfalls in geologic mapping within plays: A case study from the Three Forks Play in the Williston Basin." *Search and Discovery* #80546. AAPG Annual Convention, Calgary, June 2016. http://www.searchanddiscovery.com/pdfz/documents/2016/80546brinkerhoff/ndx_brinkerhoff.pdf.html.

Davis, John C. 1973. *Statistics and Data Analysis in Geology.* New York: John Wiley & Sons.

Del Castillo, Enrique. 2007. *Process Optimization: A Statistical Approach*. International Series in Operations Research and Management Science, vol. 105. New York: Springer.

Demas, Alex. 2013. "USGS releases new oil and gas assessment for Bakken and Three Forks Formation." Science Features (blog). U.S. Geological Survey. May 2. www2.usgs.gov/blogs/features/usgs_top_story/usgs-releases-new-oil-and-gas-assessment-for-bakken-and-three-forks-formations/.

Dubrule, Olivier. 2003. *Geostatistics for Seismic Data Integration in Earth Models: 2003 Distinguished Instructor Short Course*. Tulsa: Society of Exploration Geophysicists.

Ford, Clay. 2015. "Understanding Q-Q plots." University of Virginia Research Data Services + Sciences. data.library.virginia.edu/understanding-q-q-plots/.

International Energy Agency. 2013. *Resources to Reserves 2013: Oil, Gas and Coal Technologies for the Energy Markets of the Future*. Paris: International Energy Agency. www.iea.org/publications/freepublications/publication/Resources2013.pdf.

Jin, Hiu, Michael Lewan, Stephen A. Sonnenberg. 2016. "Bakken oil-generation kinetics by hydrous pyrolysis and its testing in 1D modeling." Paper presented at the AAPG Annual Convention and Exhibition, Calgary, AB, June 19–22. Search and Discovery article 41907. www.searchanddiscovery.com/documents/2016/41907jin/ndx_jin.pdf.

Jones, Donald R. 1992. "Some basic business concepts." In *The Business of Petroleum Exploration*, R. Steinmetz (ed.). Tulsa: American Association of Petroleum Geologists, 7–12.

Kohlruss, Dan, and Kosta Stamatinos. 2014. "Preliminary bulk density mapping of the Upper and Lower Bakken Member shales of southeastern Saskatchewan: A potential indicator for oil generation and expulsion." In *Summary of Investigations 2014*, vol. 1. Saskatchewan Geological Survey, Misc. Rep. 2014-4.1, Paper A-2. publications.gov.sk.ca/documents/310/91968-A-2%20Kohlruss%20and%20Stamatinos.pdf.

Krige, Daniel G. 1951. "A statistical approach to some basic mine valuation problems on the Witwatersrand." *Journal of the Southern African Institute of Mining and Metallurgy*. 52: 119–139.

Krivoruchko, Konstantin. 2012. "Empirical Bayesian kriging implemented in ArcGIS geostatistical analyst." ESRI.com. Fall: 6–10. www.esri.com/NEWS/ARCUSER/1012/files/ebk.pdf.

Kuchinskiy, Vitaly, Keith Gentry, and Ron Hill. 2012. "Source rock evaluation technique: A probabilistic approach for determining hydrocarbon generation potential and in-place volume for shale plays." Paper presented at the AAPG Annual Convention and Exhibition, Long Beach, CA, April 22–25. Search and Discovery article 41045. www.searchanddiscovery.com/documents/2012/41045kuchinskiy/ndx_kuchinskiy.pdf.

McCain, William D. 1990. *The Properties of Petroleum Fluids*. Tulsa: PennWell.

McCammon, Richard B. (ed.). 1975. *Concepts in Geostatistics*. Berlin: Springer.

Nind, Thomas Eagleton Westwood. 1981. *Principles of Oil Well Production*. 2nd ed. New York: McGraw-Hill.

Pollastro, Richard M., Laura N. R. Roberts and Troy A. Cook. 2013. "Geologic assessment of technically recoverable oil in the Devonian and Mississippian Bakken Formation." Chapter 5 in *Assessment of Undiscovered Oil and Gas Resources of the Williston Basin Province of North Dakota, Montana, and South Dakota, 2010*. U.S. Geological Survey

Digital Data Series DDS-69-W. pubs.usgs.gov/dds/dds-069/dds-069-w/contents/REPORTS/69_W_CH_5.pdf.

Steinmetz, R. (ed.). *The Business of Petroleum Exploration*. Tulsa: American Association of Petroleum Geologists., 382.

Terry, Ronald E., and J. Brandon Rogers. 2015. "Introduction to petroleum reservoirs and reservoir engineering." In *Applied Petroleum Reservoir Engineering*. 3rd ed. Upper Saddle River, NJ: Prentice Hall.

Wilson, Hamish, Matt Luheshi, David Roberts, and Bernard Colletta. 2011. "Nova Scotia Play Fairway Analysis: Gross depositional environments and hydrocarbon prospectivity." Paper presented at the AAPG Annual Convention and Exhibition, Houston, April 10–13. Search and Discovery article 50455. www.searchanddiscovery.com/documents/2011/50455wilson/ndx_wilson.pdf.

INDEX

A

B

C

D

E

F

G

H

I

N

Q

R

S

T

U